SYNTHETIC CDOs

Modelling, Valuation and Risk Management

Credit derivatives have enjoyed explosive growth in the last decade. One of the most important assets in this industry is synthetic Collateralised Debt Obligations (synthetic CDOs). This book describes the state-of-the-art in quantitative and computational modelling of these instruments.

Starting with a brief overview of the structured finance landscape, the book introduces the basic modelling concepts necessary to model and value simple vanilla credit derivatives. Building on this the book then describes in detail the modelling, valuation and risk management of synthetic CDOs. A clear and detailed picture of the behaviour of these complex instruments is built up. The final chapters introduce more advanced topics such as portfolio management of synthetic CDOs and hedging techniques, often not covered in other texts.

Mathematics, Finance and Risk

SYNTHETIC CDOs

Modelling, Valuation and Risk Management

CRAIG MOUNFIELD

CAMBRIDGE UNIVERSITY PRESS
Cambridge, New York, Melbourne, Madrid, Cape Town, Singapore, São Paulo, Delhi

Cambridge University Press
The Edinburgh Building, Cambridge CB2 8RU, UK

Published in the United States of America by Cambridge University Press, New York

www.cambridge.org
Information on this title: www.cambridge.org/9780521897884

First published 2009

Printed in the United Kingdom at the University Press, Cambridge

A catalogue record for this publication is available from the British Library

Library of Congress Cataloguing in Publication data
Mounfield, Craig, 1969–
Synthetic CDOs : modelling, valuation and risk management / Craig Mounfield.
p. cm. – (Mathematics, finance and risk)
Includes bibliographical references and index.
ISBN 978-0-521-89788-4 (hbk.)
1. Collateralized debt obligations. I. Title. II. Series.
HG6024.A3M69 2009
332.63′2 – dc22 2008043035

ISBN 978-0-521-89788-4 hardback

Dedicated to my parents, my wife and my daughter.

Contents

Preface

This is a book about the modelling, valuation and risk management of synthetic collateralised debt obligations (or synthetic CDOs or simply CDOs for short). Synthetic CDOs are an example of a structured credit product. This is a financial product that takes targeted risk for the purpose of achieving targeted returns. Structured credit products utilise two financial engineering technologies: credit derivatives and asset securitisation. Synthetic CDOs have played an increasingly important role in the expansion of the global credit derivatives market which has grown rapidly since the turn of the century. Indeed, it is estimated that by the end of 2006 the total credit derivative notional amount outstanding was over $20 trillion (from virtually zero only a decade earlier). Increased trading volumes naturally led to market participants becoming more sophisticated (in terms of their risk/return characteristics and the strategies they employ) as well as to a commensurate increase in the complexity and subtlety of the products available. This in turn drives the evolution of the mathematical and computational models used to value these products. The objective of this book is to collate, summarise and critically assess the current state-of-the-art in quantitative and computational modelling of synthetic CDOs. The key word here is *modelling*; the book is about mathematical models and their properties. This book is not intended to provide detailed descriptions of the business and economic rationales for trading credit derivatives; there are better resources available that describe this and due reference will be given to these sources. It is meant to provide a detailed quantitative description of the modelling techniques currently employed in the marketplace for characterising synthetic CDOs.

It will be assumed that the technical level and experience of the reader is relatively high. Basic financial concepts will not be described in detail (except insofar as when such detail is necessary). Instead reference will be made to the appropriate resources. The use of financial and technical jargon will hopefully be kept to a minimum, although in a specialised, technical text such as this some jargon is inevitable. The rationale for this approach is to ensure the volume is concise and to the point. It is intended to describe just enough of the mathematical and computational modelling to enable the reader to understand the relevant issues (along with a discussion of the practical implementation considerations) and help the reader to form their own opinion as to the merits, or otherwise, of the models presented. I will consider the book to be a success if it enables readers to understand the behaviour of models and to build better versions of them. This lean approach will hopefully make the volume attractive to practitioners (who do not always have the time to study a subject in detail) who wish to understand more about the properties of the credit derivative models commonly used in the marketplace. In particular it is envisaged that the volume will be of interest to a range of different types of practitioner.

- Quantitative analysts (quants) and quant developers wanting to understand more about credit modelling and credit derivatives. The book is written with a strong emphasis on models, implementation and understanding of the model behaviour. It is therefore well suited to quants in model validation teams, for example.
- Quantitative risk managers wanting to understand better the models used for valuation, to interpret synthetic CDO risk sensitivities (e.g. spread and correlation sensitivities) and risk manage complex credit portfolios.
- Traders and product controllers seeking a better understanding of the mechanics going on in the black boxes when 'F9' is pressed (and to understand the relative strengths and weaknesses of different models).
- Structurers wanting to understand better the properties of the instruments they are using to construct strategies with specific risk/return characteristics.
- Researchers in academia looking to understand some of the practical issues surrounding the common models used in the marketplace.

The downside to this lean approach is that for less experienced readers the material may at times not give as much explanation as would be liked, or some (basic) concepts are not described fully. However, for the motivated and intelligent reader this should present not a problem but a challenge and (as the author knows from experience) the rewards in terms of deeper understanding are worth the effort.

At the beginning of a project such as writing a book one has a vision as to what the finished product will look like. The vision for this book was that it would be very much model focused, with a strong emphasis on the practical, pragmatic implementation details that are of crucial importance in a live banking environment. This means there is less focus on the 'business' topics of the economics, mechanics and structures of credit derivatives than can be found in other texts. To include this information would have detracted from the core message of models and their properties. Also, when writing a book it is necessary to make compromises and be pragmatic in terms of content. At the beginning of the project one's vision of what will be achieved is vast and expansive. By the end of the project one is simply happy to stumble across the finish line. There are occasions throughout the book when more detailed analysis of a particular model or scenario would have been very useful indeed to illustrate a particular point further, but due to time constraints was not included. On these occasions it is suggested that the reader build the models and do the analysis themselves as an exercise.

This leads into the next important point about the approach taken in the text. In the modern world of quantitative finance it is almost impossible to develop models of complex derivative trades that are wholly tractable analytically. It is therefore difficult to separate a model's mathematical description from its actual implementation. When it comes to building models suitable for use within a live investment banking environment the devil really is in the details. Full understanding of a model only comes from implementing it, analysing its properties and understanding its weaknesses. An important objective of this volume, therefore, is to provide not only the mathematical descriptions of the models, but also details of the practical implementation issues. To achieve this objective, liberal use is made of pseudo code to illustrate the implementation of an algorithm. The purpose of this code is to allow the reader to convert quickly a description of a model into the programming environment of their choice (although the author is most familiar with C++, and there may appear to be a bias towards the syntax of this language on occasion).

The volume is structured into three distinct sections. Broadly speaking Chapters 1–3 motivate the main topic, synthetic CDOs, and introduce some of the basic modelling tools necessary to describe them. Chapters 4–10 analyse the mathematical and computational modelling techniques applied to synthetic CDOs. Chapters 11–14 look at more advanced topics in the analysis of synthetic CDOs. Each of the chapters can in principle be read in isolation and each is relatively self-contained. However, there is a clear path from chapter to chapter (which reflects the author's own train of thought), particularly in Chapters 4–10. Reading each chapter sequentially will build a clearer and more coherent picture of the subject matter as a whole, but it is by no means a prerequisite.

In the first part of the book we motivate the study of synthetic CDOs by understanding their importance and usage within the broader credit derivatives marketplace. Chapter 1 provides a brief overview of the credit derivatives market in terms of instruments and introduces the crucial concepts of securitisation and tranching which are the basis of CDO technology. In this first section we also provide some of the basic mathematical building blocks necessary for later chapters. Chapter 2 describes the current market standard modelling methodologies for capturing the arrival of default risk of an obligor. This chapter also introduces the concepts and methods used for the modelling of default correlation, which as we will see is one of the most fundamental concepts in the characterisation of synthetic CDOs (and indeed any multi-name credit derivative). The first section of the book ends with a discussion, in Chapter 3, of the valuation models for the simplest and most vanilla of credit derivatives – credit default swaps or CDSs. The market for single-name default protection CDSs is extremely liquid and a good understanding of the valuation methods for these basic building blocks is a necessary prerequisite for understanding the more complex multi-name products.[1] For a reader already conversant with single-name credit derivatives, the material in Chapters 1–3 will be familiar. Indeed these chapters are only included in order to provide a reference guide to the concepts underpinning the rest of the book.

The second part of the volume, Chapters 4–10, which is its mathematical and computational core, focuses specifically on the valuation and risk analysis of multi-name credit derivatives and synthetic CDOs in particular. Chapter 4 introduces the credit indices that have emerged and evolved over the course of the last few years. The introduction and subsequent trading of these indices has provided enormous impetus to the growth of the credit derivatives market. Chapter 5 then introduces default baskets. In terms of materiality, default baskets are a very small fraction of the overall structured credit marketplace. However, they are the simplest form of multi-name credit derivative and an understanding of their valuation and risk sensitivities can provide substantial insight into the behaviour of more complex synthetic CDOs.

Chapters 6 through 8 develop and analyse the core mathematical models for valuing synthetic CDOs. Chapter 6 describes a number of different methodologies for valuation and, in particular, introduces the current market standard valuation model, the so-called normal copula model. Chapter 7 investigates the fundamental behaviour of the model as certain key parameters are varied systematically. As will be seen in this chapter, the phenomenology of the model is relatively complex and subtle. Chapter 8 analyses the risk sensitivities of the standard market model to variations of input parameters. More importantly this chapter discusses the different risk sensitivity measures such as credit spread 01 (CS01) and value-on-default (VoD) that are necessary to capture and characterise the risk inherent in synthetic CDOs.

The next chapters look at the implications for the standard market model that standardised tranches and the development of a liquid market have had. Initially the market for synthetic CDOs was relatively illiquid and deals were done on a bespoke basis. The introduction of standardised credit indices and the subsequent development of a market for trading tranched exposures to slices of the index provided enormous impetus to the liquidity and volume of trades in single-tranche synthetic CDOs (STCDOs). Eventually the market became sufficiently liquid to allow transparent price discovery for the prices of these standardised index tranches. At this

[1] The main focus of the book is synthetic CDOs. Therefore we will not spend a great deal of time talking about CDSs and other credit derivatives – there are better texts available that describe these products in great detail.

point the role of the standard model changed; it became a mechanism whereby market participants could express and trade their views on default correlation. Chapter 9 introduces the concepts of implied and base correlations that have been developed to capture implied pricing information from market observed prices. As the prices of instruments become transparent in the open market it is crucially important for the standard model to be able to reproduce these prices accurately. Chapter 10 describes some of the different methodologies that have been developed to allow calibration of models of synthetic CDOs to market observed prices (the so-called 'correlation skew').

The final part of the volume, Chapters 11–14, looks at more advanced topics in the characterisation and analysis of synthetic CDOs. Chapter 11 introduces a number of exotic CDOs. Examples include CDOs with asset backed securities as the underlying pool of obligors as well as CDOs with CDOs as the assets in the underlying pool (so called CDO squareds). Correlation trading is the term used to refer to trading strategies designed to exploit the risk/return characteristics of portfolios of CDO tranches. Chapter 12 analyses the risk/return characteristics of a number of popular CDO trading strategies. Chapter 13 considers extending the models developed thus far for a single-tranche position to a portfolio of tranches and assesses how the risk in the tranche portfolio can be quantified and controlled.

Finally, a natural extension of analysing the static (in time) performance of CDO trading and hedging strategies is to look at the through life performance of the trading strategy. In the pricing of simpler derivatives, the value of the derivative is equal to the cost of the dynamic hedging strategy. If a hedging strategy is good at capturing all the risks a position is exposed to then the overall P/L generated from the process of selling the derivative instrument and rebalancing the hedging portfolio as the market risk factors evolve should be small. If the hedging strategy is not adequate there will be significant P/L leakage. Chapter 14 sets up and analyses a simple hedging simulation of synthetic CDO tranches. This chapter is more speculative in nature than previous chapters as it represents the cutting edge of technology applied to the analysis of complex derivative securities.

Acknowledgements

A book is never written in isolation, and it is a pleasure to acknowledge the contribution that a number of individuals have made to the current text. I would like to thank all the people I have worked with in the Model Validation and Risk Management teams of Credit Suisse and Barclays Capital as well as my co-workers at Cheyne Capital Management. A lot of the experience that is encapsulated in this text is a direct result of day-to-day interactions with my colleagues at these institutions. In particular, I would like to thank Dr Niclas Sandstrom of Barclays Capital and Dr Andrea Petrelli of Credit Suisse for their detailed reading of the original manuscript, and for making numerous suggestions as to how it could be improved.

I would also like to thank my editor at CUP, David Tranah (and all the other staff who have contributed to the bringing to fruition of this project), for providing me with an opportunity to write this book. Finally I would like to acknowledge the contribution of my Ph.D. supervisor Professor Sir S. F. Edwards of the Cavendish Laboratory, Cambridge. The scientific training I received under his tutelage has proven to be of enduring value throughout my career. I hope this text reflects some of what I learnt from him.

1

A primer on collateralised debt obligations

Credit – Derived from the Latin verb credo meaning 'I trust' or 'I believe'.

1.1 Introduction

In this book we will introduce and describe in detail synthetic collateralised debt obligations (or synthetic CDOs for short). Synthetic CDOs are a sophisticated example of a more general asset class known as credit derivatives. In their simplest form credit derivatives facilitate the transfer of credit risk (the risk that a counterparty may fail to honour their outstanding debt obligations such as paying coupons or repaying principal on bonds they issued) between different counterparties to a trade. The rationale for trading credit derivatives is to allow this risk to be transferred efficiently between counterparties, from those who are unwilling or unable to hold it, to those who want it. This chapter will introduce some of the important credit derivative products that will be analysed in detail later in the book. The chapter will also introduce the financial engineering concepts that underlie synthetic CDOs.

Section 1.2 introduces the concepts of securitisation and tranching. These are the key financial innovations that underpin CDOs and indeed much of structured finance technology. Section 1.3 then provides an overview of some of the most common credit derivative instruments. These include credit default swaps, credit indices and most importantly synthetic CDOs. The key features of the different instruments will be described and some discussion given of the motivations for trading them (although the level of detail of this final point is by no means exhaustive since there are other resources available which already extensively cover this material [Das 2005, Kothari 2006, Rajan *et al.* 2007]). Finally in Section 1.4 we briefly summarise the key points introduced in the chapter and set the scene for the remainder of the book.

1.2 Securitisation and tranching

In this section we provide an overview of the concepts of securitisation and tranching (a very detailed reference on this topic is Kothari [2006]). These are the fundamental financial engineering techniques that underpin CDOs and indeed most of structured finance. We motivate the discussion of securitisation by considering a simplified model of a bank's business.

The business of banks is to invest money and speculate with the expectation of making a positive return on their investments. They will, for example, provide loans or extend lines of credit to corporate entities for them to invest in expanding their business. In return for these loans the corporate pays interest to the bank and at the maturity of the loan repays the initial principal back (or alternatively the principal is paid back gradually over time). The risk the bank runs is that, for one reason or another, they will not get back the income due to them from the periodic coupons or their original investment (return of principal). For example, if the corporate were to go into bankruptcy or administration due to poor management or a global recession, it is unlikely the bank would receive all of their investment back.

The key component in the whole of the global financial system is liquidity (as was painfully apparent during the latter half of 2007 – a good history of financial crises past can be found in Kindleberger and Aliber [2005] and Bookstaber [2007]). Banks need cash in order to survive day-to-day. If all of the loans that a bank has made were to go bad simultaneously, the income the bank receives from this business would evaporate, forcing them to raise their costs of borrowing in other areas to recoup some of the lost income (in turn putting pressure on other parts of the economy such as consumer spending). Or worse, the bank could go out of business. In order to mitigate against the risk of loans going bad, banks are required by their regulatory bodies to hold capital against their investments. For example, if it was assumed that loans on average default at a rate of 5% per year the bank may be required to hold in readily available assets (not illiquid securities such as retail mortgages) a total of 8% of the value of their book. To a bank seeking the maximum possible return on their capital to keep shareholders happy this regulatory capital is dead money. Any means for reducing this amount is most welcome.

Unfortunately investments such as loans to corporate entities, mortgages to individuals, automobile loans, credit card receivables, home equity loans etc. are very illiquid assets. There is no secondary market for actively trading individual loans in the same way that there is for trading, for example, shares in IBM. It is difficult therefore for the bank to do anything with these assets. This is where the concept of securitisation enters. The basic concept of securitisation is to bundle up large numbers of the illiquid securities (for example pooling many thousands of

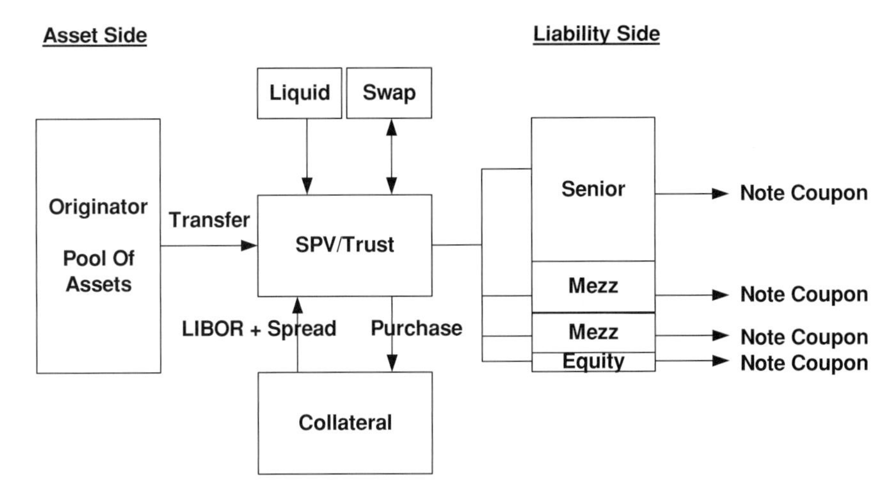

Figure 1.1 Securitisation of a pool of illiquid assets into tradable securities via the mechanism of an SPV. See the text for a full discussion.

mortgage commitments to individual domestic customers) into a new 'super' security. Figure 1.1 shows this schematically.

In this figure we have on the left-hand side the originator of the transaction (for example the bank). Let us assume the originator has a pool of illiquid assets which they own and wish to securitise. For example this might be a large number of corporate loans which are currently sitting on the balance sheet eating up regulatory capital. To securitise these assets the originator will physically transfer the ownership of these assets to a bankruptcy remote special purpose vehicle (or SPV, sometimes also referred to as the Trust). The SPV in essence purchases the assets from the originator. The funds for this are provided by the note investors, as described below, because the SPV has no funds of its own. From the originator's point of view the future (and potentially uncertain) cashflows from the assets have been transformed into an immediate cash payment, which can be beneficial to the originator's liquidity. The value of this cash payment is presumably the fair value of the expected future cashflows. The fundamental problem in mathematical finance is to develop realistic models for estimating the value of these future cashflows.

The SPV is a separate entity and most importantly is bankruptcy remote from the originator. This means that if some of the assets in the pool default, it will have no impact upon the originator (since these assets no longer sit on their balance sheet). Conversely, if the originator itself defaults it has no impact upon the SPV (and the notes that the SPV issues). Because the assets have been physically transferred the originator no longer has to hold regulatory capital against them, thereby freeing up the aforementioned '8%' for further investment in new business opportunities. Regulatory capital relief was one of the initial motivations behind securitisation.

The effect of this transfer of assets upon the underlying collateral (the corporate loans or individual mortgages) is minimal; the loans still have to be serviced, meaning that the SPV receives coupon payments (typically LIBOR plus a spread) and principal from the loans. However, it is the SPV (not the original owner) that will now be sensitive to any interruption to these cashflows due, for example, to defaults in the underlying pool. To facilitate all this, the role of the servicer (often the originator) in Figure 1.1 is to manage the collection and distribution of payments from the underlying pool (distributed to where we will now describe).

So far the discussion has focused on the 'asset' side of the structure. We now discuss the 'liability' side and introduce the concept of tranched exposures. The assets in the pool pay their owner income. The assets in turn can be used to fund further debt obligations, i.e. bonds or notes. The next step in the securitisation process is to sell the rights to the cashflows that the SPV is receiving (using these asset cashflows as security for the new debt to be issued). However, rather than selling the rights to individual cashflows or loans, the SPV sells exposure to a particular slice, or *tranche*, of the aggregate cashflows from the entire pool. For example, if the collateral is composed of 100 loans each of \$10 m then the total notional amount of loans issued is equal to \$1 bn. Each individual loan will pay a coupon of LIBOR plus a certain spread. The originator slices up this capital into a series of notes of sizes (notional amounts) \$800 m, \$100 m, \$70 m and \$30 m (for example). Each of these notes pays a coupon of LIBOR plus a spread based on the (aggregated) notional of that note. For example, the note with a notional of \$800 m may pay an annual coupon of 30 bps over LIBOR quarterly. Hence each coupon payment is (roughly) equal to $\$800\text{ m} \times (\text{LIBOR} + 30\text{ bps}) \times 1/4$. The investors in the notes pay the principal upfront, which is used to fund the purchase of the assets in the collateral pool, in return for receiving the periodic coupons and principal redemption at maturity. The risk, of course, to the investors is that the assets on the asset side do not deliver the expected returns (due to default, prepayment etc.).

The tranches are named according to their place in the capital structure and the legal seniority that the notes associated with the tranches have in terms of distribution of payments from the SPV. The most senior tranches have the first legal claim to the aggregate cashflows from the collateral pool and are referred to as the 'senior' tranches. The next most senior tranche has the next claim (typically the tranches in the middle of the capital structure are referred to as 'mezzanine' or mezz), all the way down to the most junior note at the bottom of the capital structure which is referred to as the equity tranche (or residual or first-loss piece). In the example shown in Figure 1.1 the capital structure has a senior tranche, two mezz tranches (typically referred to as junior and senior mezz) and an equity tranche. The (notional) sizes of the tranches are arranged so that the senior tranches have the largest notional and the equity tranche has the smallest amount ($800 m and $30 m respectively in the example given).

In general the income from the collateral pool is allocated down the capital structure starting with the most senior notes and working their way down to the most junior. Losses on the other hand are allocated from the bottom up. For example, if one of the assets in the pool defaults and 40% of the notional amount is recovered (leading to a loss of $10 m × (100%–40%) = $6 m) it is the equity tranche that is impacted first. This results in a reduction of the notional amount of the equity tranche from $30 m to $24 m, reducing the payments that the equity note holder receives. In addition to this, going forward the asset pool now has less collateral and will therefore make fewer coupon payments. This leads to less cash being fed into the top of the capital structure, meaning less for the junior note investors once all the senior liabilities have been met.

The tranches are also rated by an external rating agency such as Moodys, S&P or Fitch. One of the upfront costs of securitising a pool of assets is the fees paid to the rating agency to provide a rating for the issued liabilities. The rating of a note is determined by the level of losses that can be sustained by the collateral on the asset side before the note cashflows on the liability side are impacted. Obviously the equity tranche is immediately impacted by losses and is therefore the riskiest tranche. For this reason it is typically unrated, and is often held by the originator of the deal (as a sign of confidence to investors that the assets in the underlying pool do not represent a moral hazard). To compensate the equity tranche holder for the enhanced risk they are taking on, the spread on this note is typically much larger than that on more senior tranches.

More senior tranches have a greater layer of protection (subordination) and so warrant higher ratings. It is important to note that a pool of assets that individually have poor ratings can, when securitised (with a priority of payments from senior to junior liability), result in new notes which have substantially better credit quality. This immediately broadens the appeal of the notes issued by the SPV to a whole range of new investors. For example, pension funds may be prohibited from investing in assets that are rated BBB due to their default risk (but which have a substantially enhanced yield compared to say AAA rated assets making them attractive to investors who are prepared to take on the risk). But a pool of BBB assets that are securitised and reissued as a series of notes including an AAA rated one is a different matter (the AAA rating being awarded based on the level of subordination that this note has relative to more junior notes). If the original BBB rated assets perform well then the pension fund benefits from this; on the other hand if the BBB rated assets do not perform well and default, the subordination provided by the equity and mezz tranches insulates the AAA notes from this. Everyone's a winner. That is, of course, unless large fractions of the underlying collateral start to default. For example, if all the underlying collateral were composed of US subprime mortgages which suddenly reset from a low teaser rate to 10%, this might have an impact on the returns of the notes.

One practical consideration of importance is the actual process of building up the collateral on the asset side. It is unlikely that an SPV will simply be able to go out and buy all of the collateral at a single instant in time. It is much more likely that the collateral pool will be assembled over an extended period as and when suitable assets that the manager of the structure deems fit to include in the pool become available. This is known as the ramp-up period and can last for several months. This represents a potential risk to the manager as they have to purchase and warehouse all of these assets until the structure is ready to sell on to investors. During the ramp-up period market conditions can change adversely, leading to the manager holding collateral which is not as attractive as initially anticipated. A solution to this ramp-up problem is provided by the use of credit derivative technology to construct the exposures to the assets synthetically, without actual physical ownership (more on this later). Another practical difficulty with the process described so far is that there is unlikely to be much standardisation amongst the type of collateral in the underlying pool. This means that for the types of structure described there is unlikely to be a highly liquid secondary market.

Finally there are two other components of the securitisation structure that need explanation. The role of the swap counterparty in Figure 1.1 is to provide a macro hedge against interest rate and FX rate fluctuations. There is also a liquidity provider. One of the less obvious risks of the structure described is mismatches in the timing of cashflows. For example, all of the assets on the asset side may pay coupons semi-annually, but the notes issued by the SPV may be quarterly. This would lead to short-term liquidity problems for the SPV in meeting its liabilities. To provide protection against this the liquidity provider (which may for example be the originating bank) will give the SPV lines of credit that it can draw down on, on an as-and-when needed basis.

1.3 Credit derivative products

In the previous section we described in quite general terms securitisation and tranching. In this section we discuss the application of these concepts to cashflow and synthetic CDOs. We also briefly describe some of the other important credit derivative products in the marketplace. More detailed business and economic descriptions of many of the products described in this section can be found in, for example, Gregory [2003], Das [2005], Chaplin [2005] and Chacko *et al.* [2006].

1.3.1 Credit default swaps (CDSs)

CDSs are the simplest example of a single-name credit derivative [Gregory 2003, Das 2005, Rajan *et al.* 2007]. The principal motivation of a credit derivative is to transfer credit risk (risk of default on outstanding obligations of a specified reference entity) between investors. A credit derivative will therefore usually have three counterparties to the trade: the counterparty wishing to purchase protection, the counterparty willing to sell protection and the reference entity to whom the bought and sold protection refers. For example counterparty ABC may own bonds issued by a separate reference entity C. ABC might be concerned about C defaulting (meaning ABC would receive no further coupons or its principal back if C did default) and may want to purchase protection against this risk. This protection is purchased by entering into a bilateral trade with counterparty XYZ who is willing to provide protection in return for a fee. A CDS provides the legal and financial mechanisms to achieve this transfer of risk.

Reference counterparties in the CDS market can include corporate entities as well as sovereign states (allowing protection to be purchased against a sovereign defaulting on its debts – this sort of protection is particularly popular for sovereigns in emerging markets where geopolitical risk can be a significant factor). The type of reference obligor asset that protection is bought or sold on has also evolved over time. Originally CDSs referenced the plain bonds of the reference asset. This has grown to include leveraged loans (LCDS) as well as asset backed securities (ABSCDS) as the underlying assets. CDSs are usually quoted on a spread basis, which is the coupon rate that is applied to the periodic protection payments. The par CDS spread is the spread (given the prevailing market conditions) which gives a fair value of the CDS at contract inception of zero. Protection is purchased for a specified period of time. During this period the protection purchaser makes periodic fee payments to the protection seller. These payments continue until the reference entity defaults or the protection period expires. If the reference entity defaults, subsequent coupon payments cease and the protection seller makes a contingent payment to the protection purchaser to compensate them for any loss. The contingent payment is a fraction of the notional amount of protection purchased. The fraction is termed the recovery rate and is determined in the market (by a dealer poll) at the time of the default.

As the credit derivative market has grown the uses of CDSs have evolved. They are now used as much for speculation and relative value trading (playing the default risk of one obligor off against another) as for providing long-term protection against the risk of a particular obligor defaulting. One of the important developments has been the growth of the market for trading protection over different time horizons. Initially, protection was purchased for a period of, typically, five years. As the market grew, investor demand for different time horizons led to the emergence of contracts specifying protection for maturities ranging from a few months up to ten and more years. As with the bond market, this introduced an additional degree of freedom that investors can express a view on: the likelihood of default over a certain time horizon. For example, a corporate that is subject to a private equity buy-out might be viewed by the market as having a higher long-term default risk than short-term. This is because the buy-out may typically be financed by the corporate taking on long-term debt (two-thirds of the buy-out cost is normal). Its liabilities in the short term are therefore less onerous than in the long term. Conversely, a whole sector may be perceived as having significant short-term default risk. For example, banks experiencing short-term liquidity problems might be viewed as a short-term risk, but not long term (if they survive the short term, they will go from strength to strength).

Having a term structure of CDSs also allows for investors to implement trading strategies based on the relative dynamics of different CDS maturities. This is analogous to what is observed in the interest rate market where interest rates are set for borrowing over a specific time horizon. Examples include so-called curve steepeners and flatteners [Rajan *et al.* 2007] where opposite trades are placed at different ends of the term structure of par CDS spreads.

Variations on the basic CDS trade have also appeared over time. Some of these variations are now briefly described.

1.3.1.1 Forward starting CDSs

A forward starting CDS is a CDS where the protection (purchased or sold) is specified to begin at a future point in time.

1.3.1.2 Credit default swaptions

Options on CDSs, or CD swaptions, are an important class of credit derivative because they allow investors to speculate on the volatility of CDS spreads. A CD swaption gives the holder of the option the right to enter into a CDS at a future date if the prevailing par spread at that time is such that the option is in the money. CD swaptions can in principle be of European, American or Bermudan exercise variety [Hull 1999, Wilmott 2000]. More details about the mechanics and strategies for trading CD swaptions may be found elsewhere [Rajan *et al.* 2007].

1.3.1.3 Recovery rate plays

For a plain, vanilla CDS the protection purchaser receives a payment upon default of the recovered amount of notional (assuming cash settlement for the moment – the different settlement mechanisms will be discussed in Chapter 3). The amount of notional recovered is a function of the prevailing market conditions at the time the payment is due. It is usually determined by a dealer poll (taking an average of the quotes received, having stripped out the highest and lowest quotes). Although this process is (relatively) transparent it does introduce an uncertainty into the amount that the protection purchaser will actually receive since the quotes provided by dealers will depend on a lot of different factors. Recovery rate strategies can be used to express outright views on recovery rates or to fix recovery rates at a desired level.

As the name suggests, the *recovery rate lock* is a contract that enables investors to 'lock-in' a specified recovery rate. The recovery rate lock was released in May 2006 as a specific contract by ISDA and is a product for trading views on recovery rates. Prior to the recovery rate lock, two separate CDS contracts (a standard CDS and a digital CDS – where the actual recovery rate paid in the event of a default is fixed at contract inception) were needed to express a similar view on recovery rates. The economics of the recovery rate lock and dual CDS position are the same, but the recovery rate lock is a single contract. In a recovery rate lock there is an agreed fixed recovery rate set at contract initiation. There are no upfront payments or periodic coupon payments. The only cashflow in the contract is the payment in the event of a credit event. The contract is physically settled. The lock buyer is the protection seller; they want the recovery rates to increase. The lock seller is the protection buyer and they want recovery rates to decrease.

1.3.1.4 Constant maturity CDS (CMCDS)

In a vanilla interest rate swap product the two counterparties swap a periodic stream of cashflows. One stream of cashflows may be based on a fixed coupon rate, and the other on a coupon rate which is a function of the prevailing LIBOR rate. A variation of this is a constant maturity swap (CMS). In a CMS the coupon rate for the floating leg payment is based on the prevailing par swap rate of a swap with a constant maturity at each coupon date [Hull 1999]. For example, a CMS may have a coupon based on the 10 year swap rate. At the first coupon date, say after 3 months, the par swap rate of a swap with a maturity of 10 years (from the 3 month point) will be determined and used as the coupon rate for the next floating leg. At the next coupon date at 6 months, the par swap rate for a swap of maturity 10 years (from the 6 month point) is determined and is used as the coupon rate for the next floating leg payment, and so on.

A CMCDS is a similar concept [Pedersen and Sen 2004, Das 2005, Rajan *et al.* 2007]. However, instead of a par swap rate, the rate that is determined is the par CDS spread for a CDS of a specified (constant) maturity. The coupon spread that the protection purchaser pays therefore changes (resets) at each coupon payment date. The CMCDS references an obligor's debt just like a normal CDS (popular CMCDS trades reference sovereign debt) or can reference a credit derivative index.

If there is a credit event during the lifetime of the contract, the contract terminates like a CDS with either cash or physical settlement.

The buyer of protection pays periodic coupons based on the current par spread (determined at each reset date). At the contract inception a participation rate (less than unity) is agreed. The participation rate is the multiple of the prevailing par spread which is actually paid at each coupon date. The buyer of protection is taking the view that the par CDS spread on the credit will increase by less than the spread implied by existing forward rates. If the spread remains low, then a low rate will continue to be paid at each reset date (and the protection buyer receives protection for a cheaper rate than would be currently available in the market). If the spread is higher than the expected forward value, the buyer of protection will have to pay a larger coupon for the protection. The initial participation rate reflects the steepness of the credit curve at contract inception. If a CMCDS position is combined with an offsetting CDS referencing the same obligor, then the overall position is default neutral. However, the spread payable allows investors to take curve and directional spread exposures to the obligor having isolated and removed the default risk.

1.3.2 Default baskets

A natural extension of a vanilla CDS is an instrument which provides protection against not one but a basket of obligors. Default baskets provide this [Chaplin 2005]. For example, an automotive manufacturer relies on different suppliers to provide the raw materials from which cars are constructed. The automotive manufacturer may want to purchase protection against any of the suppliers defaulting. This could be achieved by purchasing individual CDSs referencing each of the suppliers. Alternatively a default basket could be constructed composed of all the suppliers, which paid out on the first default of any of the individual constituents. Default baskets are leveraged positions with respect to a single-name CDS (they generate a higher yield than an equivalent – in notional terms – position in a single-name CDS). This is because the investors are exposed to a specific fraction (e.g. second default) of the default risk of the pool of obligors. Basket sizes are usually small (of the order of 3–20 obligors) with clip sizes typically $10–50 m. They are usually unfunded products entered into by both counterparties at par (meaning that no initial exchange of cashflows takes place). Funded, or note variants (where the investor pays a principal amount upfront which they receive back at maturity or upon termination of the contract) are also traded, but these are not as common.

Higher-order extensions to default baskets, such as second-to-default, are also possible variations on the basic theme. Some baskets can also be extendable/cancellable and with digital payoffs. However, by far the most common form of contract is the first-to-default basket. Default baskets also tend to be bespoke instruments. That is, each default basket provides protection on a unique pool of obligors. This is because baskets are often used to provide protection against particularly worrisome credits in a larger synthetic CDO. For example, a hedge fund may sell protection on an equity tranche of a bespoke, synthetic CDO. If they have particular concerns about a small group of obligors within the pool (for example, obligors in the banking sector), they may consider buying protection on a small basket (a typical basket size is five obligors) referencing these obligors. This provides protection against default of these names, but at the expense of reducing the positive net carry for the hedge fund (since the basket protection requires some funding).

1.3.3 Credit indices

Credit indices are another natural extension of the CDS concept [Rajan *et al.* 2007]. Like default baskets they are a product that provides exposure to a portfolio of assets, in this case the CDSs of a reference portfolio of obligors. The crucial feature of credit indices is that they are constructed, traded and managed according to transparent and standardised rules. This degree of standardisation has proven absolutely pivotal in the development of the credit derivative market. The first-generation credit products were highly bespoke, requiring not only traders and structurers to make the deal, but also legal and documentation experts to ensure that there was a clear understanding between the counterparties as to what constituted a default event, what suitable *pari passu* deliverable obligations were etc. This meant that it was not possible to transact large volumes of deals. Standardisation of trades obviously is a key facilitator of liquidity.

The credit indices in their current incarnation were introduced in 2004. There are two broad families of indices: the CDX family based on North American investment grade corporates and the iTraxx family based on European and Asian investment grade corporates. The indices are managed and administered by an independent calculation agent, MarkIt [MarkIt] who provide the transparent construction and trading mechanics of index products.

Unlike a single-name CDS which terminates on the occurrence of a single default, an index CDS does not terminate. Instead the protection seller compensates the protection purchaser (as in a normal CDS), but the contract then continues with the defaulted obligor removed from the pool. The protection purchaser continues to make coupon payments with the same coupon spread, but based on a reduced notional amount (the reduction corresponds to the notional amount of the defaulted obligor). This continues to the overall maturity of the contract.

Index products are very good for rapidly expressing views, both long and short, as to the macro state of the credit markets. They are also relatively effective for hedging other credit derivative exposures. Index products are followed not just by those actively engaged in the credit markets, but also by market participants in other sectors. Specifically, the common indices that are used to gauge the mood of the structured finance and credit markets include the following:

- the iTraxx and CDX indices (main and crossover indices) to gauge corporate credit quality;
- the ABX indices (see Chapter 11) to gauge ABS referencing home equity collateral quality;
- the LCDX (an index of 100 LCDS contracts) to gauge the corporate loan market.

1.3.4 Collateralised debt obligations (CDOs)

CDOs and synthetic CDOs in particular are the most prominent and important example of multi-name credit derivatives. CDOs were first developed in the 1980s as a financing vehicle that allowed an owner of a pool of assets to finance the purchase of a pool of assets on a non-recourse basis at tighter spreads than were paid by the underlying assets [Gregory 2003, Das 2005, Kothari 2006].

From the original bespoke deals that were arranged individually between originator and investor (see the above references for examples of early deals and their economic rationale) they have evolved into a sophisticated and liquid asset class in their own right, particularly the single-tranche CDO (STCDO) market. The introduction of standardised credit derivative indices and subsequent trading of tranched exposures to portions of the portfolios' losses has resulted in the appearance of correlation as a new asset class to trade (although correlation is only the mechanism traders use to communicate prices and assess the relative value of different instruments, the fundamental tradable is still price).

A CDO is a specific example of the general securitisation structure shown in Figure 1.1. To characterise a CDO we need to specify the following (in addition to standard contract conventions such as holiday calendars, day-count conventions, calculation agents etc.):

- the composition of the underlying asset pool (type of asset, maturities, notional amounts, coupons etc.);
- the capital structure of the issued liabilities (the tranche structure);
- the maturity of the CDO;
- the priority of payments to the investors (the cashflow waterfall);
- the allocation of losses on the asset side to the investors on the liability side;
- whether the deal is managed or static (can names be substituted in and out of the underlying pool).

By an appropriate choice of these parameters the issuer and investor can tailor the benefits of the structure to fit their risk/return requirements. A big motivation for structuring and issuing CDOs is regulatory capital relief and ratings arbitrage. Typical regulatory regimes require a CDO investor to hold 100% capital against equity tranche exposure and of the order of 8% against exposures to other tranches. This is to be compared with the requirement to hold 8% against the entire pool of assets which has been securitised. The one-for-one holding of capital against equity exposures reflects the enhanced risk of the equity tranche which is a highly leveraged exposure to the overall pool of assets (it is leveraged because it is a thin slice of the overall portfolio losses, meaning the holder of the tranche is highly exposed to the first losses in the underlying pool). This leverage enables the originator to enhance their yield on the underlying assets by securitising the assets and purchasing back the equity tranche. This facilitates ratings arbitrage.

In addition to this form of arbitrage, it is also possible that the securitisation and tranching process can lead to a beneficial spread differential between the asset and liability sides. For example, the (notional) weighted average spread of the liabilities may turn out to be less than the spread received from the assets. The equity tranche holder can therefore earn a positive carry simply by arranging the transaction. In order to achieve this, however, the assets have to be chosen carefully (in terms of their ratings etc.) as does the structure of the liabilities. Why might this be possible? One possibility is that CDO liquidity is less than that of the corporate bond market, meaning that CDO investors will require an enhanced spread to compensate for the liquidity risk. CDOs are also leveraged instruments meaning that mark-to-market volatility is higher compared to straight corporate bonds, again leading to investors demanding an increased spread to compensate for this.

1.3.4.1 Cashflow CDOs

A cashflow CDO has the following general characteristics [De Servigny and Jobst 2007, Rajan *et al.* 2007]:

- an asset pool drawn from a potentially diverse range of collateral (see below);
- a tailored, highly bespoke capital structure;
- a long legal maturity;
- a detailed cashflow waterfall that sets out how the interest, principal and recovery amounts generated by the asset pool are propagated through the issued liabilities from the top of the capital structure down to the bottom (including appropriate collateral and interest coverage tests to protect senior note holders from losses in the asset pool);
- a detailed specification setting out how losses on the asset side are propagated up from the bottom of the capital structure;
- typically cashflow CDOs are actively managed deals where the portfolio/asset manager can choose to substitute collateral dynamically in and out of the asset side. For the benefit of their investment expertise and market insight the manager will charge a fee. The indenture of the deal will set out the terms and conditions for substitution and reinvestment.

The securitisation example discussed in the previous section refers primarily to a cashflow CDO. Cashflow CDOs are a natural extension of asset backed securitisation (ABS) technology. The most common cashflow CDOs are constructed from underlying collateral which is the following:

- corporate loans (a collateralised loan obligation or CLO);
- corporate bonds (a collateralised bond obligation or CBO) although the popularity of these products has diminished somewhat;
- senior or mezzanine tranches of existing asset backed securities (ABS), known as ABSCDOs. Suitable ABS collateral can include residential or commercial mortgages, home equity loans, automobile loans, student loans, credit card receivables etc. (even future gate receipts from a football stadium can be securitised in order to raise cash to finance other projects – the risk to the investors being that the team is not successful);
- combinations of all the above collateral types.

The complexity of cashflow CDOs arises principally from the diversity of collateral that can be incorporated in the underlying pool (as described above). An additional complexity arises if a note issued by the master CDO is backed by collateral which is itself securitised providing tranched exposures to other assets, which in turn might be securitised and tranched etc. That is to say, there may exist multiple layers of securitisation and tranching within the overall CDO structure. CDO managers and investors need to understand very clearly what the underlying collateral they are purchasing exposure to is. It is very easy to purchase assets which have deeply embedded exposures to all sorts of unwanted asset classes. This also raises the possibility that the same underlying assets might appear multiple times in different pools all referenced by the master CDO.

Therefore, on the one hand CDOs, being portfolio products, provide investors with diversification (compared to a position in, for example, a single corporate bond). On the other hand, if the same assets appear in multiple securitisations the degree of diversification might be far less than expected or desired. Indeed if there are a significant number of assets which appear in multiple pools all referenced by the same master CDO, 'cliff' risk becomes important. This refers to the possibility that a master CDO might have, on paper, a large degree of subordination and thus appear to be well protected against default. This may indeed be the case. On the other hand if the same assets appear in different pools, it may be that it only takes one or two particularly highly connected obligors to default before all the subordination is rapidly eaten away. As a consequence of this, a very important aspect of risk managing cashflow CDOs (including ABSCDOs) is constant monitoring of the composition of the underlying collateral to identify hot-spots of asset-class concentration (drilling right down as far as reasonably possible to the atomic level of individual exposures).

Another complex issue in the management of cashflow CDOs is the timing of cashflows. If, on the asset side, there is a rich mixture of different types of collateral, it is highly likely that the different assets will have coupons which pay with different tenors leading to a cashflow profile that is relatively stable over time. In an extreme example, however, all of the collateral on the asset side might pay coupons semi-annually. On the other hand, the issued liabilities may require quarterly payments. This means that there is a funding mismatch between the cash the SPV has to pay to meet its liabilities, and the cash it receives from the asset side. This is an issue that has to be monitored on a daily basis to ensure that the SPV has enough liquidity from its long-term investments to meet its short-term liabilities. To mitigate against this risk, cashflow CDOs usually have access to short-term revolving credit facilities (from banks) to provide emergency liquidity.

Other types of risk that cashflow CDOs are sensitive to stem principally from interruptions to the stream of cashflows on the asset side of the structure. Interruptions can occur due to defaults amongst the collateral or amortisation of the asset notional due, for example, to prepayments of loans (residential or commercial mortgages, automobile loans, home equity loans etc.). When interruptions to the asset cashflows occur this impacts the rated liabilities because there is now less cash being fed into the top of the waterfall.

1.3.4.2 Synthetic CDOs

Cashflow securitisation physically removes assets from the originator's balance sheet. In some cases, however, selling assets can be detrimental to a relationship with a client. Selling a corporate loan to an SPV is hardly a sign of confidence or respect for that corporate, or the value the originator places on doing business with them. Synthetic securitisation can alleviate this problem by generating the exposure to the obligors synthetically, without the physical transfer of ownership of assets. Synthetic CDOs utilise credit derivative technology to achieve this [Joannas and Choudhry 2003, De Servigny and Jobst 2007, Rajan *et al.* 2007].

Synthetic CDO issuance was, like cashflow CDO issuance, initially driven by regulatory capital relief motivations (capital relief is achieved synthetically via the use of credit derivative technology). The next evolution of the market was full capital structure synthetic CDO deals (which in turn helped the development of the CDS market). These positions required relatively simple post-trade risk management. Single-tranche CDOs were the next step in the evolution of the product. These products allowed dealers to retain and hedge individual slices of the capital structure that could not be placed in the marketplace. These products were also attractive to new types of investors, in particular hedgies wanting leveraged exposure to a particular slice of the tranche.

A synthetic CDO (also sometimes referred to as a collateralised swap obligation or CSO) has the following general characteristics.

- An asset pool composed solely of vanilla CDSs each referencing an individual obligor (the pool can be bespoke or standardised).
- In some cases a tailored, highly bespoke capital structure; in other cases (STCDOs) the capital structure is standardised.
- A relatively short legal maturity and matching of the maturities of the assets (CDSs) and liabilities (notes).
- A simple cashflow waterfall structure. Compared to cashflow CDOs, synthetic CDOs have very simple cashflow mechanics, only requiring the management of default losses (and not having complex waterfall structures). Each note pays a coupon based on the remaining notional of the note's tranche.
- Losses on the asset side are propagated up from the bottom of the capital structure (and also sometimes the recovered amounts are amortised from the most senior tranche downwards). Each loss reduces the tranche notional until all of the tranche's notional is exhausted. After this point subsequent losses have no impact upon the tranche.
- Cashflow mechanisms can be on an all-upfront basis (the only coupon cashflow is at contract inception), an all running coupon basis (there is no upfront payment, but there are periodic coupon payments) or a mixture of the two (combining an upfront payment with periodic coupons).
- The composition of the asset pool can be either static or dynamic, depending principally upon who the investor is (hedge funds for example are big users of managed tranches since it is their business to invest funds and dynamically manage the investment to achieve the best return, whereas investment banks tend to be users of static tranches to hedge bespoke positions).

Synthetic CDOs are created by applying credit derivative technology to portfolios of assets. They are portfolios of CDSs whose cashflows are tranched and these tranches are sold to investors. They facilitate the transfer of risk synthetically, without physically transferring the assets off the originator's balance sheet (thereby removing the potential problem with client management). In terms of regulatory capital relief, synthetic CDOs do act as a hedge for some of the credit risk. This allows some capital to be reduced, but the reduction is not as great as for cashflow CDOs where the assets are physically taken off the balance sheet. Because the assets are not physically purchased the ramp-up, availability of collateral and warehousing issues experienced with cashflow CDOs are greatly alleviated, although not removed completely.

The simplest rationale for investing in synthetic CDOs is to express a view on the default risk of the obligors in the underlying portfolio (purchasing or selling protection accordingly). A counterparty who is concerned about default risk can purchase protection from another counterparty willing to sell them the protection.

Conversely a yield hungry hedge fund (whose portfolio managers get paid to take risks and outperform the market) might be willing to sell protection on a portfolio of obligors if their view of the default risk is sufficiently sanguine. If the deal is also managed it gives the manager additional freedom to dynamically replace poorly performing assets in the pool. Other uses for synthetic CDOs include relative value trading (the CDO may have a greater yield than an equivalently rated bond), expressing leveraged views (both spread leverage and leveraged exposures to the slices of tranche losses) and hedging of other exposures.

Initially synthetic CDO issuance was small compared to that of cashflow CDOs. However, as the credit derivative market (for CDSs) has grown, synthetic CDO issuance has outgrown that of cashflow CDOs. Part of the growth of the CDS market has in fact been driven by the synthetic CDO market which needs large volumes of CDSs for hedging purposes. The synthetic CDO market is very dependent on the CDS market for its operation.

1.3.4.3 Single-tranche CDOs (STCDOs)

One of the disadvantages with bespoke CDOs (either cashflow or synthetic) is that it takes time to ramp up the underlying collateral pool. This is even true for synthetic CDOs despite the fact that the exposure to the underlying obligors is generated synthetically via credit derivatives. The problem is that the protection must still be purchased or sold, meaning there needs to be a counterparty on the other side of the trade willing to do the deal at an acceptable market level, and that all of the tranches must be placed into the market near simultaneously. The protection seller then has to hedge the relevant exposures separately.

An elegant solution to this problem is provided by the standardised indices which have clear and transparent rules for construction, modification and trading (standardisation of course boosts liquidity by making trading between counterparties easier). The composition of the indices is also static through time (except at the roll dates when the portfolio is updated). Tranched exposures on these indices have standardised

- pool composition (the index),
- attachment/detachment points,
- maturities.

An index tranche provides investors with exposure to a slice of the loss distribution of the credit index portfolio. Greater standardisation of product has led to greater liquidity since investors are familiar with the terms and conditions of the deal leading to less difficulty in trading and encouraging more trading etc. An STCDO trades only a single tranche of the full capital structure. STCDOs therefore have

Table 1.1 *Standardised tranches defined on the iTraxx main index*

Tranche	Attachment point (%)	Detachment point (%)	Attachment point ($ m)	Detachment point ($ m)	Tranche notional ($ m)
Equity	0	3	0	37.5	37.5
Junior mezz	3	6	37.5	75	37.5
Senior mezz	6	9	75	112.5	37.5
Senior	9	12	112.5	150	37.5
Super senior	12	22	150	275	125

Table 1.2 *Standardised tranches defined on the CDX main index*

Tranche	Attachment point (%)	Detachment point (%)	Attachment point ($ m)	Detachment point ($ m)	Tranche notional ($ m)
Equity	0	3	0	37.5	37.5
Junior mezz	3	7	37.5	87.5	50
Senior mezz	7	10	87.5	125	37.5
Senior	10	15	125	187.5	62.5
Super senior	15	30	187.5	375	187.5

the advantage that they allow a very targeted exposure to the portion of the capital structure that the investor wishes.

Standardised tranches are defined on both the iTraxx and CDX main indices. The attachment/detachment points are shown in Tables 1.1 and 1.2. Each obligor in the index has a notional of $10 m and there are 125 obligors in each index. To hedge against 'small' spread movements, index tranches trade with an opposite index position (of size delta $\times$ tranche notional). The price that is quoted is known as the delta exchange price (see Bluhm and Overbeck [2007] for an example of the mechanics of delta exchange).

STCDOs based on the standardised indices are almost always static, where the underlying asset pool composition does not change over time. Investment banks are big users of STCDOs for hedging bespoke exposures. The bespoke exposures are typically with respect to investors such as hedge funds who will want to construct bespoke STCDOs to express a view on the behaviour of the credit markets. Bespoke synthetic CDOs enable customisation in terms of

- structure of protection payment coupons (fixed, floating or zero-coupon),
- funded (note)/unfunded (swap) form,
- static or managed portfolio and the composition of that portfolio,
- custom yield/leverage,

- tranche size and subordination,
- principal protected structures (to ensure that investors receive their original investment back).

If the investment bank is on the other side of the trade they will want to hedge their exposure. Slicing up the indices and trading only specific slices of the risk solves the problem of ramping up the underlying portfolio. However it introduces new complexities:

- correlation sensitivity (later chapters will describe in detail how correlation impacts STCDOs, but for the moment we simply observe that STCDOs facilitated the introduction of correlation as a new financial variable to trade);
- how to hedge the exposure to the bespoke tranche.

This latter point is one of the most important issues in the STCDO market. One of the major (if not the major) modelling challenges in the synthetic CDO world is determining how to value bespoke tranches consistently with standardised tranches (the bespoke may only differ from the standard tranche by a few obligors in the underlying pool, non-standard attachment/detachment point etc.). If the bespoke tranche had a set of properties that were equivalent to an index tranche, then it would be easy to hedge the position. However, it is unlikely this will be the case. Therefore to hedge the bespoke exposure the investment bank must try and 'map' the bespoke to an appropriate index position. Prices of bespokes are mapped to index products because these are the only readily observable prices in the marketplace.

In terms of risks, synthetic CDOs do not have the prepayment type of risks that are common to cashflow CDOs since the only early termination mechanism for a CDS is default. The main risks to synthetic CDOs arise from the spread movements of the underlying CDSs as well as the possibility of outright obligor default. In addition to this STCDOs introduce another very important risk sensitivity: correlation (because a STCDO exposes the investor only to a certain fraction of a pool's loss exposure and the pool's loss distribution is heavily dependent on the correlation between the constituents).

Cashflow CDOs are typically buy and hold investments, where the exposure is usually long only (i.e. the investor buys the notes issued by the SPV). Synthetic CDOs on the other hand, being highly liquid, allow for investors to take long and short positions (selling or buying protection respectively). They also allow for unleveraged (index) positions or leveraged (tranche) positions to be taken. The synthetic CDO market principally consists of a flow business based on the trading of standardised index tranches. These trades are used for relative value trading as well as for hedging certain exposures. More importantly there is trading in bespoke synthetic CDO tranches. These bespoke tranches may have non-standard attachment/detachment points, maturities or underlying pool composition (or all three).

1.3.4.4 Options on index tranches

Recently we have seen the introduction of derivatives which have synthetic CDOs as their underlying security. Examples include forward starting CDOs and options (European, American and Bermudan) on synthetic CDOs. Options on tranches are options on the spread of the tranche and allow investors to trade the volatility of the tranche spreads. A put is defined as the right to buy protection (selling the risk to another counterparty), and a call is defined as the right to sell protection (buying the risk).

1.3.4.5 Tranchelets

Tranchelets are very thin tranches on a given portfolio. Because the tranche width is smaller than usual, tranchelets provide an even more granular and leveraged exposure to the portfolio losses.

1.3.4.6 Zero-coupon equity tranches

In a zero-coupon equity tranche there are only two cashflows: one at trade inception and one at maturity (it is therefore an all-upfront cashflow mechanism). This is similar to a zero-coupon bond. The payment at maturity depends only on the notional losses incurred by the tranche during the lifetime of the tranche (but does not depend upon the timing of the losses). The structure can provide a high internal rate of return with a payout whose timing is known (as opposed to a normal synthetic CDO structure where the timing of contingent payments is unknown). The high rate of return stems from the fact that the initial upfront payment prices in the full expected losses of the portfolio (upfront). If this amount is higher than the actual realised loss then the investor benefits at the maturity of the tranche.

1.3.4.7 Structured investment vehicles (SIVs)

A SIV is not technically a credit derivative, it is more correctly described as a structured finance vehicle. However, the technology and investment rationale for SIVs is quite similar to that described for CDOs so it is worth digressing for a moment to mention this type of vehicle. It is also true that SIVs have achieved a certain amount of notoriety due to their role in the credit crunch of 2007. A very good description of SIVs can be found in De Servigny and Jobst [2007].

Like an SPV, a SIV is a structured finance vehicle that is bankruptcy remote from the originating institution. Also in common with an SPV, a SIV issues rated liabilities (notes) and uses the proceeds from investors to purchase assets (collateral). The core investment rationale of a SIV is that it will purchase high-yielding, long-term assets and fund these purchases with short-term, lower yielding paper. The SIV manager will therefore be able to make a profit on the difference between the monies received from the assets and the monies paid to investors to service the debts. In essence the SIV is taking long-term exposure (risk) to high-grade assets and funding it with short-term debt. SIVs are typically buy-and-hold investment vehicles. Active trading of assets by the portfolio manager is only undertaken when assets mature and the proceeds need to be reinvested (and of course trading is undertaken when investors initially invest in the issued liabilities). SIVs usually do not have a well-defined lifetime, instead simply continuing to operate indefinitely (although so-called SIV-lites have a well defined legal maturity beyond which the SPV ceases to operate).

The collateral purchased will typically have high credit ratings (usually AAA to A with little BBB rated exposure). A broad range of collateral can be purchased, similar to the types of collateral that appear in cashflow CDOs. Some SIVs also have synthetic credit derivative exposure. Assets are typically well diversified according to type, rating, sector and geography. Strict concentration limits (for example, limiting the geographical concentration in any one region) are specified as part of the SIVs operating manual. The capital model (see the following paragraph) will impose penalties for breaches of these concentration limits. For example, if a rating concentration is breached the SIV will have to liquidate assets in order to re-establish the appropriate capital buffer.

Because the assets are highly rated, the default risk to the collateral on the asset side is expected to be low. When an asset matures the income stream from that asset disappears. Therefore it is necessary to replace the assets that are rolling off with new assets in order to ensure there is enough income coming in to service the liabilities. Similarly, short-term liabilities must be rolled in order to fund new asset purchases. SIVs dynamically manage the debt that they issue. In particular, existing debt can be rolled and new debt issued at a very high frequency (weeks). This provides a constant stream of paper. Typically the principal amounts raised from the liabilities will be used to fund the purchase of (longer maturity) assets on the asset side. Usually in a SIV there are two tranches of liabilities. Senior liabilities are rated AAA and are issued in several classes. Capital notes are the mezz piece and were originally one tranche. But the capital notes can also be tranched into a rated and unrated (first-loss) piece. SIVs therefore incorporate the securitisation and tranching technologies that are the basis of the CDO world.

An important role of the SIV manager (in addition to choosing the collateral to purchase and the liabilities to issue) is to manage the mismatch between assets and liabilities and the consequences of a temporary shortfall in liquidity. Because of this mismatch, temporary liquidity shortfalls must be managed with internal/external liquidity in the form of bank lines, breakable deposits, committed repos, put options and liquid assets. A SIV that is backed by a major financial institution is likely to be able to call upon that institution for funding in times of liquidity crises. Clearly one of the key risk management tasks for a SIV is to monitor the peak liquidity needs (i.e. the net of the liabilities that must be met and the income that will be received) over a specified time horizon in the future. These liquidity tests should be run daily and enable quantification of what proportion of the asset pool needs to be in liquid form (to cover any short-term liquidity problems).

The value of the collateral will be sensitive to the credit quality of the assets (ratings and credit spreads), recovery rates as well as interest rates and FX rates. It is important to note however that SIV exposures are usually hedged at an individual trade level. That is, for each exposure purchased for the asset side, the SIV manager will enter into an offsetting position to neutralise the SIV's overall position to day-to-day fluctuations in interest rates and FX rates. However, the value of the assets can of course decline over time irrespective of how they are hedged. For example, if the assets are ABS of home equity loans then rising defaults and delinquencies amongst the individual borrowers will negatively impact the value of the security. If the SIV needs to liquidate assets in order to cover short-term liquidity requirements, the mark-to-market value it receives for them may be less than what is needed to repay the liabilities. Therefore an SIV needs to be equipped with equity to ensure it can meet its liabilities in the event of credit losses and mark-to-market losses. This is achieved by issuing capital notes which are first-loss in nature (so that the note holders absorb the expected losses from the asset pool). The size of the first-loss piece should be sufficient to cover the expected losses of the assets.

Because of the SIV's need to continuously roll the notes it issues, the liquidity of the short-term commercial paper (CP) market is a key factor in determining the SIV's operational effectiveness. The CP market typically consists of notes with maturities of 60–90 days. In a perfect world there would be no defaults and there would be an endless supply of CP (hence when an asset reached its maturity, it could be immediately replaced with a similar asset and short-term paper could be continuously rolled). Under these circumstances there would be no need to keep resources to cover liquidity and asset shortfalls. In order to mitigate against potential difficulties SIVs are required to hold capital against such eventualities. The capital buffer can be computed either by a matrix based or by a simulation based approach. The matrix based approach to calculating the amount of capital required calculates the capital required on an asset-by-asset basis. When debt is issued and used to buy an asset, a capital charge is associated with the asset which is a function of the rating and tenor of the asset. The capital charges for each asset (and asset tenor) are estimated from historical data. The total capital charge is the sum of all the individual asset charges. Alternatively, Monte Carlo simulation methods can be used to compute the capital charge. In this approach the future (correlated) values of all the assets are simulated and losses computed. From the expected value of losses calculated the appropriate capital charge can be determined. The simulation approach is more sophisticated than the matrix based approach (and can incorporate portfolio diversification effects), but is more difficult to implement and requires greater computational resource (the level of detail of the simulations can be quite significant in terms of the number of different market variables to simulate and assets to re-price).

The risk sensitivities of a SIV are broadly comparable to the risks of the collateral on the asset side. Outright default and prepayment are common sorts of risks. In principle SIVs should be market neutral (because of the micro-hedging) so even if the market values of assets decline, as long as they do not default and they continue to pay their coupons there should be no risk (because the SIV will simply hold the asset until maturity). However some SIVs have operating modes which are triggered by the market value of the assets under management. Therefore the mark-to-market value must also be constantly monitored. As we have also discussed, in addition to these risk factors quantitative risk management of a SIV should additionally focus on capital adequacy, market neutrality and liquidity management. Given the diversity of collateral and the complexity of the basic SIV structure, risk managing SIVs is probably as hard as it gets in terms of the current state-of-the-art in structured finance.

This is the theory of SIVs. During the ongoing fallout from the problems in the American mortgage market during 2007 SIVs were particularly hard hit. This is because the CP market effectively dried up during the summer of that year. In particular, buyers of CP were reluctant to invest because of concerns over the quality of the paper being issued and the quality of the underlying collateral backing this paper. This in turn stemmed from events earlier in the year when the market began to recognise gradually the extent of the problems in the American home loan market (one of the big concerns being the extent to which people had been able to obtain mortgages without the means to repay them). In addition to this a number of hedge funds which were heavily invested in the ABS world blew up in the summer of 2007, further spooking the whole market. When investors did purchase CP they demanded significant risk premiums as mitigation against the possibility they may not get their investment back. As liquidity ground to a halt, supply of assets exceeded demand leading to downward pressure on prices. All of this led to the SIVs experiencing severe difficulties in rolling (reissuing) their short-term debt. Effectively investors were not willing to re-finance the vehicles.

1.4 Chapter review

In this chapter we have introduced the financial engineering concepts of securitisation and tranching that are at the core of modern structured finance and structured credit in particular. The core function of credit derivatives is to facilitate the transferral of (credit) risk from those who do not want to bear it to those who are willing to. Basic CDSs achieve this objective quite well. But as this chapter has shown, the structured credit market has evolved far beyond this basic objective. Such financial innovation to transfer risk may in fact work too well and disperse risk far and wide throughout the global financial system. The risk then is that problems in one particular market may impact seemingly unrelated other markets.

We have also discussed the application of securitisation and tranching technologies to CDOs, and synthetic CDOs in particular, as well as briefly describing some of the primary motivations for issuers and investors in CDOs alike. The level of detail provided in this chapter in this regard is by no means exhaustive. There are other reference texts that go into substantial detail describing the underlying financial motivations, the economic benefits of securitisation and examples of groundbreaking deals (see the references referred to in the text for examples). There are also other resources that provide up-to-date statistics on the size and growth of the credit derivatives market (for example the news service agencies such as Bloomberg and Reuters and the rating agencies). It is recommended that the reader consult these resources for more 'business' focused detail where necessary (reproducing the material here would add little to this text). This chapter's goal was simply to introduce the concepts we will need later in the book.

Credit derivatives are instruments whose cashflows depend upon the 'living' state of the reference obligors. One of the most important elements to model therefore is when do obligors default. This is the subject of the next chapter. After this we then start the process of analysing specific products, starting with credit default swaps, credit indices, default baskets and then finally synthetic CDOs. These topics constitute Chapters 1–10 of the text. Once the basic products and their properties are understood, we will then, in Chapters 11–14, analyse some more complex applications of the underlying technology introduced in Chapters 1–10.

2
Modelling of obligor default

2.1 Introduction

What constitutes an obligor (corporate or sovereign entity) default? How do we model and quantify it? These are two simple questions. The first question can be answered (at least from a quantitative analyst's perspective) by stipulating what constitutes a default from a legal perspective (such as failure to meet a debt obligation, legal bankruptcy or restructuring of a company) and enshrining this in legally enforceable terms and conditions of a contract (specifically the types of events which will trigger contingent cashflows). The second question, which is the subject of this chapter, is more difficult to answer.

In this chapter we introduce the standard market models for characterising obligor default. The fundamental modelling assumption is that default events happen randomly and at unknown times which cannot be predicted deterministically. Default events must therefore be modelled using the powerful machinery of probability theory. For modelling the default of single obligors, the relevant quantity is the default time. The statistical properties of this random variable are postulated to be governed by a Poisson process. Poisson processes occur throughout the mathematical and physical sciences and are used to model probabilistically events which are rare and discrete. Modelling default times makes sense since credit derivatives have payoffs that are a function of the timing of defaults of the underlying reference obligors. Unfortunately, introducing discontinuous jumps in variables (such as credit spreads) renders a lot of the machinery of traditional no-arbitrage financial mathematics ineffective. This is because adding jump components to diffusion based models for spread dynamics makes the market incomplete. As a consequence it is no longer possible to hedge contingent claims dynamically (essentially this is because the unpredictable nature of the instantaneous jump means the hedge cannot be promptly rebalanced).

There are broadly two different approaches to modelling default risk: reduced form and structural modelling approaches. Reduced form models use dynamics derived from rating information (full rating state information or binary default/no-default information) to characterise defaults. In the reduced form modelling methodology, default is an unpredictable event which is governed by a particular statistical process (in this case a Poisson process).

Structural models of credit risk assume that default events are closely linked to the balance between the assets and liabilities of an obligor. Both of these quantities are time varying, and default is assumed to occur when the value of the assets falls below the value of the liabilities. Structural models require a lot of obligor-specific data in order to parameterise the model correctly. Reduced form models on the other hand dispense with a lot of these difficulties by ignoring the specific details of an obligor's default mechanism (essentially sweeping all the details under the carpet into a single – hazard rate – parameter which is calibrated to market data). In reduced form models default is unpredictable and is driven by a stochastic jump (Poisson) process. The hazard rate model is introduced in Section 2.2.

When considering the default of multiple obligors (for example the obligors referenced in a synthetic CDO pool), an important new complication arises: correlation amongst defaults. Corporate entities never exist in isolation from other corporate entities or from the recipients of their services (their customers). It is reasonable to expect that there is a degree of coupling between the fortunes of different obligors. The market standard methodology for modelling the co-dependence amongst a group of risky obligors is to use copula functions. These will be introduced in Sections 2.3 and 2.4 and some simple examples of the most commonly used copulas in financial applications will be discussed. In Section 2.5 we discuss an extension of the simple hazard rate model to the case of modelling obligor survival not as a single binary event, but as a series of transitions between different rating states. The technology introduced in this section will prove to be useful in some more advanced models described in Chapters 11–14. Finally in Section 2.6 we summarise the important points that have been introduced in this chapter.

2.2 Modelling single-name default as a Poisson process

2.2.1 The hazard rate model

As described in Chapter 1, credit derivatives are contracts that have payoffs that are dependent upon the occurrence of obligor related trigger events, typically defaults (or other extreme events such as debt restructuring). The valuation of these contracts therefore requires some means of modelling how and when obligor default occurs. In this section we introduce the market standard model for capturing the occurrence of individual instances of obligor default. For more detailed expositions of the financial mathematics underpinning the models introduced in this section see, for example, Giesecke [2004] or Bielecki and Rutkowski [2002] and Bielecki *et al.* [2006a, 2006b]. The hazard rate model for obligor default is based on the assumption that default is a random event and that it is characterised by a Poisson process [Lando 1998, Duffie and Singleton 1999]. A recent extension to the fundamental modelling framework is described in Giesecke [2007].

To begin we introduce a random counting process. A counting process, in the context of credit default risk, is a non-negative strictly increasing integer-valued (stochastic) process that records the number of events that have occurred in the specified time horizon. Consider a counting process $N(t)$ such that $N(0) = 0$ (initially no defaults) and $N(u) < N(t)$ for times $u < t$ (this is a strictly increasing number of defaults – once an obligor has defaulted it is a terminal event for that obligor) and $N(t) \geq 0$ for all t. This counting process has a (possibly stochastic) intensity $h(t)$ representing the expected rate of arrival of counted events per unit time, i.e. the expected number of defaults per unit time. The time of default, τ, of an obligor is defined as the first jump in the counting process, i.e. $\tau = \inf\{t \in \mathbb{R}^+ | N(t) > 0\}$ (the first time when an event occurs). The survival probability over the interval $[0, T]$ is therefore $S(0, T) = P[N(T) = 0|\Im_0]$ where the filtration $\Im_t$ represents the information knowable at time t. The default probability over the same time period is given by $Q(0, T) = 1 - S(0, T)$.

In order to progress it is necessary to make some assumptions as to what the physical process generating defaults is. The fundamental assumption we make is that defaults are modelled as a Poisson process and that default occurs at the first jump of the counting process. Poisson processes appear frequently in the natural and mathematical sciences and are used to model the statistical properties of events that are rare and independent (for example radioactive decay). Corporate default events are certainly rare (compared to other types of corporate event such as reporting earnings) and the occurrence of default events for an individual obligor can be considered to be independent of other events.

The counting process $N(t)$ is therefore assumed to be a Poisson process which takes integer values 0, 1, 2, . . . etc., and the default/jump times when the process increments by unity are $\{\tau_i\}$. Let us assume that the intensity of the process $h(t) = h > 0$, i.e. the default intensity is positive and homogeneous (non-time-dependent, therefore defaults occur at a constant rate, for example three expected defaults per year, i.e. $\mathbf{E}[dN(t)] = h dt$). The probability of a default in the time interval $[t, t + \Delta t]$ where $\Delta t \ll t$ is given by $P[N(t + \Delta t) - N(t) = 1] = h\Delta t$, and the probability of no jump in this interval is $P[N(t + \Delta t) - N(t) = 0] = 1 - h\Delta t$. If we assume that jumps in non-overlapping time intervals are independent (i.e. the process is Markovian) then we have that

$$P[N(t+\Delta t) - N(t) = 0] = 1 - h\Delta t,$$
$$P[N(t+2\Delta t) - N(t) = 0] = (1 - h\Delta t)^2,$$

and so on. We now subdivide the interval $[t, T]$ into n partitions of equal size Δt such that $n\Delta t = T - t$. In each period Δt there are no jumps with probability $1 - h\Delta t$. Therefore over the n intervals in $[t, T]$ the probability that there are no jumps is given by

$$P[N(T) - N(t) = 0] = (1 - h\Delta t)^n \rightarrow \mathrm{e}^{-h(T-t)}$$

as $\Delta t \rightarrow 0$ in the limit of continuous time.

Consider the probability of one jump (default) in the interval $[t, T]$. If there is one default/jump this has a probability $h\Delta t$ and there are n possible steps at which to have exactly one jump. If there is one jump at a particular step there are $n - 1$ steps in which there was not a jump with a total probability of $(1 - h\Delta t)^{n-1}$ (since all the increments are independent). The probability of one jump in $[t, T]$ is therefore

$$P[N(T) - N(t) = 1] = n(h\Delta t)(1 - h\Delta t)^{n-1} \rightarrow (T - t)h\,\mathrm{e}^{-h(T-t)}$$

as $\Delta t \rightarrow 0$. The probability of n jumps is therefore

$$P[N(T) - N(t) = n] = \frac{1}{n!}[h(T - t)]^n\,\mathrm{e}^{-h(T-t)}.$$

This is the Poisson process which is used to characterise rare or discretely countable events. All of the increments/events are independent. The survival probability over the interval $[t, T]$ corresponds to the case $n = 0$ when there are no jumps and is given by $S(t, T) = \mathrm{e}^{-h(T-t)}$.

The Poisson process has no memory (it is Markovian). Hence the probability of n jumps in $[t, t + s]$ is independent of $N(t)$, the number of jumps that have occurred in the interval $[0, t]$, and of its history prior to t. Therefore a jump is no more likely at time t if in the preceding periods there has or has not been a jump.

We now let the default intensity become a deterministic function of time $h \rightarrow h(t) > 0$, i.e. the default trigger is an inhomogeneous Poisson process. The probability of no jumps/defaults in $[t, t + \Delta t]$ is $P[N(t + \Delta t) - N(t) = 0] = 1 - h(t)\Delta t$ and the probability of no jumps in the interval $[t, T]$ (subdividing the region into n intervals) is

$$\ln P[N(t+\Delta t) - N(t) = 0] = \ln \prod_{i=1}^{n} (1 - h(t + i\Delta t)\Delta t),$$

implying that

$$P[N(T) - N(t) = 0] \rightarrow \mathrm{e}^{-\int_t^T h(s)\mathrm{d}s}$$

as $\Delta t \rightarrow 0$. Similarly the probability of n jumps in the interval $[t, T]$ is

$$P[N(T) - N(t) = n] = \frac{1}{n!}\left(\int_t^T h(s)\mathrm{d}s\right)^n \mathrm{e}^{-\int_t^T h(s)\mathrm{d}s}.$$

The survival probability over this interval ($n = 0$) is

$$S(t, T) = \exp\left\{-\int_t^T h(s)\mathrm{d}s\right\},$$

and the density of the time of the first jump given that there has been no jump in $[0, t]$ is

$$h(t)\exp\left\{-\int_0^t h(s)\mathrm{d}s\right\},$$

i.e. the probability of surviving from 0 to t multiplied by the expected default rate in the period $[t, t + \Delta t]$.

Finally we can now let the hazard rate process itself be stochastic (a Cox process). Now the time of the default event is stochastic, as is the conditional probability of observing a jump over the interval $[t, T]$. This means

$$P[N(T) - N(t) = n] = \mathbf{E}\left[\frac{1}{n!}\left(\int_t^T h(s)\mathrm{d}s\right)^n \mathrm{e}^{-\int_t^T h(s)\mathrm{d}s}\right].$$

A Cox process is a generalisation of a Poisson process to include stochastic intensity dynamics. To specify the model fully we need to postulate some form for the stochastic dynamics of the hazard rate, for example, $\mathrm{d}h(t) = \mu(t)\mathrm{d}t + \sigma(t)\mathrm{d}W(t)$.

2.2.2 *Valuation of risky cashflows*

In the framework of the hazard rate model the price at time 0 of a risky bond (i.e. a bond issued by an obligor which is subject to default risk) $\bar{B}(0, T)$ which pays 1 at time T is given by

$$\bar{B}(0, T) = \mathbf{E}\left[\mathrm{e}^{-\int_0^T r(s)\mathrm{d}s}\mathbf{1}_{N(T)=0}\right]$$

where $r(t)$ is the (spot) risk-free interest rate at time t. That is, the value is equal to the expectation under an appropriate risk-neutral measure [Baxter and Rennie

1996] of the discounted cashflow multiplied by the probability of there being no defaults in the interval $[0, T]$. If we assume that the dynamics of the process for the default-free interest rate does not depend upon the arrival of the default of the obligor (the two processes are uncorrelated) and that the hazard rate process is deterministic (but time varying) we can separate out the calculation of the two expectation values and this becomes

$$\bar{B}(0,T) = B(0,T)\exp\left\{-\int_0^T h(s)\mathrm{d}s\right\},$$

where

$$B(0,T) = \exp\left\{-\int_0^T r(s)\mathrm{d}s\right\}$$

is the risk-free discount factor and $\mathbf{E}[\mathbf{1}_{N(T)=0}] = P[N(T) - N(0) = 0]$ is the survival probability over the interval $[0, T]$.

In order to proceed further and actually calculate expectation values, it is necessary to postulate a suitable process for the dynamics of the risk-free interest rate and the hazard rate. If the choice of dynamics is sufficiently simple it will be possible to evaluate the expectation values analytically. For example, if the hazard rate is assumed to follow a Hull–White type model the dynamics of the hazard rate will be governed by

$$\mathrm{d}h(t) = [\theta - ah(t)]\,\mathrm{d}t + \sigma\,\mathrm{d}W(t)$$

where θ is the long-run mean-reversion level of the hazard rate, a the speed of mean-reversion and σ the volatility. In this case there are analytic solutions to the stochastic differential equation (including negative hazard rates – clearly not a feasible scenario) [Brigo and Mercurio 2001]. For more complex payoffs Monte Carlo simulation methods can be used to evolve the hazard rate. Indeed all of the powerful techniques developed in the interest rate world can be brought to bear on pricing instruments using an appropriate choice of dynamics. See Schonbucher [2003], and De Servigny and Jobst [2007] for a good discussion of different stochastic models for hazard rates and valuation models for risky cashflows.

Finally we consider a slightly more complex situation where we have a payoff of 1 at time T_i if and only if there is a default in the interval $[T_{i-1}, T_i]$. The value of this claim is given by

$$\mathbf{E}\left[\exp\left\{-\int_0^{T_i} r(s)\mathrm{d}s\right\}\right]\mathbf{E}\left[\mathbf{1}_{N(T_i)=0} - \mathbf{1}_{N(T_{i-1})=0}\right]$$

(using the independence assumption). In the hazard rate framework we can compute

$$\mathbf{E}[\mathbf{1}_{N(T_i)=0}] = S(0, T_i) = \mathrm{e}^{-\int_0^{T_i} h(s)\mathrm{d}s},$$
$$\mathbf{E}[\mathbf{1}_{N(T_{i-1})=0}] = S(0, T_{i-1}) = \mathrm{e}^{-\int_0^{T_{i-1}} h(s)\mathrm{d}s},$$

and therefore value the cashflow. This result is the basis for valuing the contingent leg in vanilla CDS contracts.

2.3 Modelling default correlation – fundamental concepts

The previous section introduced the market standard model for capturing the default behaviour of a single obligor. This is perfectly adequate for valuing credit derivatives whose payoff is only contingent upon the default state of a single obligor. However, when considering instruments which have payoffs dependent upon the joint behaviour of multiple obligors it is necessary to introduce mechanisms for capturing synchronisation or *co-dependence* between obligor behaviour. In particular, quantities such as the overall portfolio loss are not simply a straight sum of the losses to each individual defaulting obligor due to the interactions between them. Default correlation modelling is one of the most important components in assessing portfolio credit risk, and pricing instruments dependent upon a portfolio of risky obligors. In general, defaults can occur due to idiosyncratic effects (an obligor defaulting independently of the behaviour of other obligors), industry-specific or systemic effects (where large numbers of obligors default near synchronously due to common macroeconomic factors which impact all obligors to a certain degree). Empirical evidence [Das *et al.*] indicates that defaults are often clustered temporally. Hence it is necessary to capture timing risk, and not just the expected number of defaults, in the modelling of default dependence. Default rates are also often observed not to be constant in time. These observations are to a certain degree simply common sense: when the economy is poor large numbers of corporate entities will default in close proximity to one another; when the economy is thriving there will be fewer defaults save for those entities that are simply badly run or suffer very specific shocks to their customer base.

A general consideration of dependency amongst obligors goes beyond simple linear correlation. Dependency includes not just pairwise synchronous co-movements between obligor values, but also the concepts of portfolio effects, temporally lagged behaviour between obligors and concepts of causality (for example a large automobile maker going bust will have a downstream impact on all the suppliers of components). Simple linear correlation alone is not capable of capturing this behaviour in a portfolio of obligors (the following argument is based on that presented in Schonbucher [2003]). Consider two obligors A and B. The probabilities of individual defaults of these obligors in isolation before time T are given by p_{A} and p_{B} respectively. The probability of A and B defaulting before T is given by p_{AB} (the probability of A defaulting conditional on B having already defaulted or vice versa). By Bayes' rule $p_{\mathrm{A|B}} = p_{\mathrm{AB}}/p_{\mathrm{B}}$ and $p_{\mathrm{B|A}} = p_{\mathrm{AB}}/p_{\mathrm{A}}$. The linear correlation coefficient is given by $\rho_{\mathrm{AB}} = \mathrm{cov}(A, B)/\sigma_{\mathrm{A}}\sigma_{\mathrm{B}}$ (where $\mathrm{cov}(A, B) = \mathbf{E}[AB] - \mathbf{E}[A]\mathbf{E}[B]$ and $\sigma_A^2 = \mathbf{E}[A^2] - \mathbf{E}[A]^2$). This is

$$\rho_{\mathrm{AB}} = \frac{p_{\mathrm{AB}} - p_{\mathrm{A}}p_{\mathrm{B}}}{\sqrt{p_{\mathrm{A}}(1 - p_{\mathrm{A}})}\sqrt{p_{\mathrm{B}}(1 - p_{\mathrm{B}})}}$$

since $\mathbf{E}[A] = p_{\mathrm{A}}$ and $\mathbf{E}[A^2] = p_{\mathrm{A}}^2$ and $\mathbf{E}[AB] = p_{\mathrm{AB}}$. For two obligors there are four possible elementary events: no defaults, A defaults, B defaults or both A and B default. The linear correlation matrix is therefore adequate to capture all of these possibilities (since there are $2 \times (2 - 1)/2$ independent entries in the correlation matrix). For three obligors there are eight elementary events: no defaults, three single defaults, three joint defaults and a triple default. It is this triple default event that cannot be captured by the linear correlation matrix specifying only the pairwise interactions between the obligors. This is because we can know the individual default probabilities of the obligors, the joint default probabilities of two obligors defaulting together and the sum of the probabilities (a total of seven pieces of information). There is one piece of information missing: the probability of all three obligors defaulting together. In general, for a portfolio of n obligors there are 2^n joint default events. However, there are only $n(n - 1)/2 + n + 1$ pieces of information available. As n becomes large this rapidly becomes a serious issue. The problem is that the linear correlation matrix only gives the bivariate marginal distributions; the full joint distribution is undetermined. Linear correlation is therefore not capable of capturing the full dependence structure in a portfolio of more than two obligors. However, it is precisely the multiple default events that are of crucial importance to capture in the context of portfolio credit derivative modelling (for example, the loss to a mezz tranche is dependent upon multiple defaults occurring before the maturity of the tranche).

Linear correlation coefficients also have other shortcomings (see Embrecht *et al.* [1999] for further details of these). For example, if $\sigma_X \to 0$ (or $\sigma_Y \to 0$) then $\rho_{XY} \to \infty$. Also if X and Y are independent random variables then X and X^2 are clearly dependent. However, $\mathrm{cov}(X, Y) = 0$ and $\mathrm{cov}(X, X^2) = 0$ implying that $\rho_{XY} = 0$ and $\rho_{XX^2} = 0$. That is, independence amongst variables implies zero correlation, but zero correlation does not imply independence amongst the variables.

To illustrate the importance of correlation in portfolio loss modelling, we can consider a simple independent default model (where the correlation is zero). Consider the behaviour of a homogeneous portfolio of n obligors, with X defaults in the portfolio over the time horizon $[0, T]$. The loss is therefore $X(1 - \delta)$ where

δ is the recovery rate (the fraction of notional recovered in the event of a default). To characterise the probability of losses (the loss distribution) it is necessary to characterise the distribution of default events.

We consider that each of the n obligors has a marginal (individual) probability of default p and that each default (trial) is independent. Because each default event is independent, the probability of $N < n$ defaults is given by

$$P(X = N) = \binom{n}{N} p^N (1 - p)^{n-N}$$

and the probability of less than N defaults is given by

$$P(X < N) = \sum_{m=0}^{N} \binom{m}{N} p^N (1 - p)^{m-N}.$$

As $n \to \infty$ the distribution of losses converges to a standardised normal distribution by the central limit theorem. Because the distribution of losses is Gaussian, the tails of the distribution will be thin. Comparing this simple model with empirical default data shows that the model does not reproduce the historical data. We therefore conjecture that it is important to capture correlations in default events in portfolio modelling.

2.4 Introducing default dependence via copulas

In this section we briefly introduce the concept of copulas and their application to credit risk modelling. Copulas have a long history in statistical modelling, but their application to financial (and specifically credit) problems is a recent event (due to Li [2000]). The speed with which they have been adopted by the marketplace has been extremely rapid and it is difficult to overstate the impact that copulas have had on credit risk modelling in particular.

The material presented here is only intended to be an introduction to the basic concepts of copulas relevant to understanding the material in the rest of the text. For a more detailed description regarding the underlying theory and applications of copulas the reader is referred to the appropriate references (a comprehensive guide to copulas applied to financial problems is provided in Cherubini *et al.* [2004], the classic text is Nelson [1999] and a good intuitive introduction can be found in Schonbucher [2003]).

2.4.1 Basic concepts

As we saw in the previous section, capturing the joint default behaviour amongst a portfolio of obligors requires more than just a simple linear correlation coefficient. Copulas have been introduced into financial modelling as a mechanism to achieve this. The basic concept of a copula is that it is a mechanism for modelling the dependency structure between a set of random variables (capturing the joint probability amongst all the random variables of an event happening, for example the joint probability that all the variables have a value less than one-half, or that three-quarters of the obligors in a portfolio default before 2 years). Each individual random variable has a univariate, marginal probability density function (and by construction a marginal cumulative distribution function). The copula function 'stitches' together all of the marginal distributions into a multivariate, joint probability density function. Because the copula joins together the marginal distributions, one of the advantages of the copula framework is that it allows the marginal distributions and the joint dependence structure to be separated and modelled in isolation (hence we can calibrate the marginal distributions independently of the choice of copula). A copula function can be extracted from an empirical multivariate distribution. Alternatively, if the marginal distributions are specified, then a new multivariate distribution can be constructed by choosing an appropriate copula function.

A copula $C(u_1, \ldots, u_i, \ldots, u_n)$ with dimension n, is an n-dimensional probability distribution function defined on the hypercube $[0, 1]^n$ that has uniform marginal distributions $\{U_i : i = 1, \ldots, n\}$. It is defined as

$$C(u_1, \ldots, u_i, \ldots, u_n) = P[U_1 \le u_1, \ldots, U_i \le u_i, \ldots, U_n \le u_n]$$

(hence it follows that $C(1, 1, 1, \ldots, u_i, \ldots, 1, 1, 1) = u_i$, i.e. the copula has uniform marginal distributions). The copula function characterises the joint, multivariate, dependence structure (probability distribution) amongst the random variables.

Probably the most useful result regarding copulas is Sklar's theorem [Nelson 1999, Cherubini *et al.* 2004]. This states that any group of random variables can be joined into their multivariate distribution using a copula. Let $\{X_i : i = 1, \ldots, n\}$ be a set of random variables. All information about these variables is contained in the cumulative joint distribution function defined as

$$F(X_1, \ldots, X_i, \ldots, X_n) = P[X_1 \le x_1, \ldots, X_i \le x_i, \ldots, X_n \le x_n] \in [0, 1]^n.$$

The marginal cumulative distribution function of each random variable is given by $F_i(x) = P[X_i \le x_i] \in [0, 1]$. Sklar's theorem states that there exists an n-dimensional copula of the multivariate distribution function $F(X_1, \ldots, X_i, \ldots, X_n)$ such that

$$F(X_1, \ldots, X_i, \ldots, X_n) = C(F_1(X_1), \ldots, F_i(X_i), \ldots, F_n(X_n)),$$

i.e. the multivariate distribution function is characterised as a (copula) function of the marginal distributions. By changing the variable $x_i \to F_i^{-1}(x_i)$ we can invert this relationship to obtain an expression for the copula function in terms of the multivariate probability density function of the random variables and their marginal distributions according to

$$C(x_1, \ldots, x_n) = F(F_1^{-1}(x_1), \ldots, F_n^{-1}(x_n)).$$

If the marginal distributions are continuous then the copula is unique. In principle there exist an infinite number of copula functions which could be used to characterise the joint-dependence structure of a portfolio of obligors. However, data on default correlations is scarce. It is prudent therefore to model the dependence structure with a copula which is as simple and parsimonious to parameterise as possible. The most commonly used copulas in financial applications are the so-called elliptical copulas, the Gaussian and Student t. Other copulas can be used and the reader is referred to Cherubini *et al.* [2004] and De Servigny and Jobst [2007] for further details of these. We only consider the Gaussian and Student-t copulas as these are by far the most common in financial applications.

2.4.2 The Gaussian copula

The Gaussian copula is the most common copula function in financial applications. The n-dimensional Gaussian copula is defined by

$$C_\Sigma^{\mathrm{G}}(u_1, \ldots, u_n) = N_\Sigma^n(N^{-1}(u_1), \ldots, N^{-1}(u_n))$$

where Σ is the correlation matrix between the normal random variables. In the two-dimensional case this copula function reduces to the standard bivariate normal distribution. The marginal distribution of the u_i is all Gaussian. The joint distribution of the u_i is therefore multivariate Gaussian.

Simulating from the multivariate Gaussian copula is straightforward using the following algorithm. We assume that the $n \times n$ correlation matrix $\mathbf{C}$ is available as an input.

1. Compute the Cholesky decomposition $\mathbf{A}$ of the correlation matrix according to $\mathbf{C} = \mathbf{A}\mathbf{A}^{\mathrm{T}}$ (a suitable algorithm for this can be found in Press *et al.* [2002]).
2. Sample an $n \times 1$ vector of standardised random deviates $\varepsilon \sim N(\mathbf{0}, I)$.
3. Correlate the random deviates ε by computing $\phi = \mathbf{A}\varepsilon$ (an $n \times 1$ vector).
4. Set $u_i = N(\phi_i)$ for $i = 1, \ldots, n$.
5. Compute $(u_1, \ldots, u_i, \ldots, u_n)^T = (N_1(t_1), \ldots, N_i(t_i), \ldots, N_n(t_n))^T$.

The problem with simulating default times is that the default times can be large for good quality obligors. Therefore simulation convergence can be slow since many simulations are required to generate enough non-zero contingent payments to make the estimators statistically reliable.

2.4.3 The t copula

Let $\mathbf{C}$ be a symmetric, positive definite matrix. Let $t_{\mathbf{C}}^{\nu}$ be the standardised multivariate Student t distribution with correlation matrix $\mathbf{C}$ and ν degrees of freedom given by

$$t_{\mathbf{C}}^{\nu}(x_1, \ldots, x_n) = \frac{\Gamma\left(\frac{\nu+n}{2}\right)}{\Gamma\left(\frac{\nu}{2}\right)} |\mathbf{C}|^{-1/2} \int_{-\infty}^{x_1} \ldots \int_{-\infty}^{x_n} \left(1 + \frac{1}{\nu}\mathbf{x}^{\mathrm{T}}\mathbf{C}^{-1}\mathbf{x}\right)^{-(\nu+n)/2} \mathrm{d}x_1 \ldots \mathrm{d}x_n.$$

The multivariate Student t copula is given by

$$T_{\mathbf{C}}^{\nu}(u_1, \ldots, u_n) = t_{\mathbf{C}}^{\nu}\left(t_{\nu}^{-1}(u_1), \ldots, t_{\nu}^{-1}(u_n)\right)$$

where $t_{\mathbf{C}}^{-1}$ is the inverse of the univariate cumulative distribution function of the Student t distribution function with ν degrees of freedom.

Simulating from the multivariate t-copula is also straightforward using the following algorithm. We assume that the correlation matrix $\mathbf{C}$ is available as an input.

1. Compute the Cholesky decomposition $\mathbf{A}$ of the correlation matrix according to $\mathbf{C} = \mathbf{A}\mathbf{A}^{\mathrm{T}}$.
2. Draw an $n \times 1$ vector of random deviates $\varepsilon \sim N(\mathbf{0}, I)$.
3. Correlate the random deviates ε by computing $\phi = \mathbf{A}\varepsilon$.
4. Simulate a random variable s from a χ_{ν}^{2} distribution with ν degrees of freedom.
5. Set $\mathbf{x} = \sqrt{\nu/s}\phi$.
6. Set $u_i = t_{\nu}(\phi_i)$ for $i = 1, \ldots, n$.
7. $(u_1, \ldots, u_i, \ldots, u_n)^T = (F_1(t_1), \ldots, F_i(t_i), \ldots, F_n(t_n))^T$ where F_i is the ith marginal distribution.

Gaussian copulas have tail-independent properties. Therefore extreme events occur almost independently in the Gaussian copula set-up. However, the t-copula has dependent extreme events. This means that the t-copula is good for sampling extreme events. Unfortunately the t-copula is symmetric meaning that there is also a high probability of zero losses.

This concludes our very brief description of copula functions. The interested reader is referred to the excellent book by Cherubini *et al.* [2004] for a more complete (mathematical) discussion of copula functions and their application to financial problems.

2.5 Rating transition methods for modelling obligor default

The standard hazard rate model presented earlier treats an obligor's behaviour as binary: either they default or they do not and the transition between the two states is abrupt and instantaneous. When it comes to sudden, outright default this is probably a reasonable approximation. However, in reality the path to default for an obligor is rarely as clear-cut as this binary choice implies. In reality an obligor will meander towards default, gradually being marked down by the market as the full extent of its difficulties begin to become apparent. In principle, if markets were truly efficient and able to process information instantaneously, all of the collective insight of the market (that is, its participants) should be reflected in the price of the obligor. This in turn should be reflected in the obligor's credit rating. An obligor whom the market perceives to have little default risk will have a high credit rating; an obligor who is perceived to be at risk will have a low credit rating.

In this section we outline a more sophisticated valuation methodology than the simple binary default model presented earlier. This methodology is due to Jarrow *et al.* [1997] (also described in De Servigny and Jobst [2007]) – referred to as the JLT methodology – and is based on rating transition mechanics for the valuation of defaultable contingent claims. The key insight in this methodology is to use the credit quality of the obligors, characterised by their rating, as the fundamental state variable used to calibrate the risk-neutral hazard rate (the model is still of the reduced form variety). An obligor's riskiness is associated with a credit rating as given by agencies such as Moody's [Moodys], S&P [S&P] and Fitch [Fitch]. The ratings given by different agencies differ. However, they all order credit risk according to the perceived default likelihood of an obligor. Without loss of generality we consider a generic set of rating states running from high rated firms (AAA) down to defaulted firms (D). For the ratings of specific agencies, please refer to their websites.

2.5.1 Formulation of the problem

Assume a complete set of rating classes $S = \{1, \ldots, K\}$ (which constitutes a complete state space). This set also includes the default state (the Kth rating state). Let the counterparty rating at time t be $R(\omega, t)$. Default is assumed to be an absorbing state with no exit. Hence the probability of migrating from default to any other state is identically equal to zero. A transition matrix $\mathbf{Q}$ defines the actual probability of an obligor changing rating over a period of time $[t, T]$ and is of the form (acknowledging default as an absorbing state),

$$\mathbf{Q}(t, T) = \begin{pmatrix} q_{11}(t, T) & q_{12}(t, T) & \cdots & q_{1K}(t, T) \\ q_{21}(t, T) & q_{22}(t, T) & \cdots & q_{2K}(t, T) \\ \vdots & \vdots & \vdots & \vdots \\ q_{K-1,1}(t, T) & q_{K-1,2}(t, T) & \cdots & q_{K-1,K}(t, T) \\ 0 & 0 & \cdots & 1 \end{pmatrix}.$$

For $i, j \in S$ the elements of the transition matrix are defined according to $q_{ij}(t, T) = P[R(T) = j | R(t) = i]$ for $t < T$, i.e. the probability of a transition from rating i at time t to a rating j at time T. Since $\mathbf{Q}$ is a transition probability matrix we require that $q_{ij}(t, T) > 0$ for all i, j and also $\sum_{j=1}^{K} q_{ij}(t, T) \equiv 1$ for any particular initial state $i \in S$. For the default state we see that $q_{KK} = 1$ and $q_{Ki} \equiv 0$ for all i.

It is important to realise that the elements of the rating transition matrix have implicit time dependence. The element $q_{ij}(t, T)$ represents the probability of a migration from rating state i to state j in the time interval $[t, T]$. Practically speaking most rating transition matrices are computed for a one year time horizon (because this is the frequency with which empirical data are collected). We note that $\mathbf{Q}(t, t) \equiv \mathbf{I}$, i.e. the equal time transition matrix is the identity matrix (no rating transitions). The probability of migration is given by $P[R(T) = r | \Im_t] = P[R(T) = r | R(t)]$ where $\Im_t$ is the filtration (information) available at time t. That is, $R(T)$ only depends upon information at $R(t)$. This is the Markov property[1] (the current rating state is only dependent on the most recent time period). Transition probabilities are also assumed to be time homogeneous[2] meaning that $\mathbf{Q}(t, T) = \mathbf{Q}(T - t)$ for all $t \leq T$.

Now consider starting in the rating state i at time t and ending in rating state j at time T. For any intermediate time $t < s < T$ the obligor can be in any one of the possible states (with an appropriate probability). Therefore the total probability of migrating from the initial to the final state is given by the Chapman–Kolmogorov equation (see Figure 2.1)

$$q_{ij}(t, T) = \sum_{k=1}^{K} q_{ik}(t, s) q_{kj}(s, T).$$

Writing this in matrix form we have $\mathbf{Q}(t, T) = \mathbf{Q}(t, s)\mathbf{Q}(s, T)$. From this relationship we can determine long time horizon rating transition events if we know shorter time horizon rating matrices. For example, if we know the Δt period transition matrix $\mathbf{Q}(t, t + \Delta t)$ then the cumulative transition probability over the period $[t, T]$ where $t < t + \Delta t < T$ and $T - t = n\Delta t$ is given by

$$\mathbf{Q}(t, T) = \mathbf{Q}(t, t + \Delta t)\mathbf{Q}(t + \Delta t, t + 2\Delta t) \ldots \mathbf{Q}(t + (n - 1)\Delta t, T).$$

The matrix $\mathbf{Q}$ is the actual transition matrix observed in the marketplace where the elements of the matrix are estimated from historical data (in the objective/historical measure). For valuation purposes, a risk-neutral transition matrix $\tilde{\mathbf{Q}}$ is needed. To

[1] Formally, consider a Markov chain of random variables $\{X_j : j = 1, \ldots, \infty\}$. The Markov property is that $P(X_{j+1} = x_{j+1} | X_j = x_j, X_{j-1} = x_{j-1}, \ldots, X_0 = x_0) = P(X_{j+1} = x_{j+1} | X_j = x_j)$.

[2] However, it is important to note that this assumption is not observed in practice, e.g. the existence of a business cycle implies that transition probabilities evolve over time.

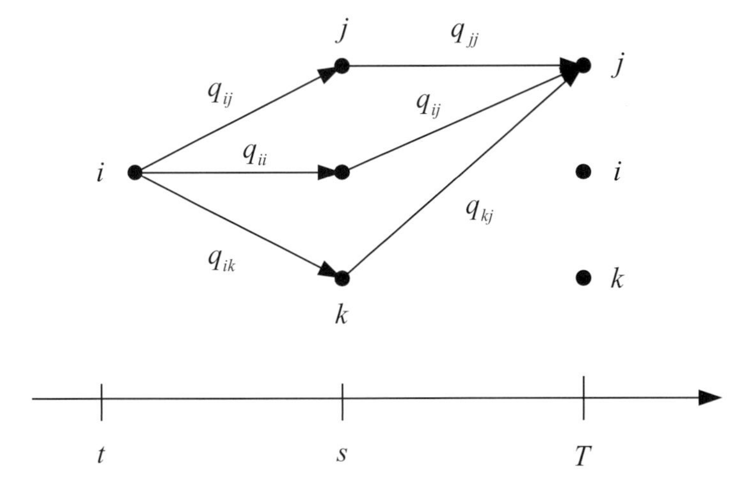

Figure 2.1 The possible paths of rating state evolution from an initial state i at time t to a final state j at time T via an intermediate time $s \in [t, T]$.

obtain the risk-neutral transition matrix, a risk adjustment is made to $\mathbf{Q}$ [Jarrow *et al.* 1997]. The risk adjustment is done in such a way that the initial term structure of risky bond yields or par CDS spreads (see later) is matched exactly. This will result in $\tilde{\mathbf{Q}} = \tilde{\mathbf{Q}}(t, T)$ being time varying (whereas $\mathbf{Q}$ is time homogeneous, i.e. $q_{ij}(t, t + T) = q_{ij}(t + T, t + 2T) = \cdots$).

We define the conditional risk-neutral transition matrix under the appropriate measure to be $\tilde{Q}(t, t + 1)$ which represents credit rating transition probabilities for the time period t to $t + 1$ as

$$\tilde{\mathbf{Q}}(t, t+1) = \begin{pmatrix} \tilde{q}_{11}(t, t+1) & \tilde{q}_{12}(t, t+1) & \cdots & \tilde{q}_{1K}(t, t+1) \\ \tilde{q}_{21}(t, t+1) & \tilde{q}_{22}(t, t+1) & \cdots & \tilde{q}_{2K}(t, t+1) \\ \vdots & \vdots & \vdots & \vdots \\ \tilde{q}_{K-1,1}(t, t+1) & \tilde{q}_{K-1,2}(t, t+1) & \cdots & \tilde{q}_{K-1,K}(t, t+1) \\ 0 & 0 & \cdots & 1 \end{pmatrix}.$$

We therefore have a term structure of conditional transition probabilities. We also have the conditions $\tilde{q}_{ij}(t, t + 1) > 0$ (if and only if $q_{ij} > 0$) and $\sum_{j=1}^{K} \tilde{q}_{ij}(t, t + 1) \equiv 1$. Therefore we require

$$\tilde{q}_{ii}(t, t+1) = 1 - \sum_{\substack{j=1 \\ j \neq i}}^{K} \tilde{q}_{ij}(t, t+1)$$

to ensure that each $\tilde{\mathbf{Q}}(t, t + 1)$ is a valid probability transition matrix. Using standard Markov chain analysis, the cumulative risk-neutral transition matrix is the product of the conditional risk-neutral transition matrices $\tilde{\mathbf{Q}}(t, T) = \tilde{\mathbf{Q}}(t, t+\delta t)\tilde{\mathbf{Q}}(t+\delta t, t+2\delta t)\ldots\tilde{\mathbf{Q}}(t+n\delta t, T)$. Hence if we can calculate $\tilde{\mathbf{Q}}(t, t+1)$, i.e. the daily default probabilities, then it will be possible to price any contingent claim.

2.5.2 Modelling rating transition dynamics

In the reduced form model of Jarrow *et al.* [1997] the rating transition process is assumed to be a time-homogeneous Markov chain (calculated under the historical measure). This will typically be defined for a particular time horizon (e.g. giving the probabilities of migrations between rating states over a one year horizon). Transition matrices for an arbitrary horizon are determined from this matrix by a generator matrix $\Lambda(t)$ which drives the transitions between different rating states. Let us assume that the time interval over which rating changes are recorded is Δt. To determine an equation for the generator matrix we will consider the limit $\Delta t \to 0$. In this case the off-diagonal elements of Λ are defined by the relationship

$$P[(R(t+\Delta t) = l|R(t) = k] = \lambda_{lk}\Delta t$$

and the diagonal elements are defined according to

$$P[(R(t+\Delta t) = k|R(t) = k] = 1 - \sum_{l \neq k}^{K} \lambda_{lk}\Delta t = 1 + \lambda_{kk}\Delta t.$$

The transition matrix over the period $[t, t+\Delta t]$ is given by

$$\mathbf{Q}(t, t+\Delta t) = \mathbf{I} + \Lambda(t)\Delta t,$$

i.e. the current rating state is perturbed by a factor (the elements of the generator matrix) proportional to the time interval. Consider the interval of time $[t, T]$ subdivided into n different grid points with spacing between nodes of $\Delta t = (T-t)/n$. Now by the Markov property of the rating transition process

$$\mathbf{Q}(t, t+2\Delta t) = \mathbf{Q}(t, t+\Delta t)\mathbf{Q}(t, t+\Delta t) = (\mathbf{I} + \Lambda(t)\Delta t)^2,$$

leading to

$$\mathbf{Q}(t, t+n\Delta t) = \left(\mathbf{I} + \frac{T-t}{n}\Lambda(t)\right)^n.$$

In the limit as $n \to \infty$ (or equivalently $\Delta t \to 0$) this becomes

$$\mathbf{Q}(t, T) = \mathrm{e}^{(T-t)\Lambda(t)}$$

where we have defined the matrix exponential via

$$e^{(T-t)\Lambda(t)} = \sum_{n=0}^{\infty} \frac{[(T-t)\Lambda(t)]^n}{n!}.$$

This is the fundamental relationship between the rating transition matrix over the interval $[t, T]$ and the generator matrix $\Lambda(t)$. Rating agencies typically publish transition matrices $\mathbf{Q}$ for rating transitions over a particular time horizon (usually one year). However, this is too restrictive for most practical purposes. It is therefore necessary to rescale $\mathbf{Q}$ via the generator matrix for application over time periods of less than one year. That is, if we know $\mathbf{Q}(0, 1)$ this implies the generator matrix can be found from $\mathbf{Q}(0, 1) = e^{\Lambda}$ where the matrix Λ is such that $\lambda_{ij} > 0$ for all $i, j \in S$ and $i \neq j$ and $\lambda_{kk} = -\sum_{l \neq k} \lambda_{kl}$.

Unfortunately, computing the log or exponential of a matrix can be problematic (the embedding problem, see Schonbucher [2003]). The rescaling procedure must ensure that the rescaled matrix also satisfies the necessary and sufficient conditions to be a transition matrix (specifically that there are no negative transition probabilities generated and that the sum of transition probabilities is equal to unity). A simple approximation is to use the power series definition of the function $\ln \mathbf{Q} = \mathbf{I} - \mathbf{Q} - \mathbf{Q}^2 - \mathbf{Q}^3 + \cdots$. This will converge quickly if $q_{ii} > 1/2$ for all $i \in S$ (which will be the case for a rating transition matrix). For more details on the computation of matrix logarithms and exponentials see Kreinin and Sidelnikova [2001] and Higham.

2.5.3 Determining risk-neutral transition probabilities

The JLT approach assumes that the relationship between $\mathbf{Q}$ and $\tilde{\mathbf{Q}}$ (real-world and risk-neutral rating transition matrices) is specified according to

$$\tilde{q}_{ij}(t, t+1) = \pi_i(t) q_{ij}$$

for $i \neq j$. Note that $j = K$ is a possible final rating state. $\pi_i(t)$ is the deterministic risk adjustment (premium) for an initial state i for the period t to $t + 1$. This is assumed to be independent of the final state j. Therefore the risk premium associated with a rating transition from state i to K (default) is the same as transitions from state i to any other state. The q_{ij} are the real-world transition probabilities which are obtained from historical data, for example. It is easy to see that

$$\tilde{q}_{ii}(t, t+1) = 1 - \sum_{\substack{j=1 \\ j \neq i}}^{K} \tilde{q}_{ij}(t, t+1) = 1 - \pi_i(t)[1 - q_{ii}(t, t+1)].$$

Therefore we have an expression for the risk-neutral transition probabilities for all states i. Since we also require $\tilde{q}_{ij}(t, t+1) \geq 0$ we have the following constraint for the risk premiums (otherwise we would have negative transition probabilities which is clearly not feasible)

$$0 \leq \pi_i(t) \leq \frac{1}{1 - q_{ii}},$$

where $1 - q_{ii}$ represents the probability of migration from the current rating state.

The risk premium corrections $\pi_i(t)$ are typically inferred from the prices of risky and risk-free zero coupon bonds. Let us assume that we have these prices for bonds of all maturities and all ratings. Denote the price of a zero coupon bond at time t, maturing at time T with credit rating j as $B_j(t, T)$. $B_0(t, T)$ is the price of a risk-free bond. If the time-to-default is τ then we have for the price of the risky bond

$$B_j(0, T) = B_0(0, T)[P(\tau > T) + \delta(1 - P(\tau > T))]$$

where δ is the recovery amount (a fixed fraction of the face value of the bond, assumed to be constant). The survival probability is $P(\tau > T) = 1 - P(\tau \leq T)$, therefore,

$$P(\tau \leq T) = \frac{B_0(0, T) - B_j(0, T)}{B_0(0, T)(1 - \delta)}.$$

If we assume that the rating transition matrix $\mathbf{Q}$ is defined for a time horizon of one year then the probability of default over this horizon is given by

$$P(\tau \leq 1) = \frac{B_0(0, 1) - B_j(0, 1)}{B_0(0, 1)(1 - \delta)} = \tilde{q}_{jK}(0, 1) = \pi_j(0) q_{jK}.$$

The risk premium adjustments are therefore given by

$$\pi_j(0) = \frac{B_0(0, 1) - B_j(0, 1)}{B_0(0, 1)(1 - \delta) q_{jK}}.$$

The risk premium adjustments are expressed in terms of the probabilities of jumping from rating state j to default. We also have to satisfy the constraint $0 \leq \pi_j(t) \leq (1 - q_{jj})^{-1}$ to ensure the risk-neutral premiums are non-negative. If q_{jK} is small relative to $B_0(0, 1) - B_j(0, 1)$ then this inequality can be violated quite readily (for example $q_{\text{AAA,Default}} \sim 0$). This has serious consequences for the use of the JLT model for pricing instruments in practice.

An alternative definition, due to Kijima and Komoribayashi (KK), of the risk premium adjustment is to assume that the relationship between $\mathbf{Q}$ and $\tilde{\mathbf{Q}}$ is defined to be [Kijima and Komoribayashi 1998]

$$\tilde{q}_{ij}(t, t+1) = \pi_i(t) q_{ij}$$

for $j \neq K$. Both the JLT and KK methods assume that the risk premium is the same independent of where the jump occurs to. Both methods require the sum of the probabilities to equal one. The difference between JLT and KK is the element used to ensure that the transition probabilities sum to one for each credit class. JLT use element i whereas KK use element K. The reason the KK method is useful is its robustness to estimation of the risk premiums. The example in JLT requires a constrained optimisation to be solved due to the small actual default probabilities of high-rated firms while the KK approach does not.

As in the JLT approach, the KK risk premium can be inferred from the prices of risk-free and risky bonds. Following Kijima and Komoribayashi [1998], the risk premiums are solved for recursively using the following formula:

$$\pi_i(t) = \frac{1}{1 - q_{iK}} \sum_{k=1}^{K} \tilde{q}_{ik}^{-1}(0, t) \frac{B_k(0, t+1) - \delta_k B_0(0, t+1)}{(1 - \delta_k) B_0(0, t+1)},$$

where $\tilde{q}_{ij}^{-1}(0, t)$ are the elements of the inverse of $\tilde{\mathbf{Q}}(0, t)$. In this formula the denominator of the prefactor $1 - q_{iK}$ is the probability of not defaulting, which is almost always approximately equal to unity. Therefore the KK approach will lead to more stable estimates of the risk premiums than the JLT approach. Note that this equation will be solved iteratively for the π, i.e. at time zero we know $\{B_j(0, 1) : j = 1, \ldots, K\}$ and hence can determine $\pi_j(0)$ etc. With the risk premium computed, the conditional risk-neutral transition matrix $\tilde{\mathbf{Q}}(t, t+1)$ can be solved for using the cumulative risk-neutral transition matrix $\tilde{\mathbf{Q}}(0, t+1)$.

Further extensions to the JLT methodology are provided by Das and Tufano [1996] (stochastic recovery rates coupled to the risk-free rate) and Arvanitis *et al.* [1999] (non-constant transition matrices).

2.6 Chapter review

In this chapter we have introduced the standard approaches for modelling individual obligor default and default dependence between obligors. The standard approach for capturing defaults in the reduced form modelling framework is to assume that default is an unpredictable and instantaneous change of state for an obligor and that the statistics of the default process are driven by a Poisson process. This is a simple and flexible framework and as we will see in subsequent chapters allows for the valuation of a wide range of credit related instruments. The key difficulty is to ensure that the hazard rate is calibrated to the available market data. This idea of default as a random process can be further refined to include more detailed ratings based information (but the basic modelling of default as a stochastic process remains the same).

When considering more than one obligor the concept of co-dependence amongst defaults becomes of paramount importance. We have briefly introduced copulas as a means whereby dependence amongst defaults can be statistically captured. The influence of copulas on credit risk modelling cannot be overstated. They have become a cornerstone on which all of the market standard valuation approaches for portfolio credit derivatives are based. It is not the intention of the current text to provide a detailed description of copula functions (there are other resources available which very adequately discuss this material in great detail). However, it was the intention to introduce the concept and their usage in a credit context.

Hazard rate modelling and copulas represent the two fundamental building blocks of credit derivative modelling. In the next and subsequent chapters we will apply these modelling concepts to valuing both single-name and portfolio credit derivatives.

3

Valuation of credit default swaps

3.1 Introduction

In this chapter we introduce and analyse the most common and vanilla of credit derivative instruments, namely credit default swaps (or CDSs). In general terms a credit default swap is a bilateral contract agreed over-the-counter (as opposed to exchange traded) by two counterparties whereby one counterparty agrees to provide protection to the other counterparty against a specified default event of a third reference counterparty (distinct from the other two). CDSs therefore allow the credit risk component of an exposure to be isolated and traded. The purchaser of credit protection trades away the credit risk but retains the market risk (in the form of mark-to-market (MtM) fluctuations). Because the protection is with respect to a single reference counterparty the generic term 'single-name' is often used to refer to CDSs (and other related credit derivatives which reference a single counterparty). This will provide a simple way of summarising the key difference between CDSs and portfolio products such as CDOs (which will be referred to as 'portfolio' products!).

CDS cashflow mechanics are introduced in Section 3.2 and the current market standard model for their valuation (based upon the hazard rate approach introduced in Chapter 2) in Section 3.3. The market standard method for the calibration of the model to market observed prices based on a bootstrapping procedure is also discussed in Section 3.4. Section 3.5 then considers risk sensitivities of the product and introduces simple measures to quantify these risks. Finally Section 3.6 reviews the important points that have been introduced in the chapter.

The development of the CDS market has been closely related to the rise in prominence of portfolio CDS products such as synthetic CDOs which construct portfolio exposure synthetically using CDSs. CDSs thus form the building block from which more complex credit derivatives have evolved. They are also among the most liquid of traded instruments. Seen from this perspective an understanding

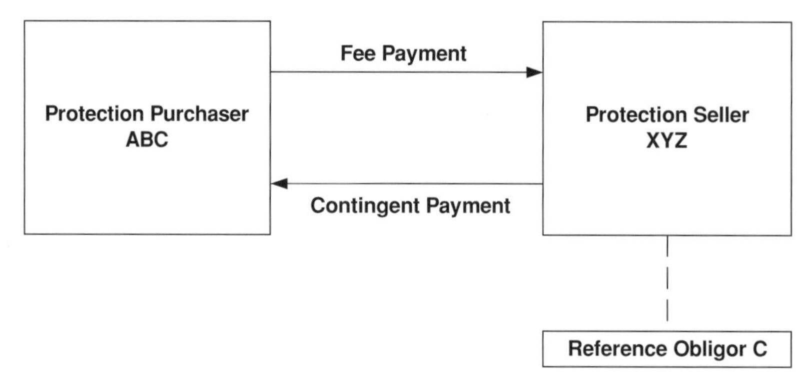

Figure 3.1 Basic mechanism of a CDS contract. Counterparty ABC purchases protection on the cashflows of the bond issued by reference entity C from counterparty XYZ. The protection seller is 'long' risk. ABC makes periodic fee payments to XYZ for the lifetime of the contract, unless C defaults. In this case the fee payments cease and XYZ compensates ABC for the loss due to C defaulting.

of CDSs is a prerequisite for an understanding of portfolio products. Much of the material in this chapter introduces concepts that will be of basic importance for later chapters.

3.2 Overview of vanilla credit default swaps

3.2.1 Description of the product

The basic mechanism of a CDS is shown in Figure 3.1. In essence a CDS is a contract between two counterparties ABC and XYZ who agree to swap a stream of payments contingent upon the behaviour of a third reference counterparty. There is also a calculation agent who is responsible for determining and distributing cashflows (this is usually one of ABC or XYZ). The essential purpose of a CDS is for one party to transfer credit risk to another counterparty (who presumably understands the nature of the risk they are taking and is able to bear that risk effectively – although *caveat emptor*). It is this transfer of risk which, it is argued by some market commentators, is the principal benefit of credit derivatives to the overall stability of the global financial system. The argument is that credit derivatives allow risk to flow from those who do not wish to bear it, to those who will (for suitable compensation). The reality is far from as clear cut as this. Indeed some investors have likened derivatives to 'financial weapons of mass destruction'.

Initially CDS contracts were motivated by a counterparty ABC, who owned a bond in a separate reference counterparty C. ABC wished to transfer the risk that C would default before the maturity of the bond (and therefore not be able to fulfil its contractual obligation to pay the coupons on the debt it had issued) to another counterparty XYZ, while at the same time retaining the right to the bond cashflows (and any mark-to-market exposure). CDSs provided the mechanism for this transfer and were therefore intended as a means to hedge default risk. However, as the market for CDSs grew in size (due to standardised contracts, terms and conditions and documentation), CDSs became a means of actively trading credit risk, thereby expressing a view on the credit worthiness of an obligor. Additionally, as it became possible to trade credit risk for different time horizons (enabling the expression of views on the credit worthiness of an obligor over a range of investment horizons) the concept of duration became important. The existence of a liquid term structure of tradable CDSs in turn opened the possibility for trades/strategies which expressed a view on the relative movements of different parts of the term structure (e.g. 10 year protection looks cheap compared to 5 year) so-called curve steepeners and flatteners (borrowing the terminology from the interest rate world).

A credit default swap has two legs. The 'fixed' fee leg is paid by the protection buyer. Usually this is a fixed fraction of the principal paid at regular intervals. Other payment arrangements, such as amortising notional amounts or the total premium paid upfront at contract inception, are possible although not common. These payments are made until a default occurs or until the specified maturity of the contract. In the event of a default, a contingent payment (the payoff of the derivative) is made. The contingent payment in case of default is also called the 'floating' leg of the swap. When a default occurs there are two common mechanisms for delivering (settling) the contingent payment: cash settlement or physical settlement

In a physically settled CDS upon the event of a default the protection purchaser delivers the reference obligation to the protection seller who pays the par amount to the protection purchaser. That is, the protection purchaser delivers the distressed debt of the reference entity (which may in fact be worthless) to the protection seller in return for the par value of the debt. The contractual specifications will specify quite explicitly what exactly the protection purchaser can deliver (in terms of the debt obligations) to the protection seller in exchange for the protection payment. Typically acceptable deliverables will be debt which is *pari passu* (at an equal level) with the original defaulted reference obligation. In a cash settled CDS the protection purchaser is compensated in the event of a default by an amount equal to $N(1 - \delta)$, i.e. notional times recovery rate (which is the fraction of the original notional amount recovered in the event of a default). The recovery rate is determined at the time of default by a dealer poll (auction).

The two principal uncertainties regarding the contingent payment are therefore the timing of the default event and the actual amount paid (reflected in the uncertainty of the recovery rate at the time of default). In some CDS contract specifications the contingent payment can be deferred to the maturity of the trade although it is more common that the contingent payment is made at the time of default at which time the contract also unwinds. Upon default there may also be an additional accrued fee payment paid to the protection seller for providing protection for part of a coupon payment period (depending upon the precise timing of the default). If there are no defaults then the fee payments continue until the maturity of the instrument. In essence a CDS has some similarities with an insurance contract. In contrast, however, in a CDS it is not actually necessary to physically own the asset the protection refers to. That is, the contract can be between two counterparties, without either of them actually owning the reference asset at all.

The identification of the credit event which triggers the unwinding of the CDS must be based on publicly available information since both parties to the trade have to agree that an event has occurred. When such an event is noted and identified (by the delivery of a 'credit event notice' and also citation of the source of the information identifying the credit event), the process of unwinding the CDS begins. Unwinding the CDS means settling the contingent payment and any accrued payment. The calculation of monies due is usually handled by the appointed calculation agent.

CDSs can be written on all types of obligors, most typically corporate entities but also including sovereign states. They can reference a wide variety of types of debt obligation (bonds, asset backed securities (ABCDS), loans (LCDS), or some other form of receivables). The limiting factors in the types of CDS contract available tend to be practical rather than technological. Finding counterparties to trade a particularly exotic form of exposure with can be problematic. In fact the entire credit derivative market is based on non-exchange-traded, over-the-counter agreements between counterparties. Initially this lack of an exchange traded basis for trading was a limiting factor in the development of the credit derivative market. However, with the introduction of standardised (ISDA) terms and conditions governing vanilla contracts the process of trading became a whole lot easier. As single-name corporate CDS trading became more liquid this in turn facilitated the development of portfolio products such as indices and synthetic CDOs. Synthetic CDOs in turn helped develop the liquidity of the CDS market by introducing additional demand for the product (as a hedging instrument).

Finally, CDSs allow investors to take leveraged exposure to the reference entity. For example an obligor may have \$1 m of outstanding debt (bonds) issued. If the obligor was to default and the recovery rate was 40%, the losses to the holders of the obligor debt would be \$600 k. If there are \$10 m of CDSs written referencing this obligor, the losses to the protection sellers would be \$6 m. In this respect credit derivatives may be transferring risk to those who wish to bear it, but the overall potential losses stemming from an obligor's default have also been greatly amplified.

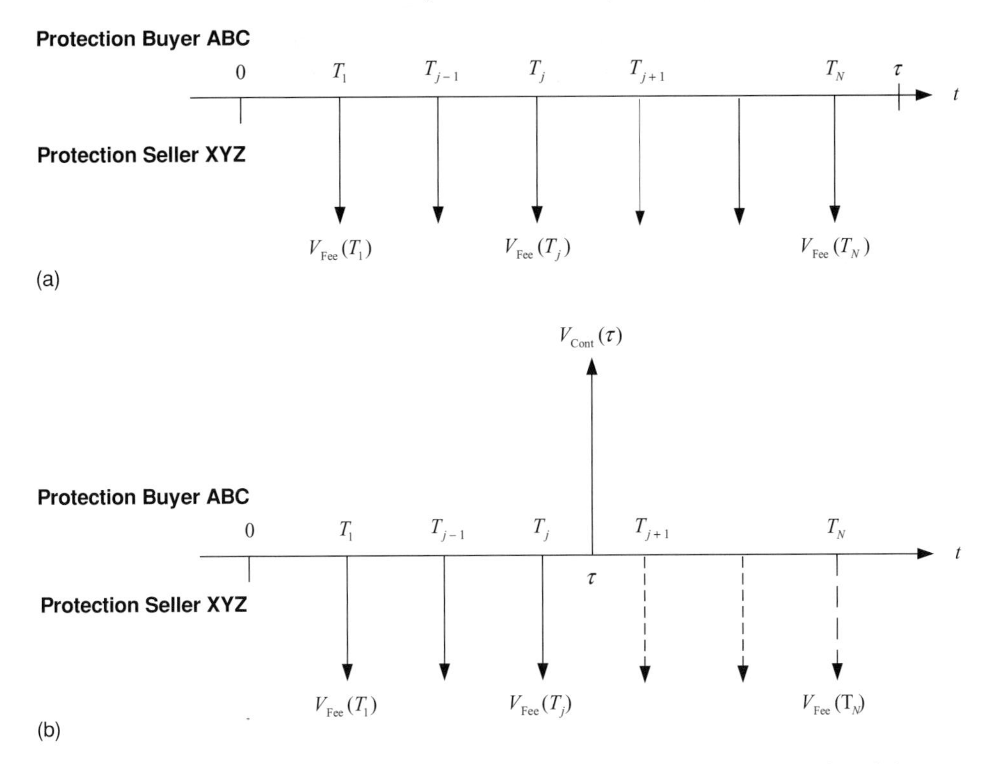

Figure 3.2 (a) Cashflows in a CDS where default occurs after the maturity of the contract. (b) Cashflows in a CDS where default occurs before the maturity of the contract.

3.2.2 Cashflow mechanics

We will now specify more precisely the cashflows in a vanilla CDS contract. In return for receiving default protection, counterparty ABC makes periodic coupon payments (based on a contractually specified tenor and notional) to counterparty XYZ. These coupon payments continue either until the maturity of the contract or until reference entity C defaults. If reference entity C defaults then XYZ makes a contingent payment to ABC (and ABC may also make an accrued fee payment to XYZ). All subsequent cashflows cease. Figures 3.2(a) and 3.2(b)[1] represent schematically the mechanics of the cashflows of a vanilla CDS. In this figure counterparty ABC is purchasing protection from counterparty XYZ (protection against a reference counterparty C).

For the purposes of illustrating the cashflows of the contract, let us assume that the default time of the reference entity is given by τ. The maturity of the contract

[1] This type of stick diagram will occur frequently throughout the book. When understanding a structured product, irrespective of its complexity, it is always beneficial to lay out diagrammatically the schedule, direction and composition of the cashflows.

is T. Figure 3.2(a) is for the case of no default of the reference entity, i.e. $\tau > T$. In this case there is no contingent default payment (and no accrued fee payment). Figure 3.2(b) has a default at $\tau \leq T$ and XYZ makes a contingent payment to ABC (who may also make an accrued fee payment to XYZ).

Typical maturities for CDS contracts are 3 months, 6 months, 1, 2, 3, 5, 7, 10, 15, 20 and 30 years. The most common maturity is 5 years (closely followed by 3, 7 and 10 years). The maturity reflects the period over which protection is sought. The term structure nature of protection allows investors to express views on the default probabilities of the reference obligors over different investment horizons. For example, an investor may feel that a particular corporate has a low likelihood of default in the near term, but a larger likelihood of default in the longer term (for example if the corporate has a large amount of debt that requires servicing over a long timeframe, but relatively little short-term debt). This view can be translated into actual protection via appropriate CDSs. High-grade CDS term structures are typically steep because the perceived default risk at short maturities is low. As the time horizon increases, the probability of default increases. High-yield debt typically has a relatively flat CDS term structure. This is because there is a high perceived default risk at all maturities (the obligor is just plain risky). Credit rating movements are also of importance in determining CDS spreads. If an obligor has a junk rating this will be reflected in the rate they must pay in order to borrow (making it more difficult for them to service their debts in the future). If an obligor has a good rating its borrowing costs will be reduced. High-grade ratings will also increase the pool of potential investors in the obligor (typically some investors, such as pension funds, cannot invest in junk debt). If it has low borrowing costs, the burden of its debt will be less and the probability of defaulting on that debt will be reduced.

The periodic coupons are paid at times $\{T_j : j = 1, \ldots, N\}$. Typically these coupons are quarterly or semi-annual. All payments for corporate CDSs occur on International Money Market (IMM) dates. These are 20th March, June, September and December. The day-count fraction between coupons is given by $\Delta_j = T_j - T_{j-1}$. Let the notional amount on which the coupons are based be given by $N(T_j)$ for the coupon payment at time T_j. In typical contracts the notional amount is constant $N(T_j) = N$. We assume that the maturity of the contract is $T = T_N$. Note that the notation here is slightly poor since we are using N for the total number of coupon payments as well as for the contract notional. It should be clear from the context what meaning is to be ascribed to the notation.

If the default occurs at some coupon date $\tau = T_j$ then the protection purchaser makes the coupon payment (since they have been receiving protection up to this point for the period $[T_{j-1}, T_j]$) and the protection seller makes the contingent payment at this time. If the default occurs in-between coupon payment dates $\tau \in [T_{j-1}, T_j]$ then the protection seller once again makes the contingent payment at time τ (the protection purchaser will also have made a coupon payment at time T_{j-1}). But the protection seller may also receive an additional fraction of a coupon payment which reflects the fact that they have been providing protection to the protection purchaser for the period $[T_{j-1}, \tau]$. This is an accrued fee payment (the contract specification will indicate if the accrued is to be paid or not and the precise timing of the payment). It is typically assumed that the accrued fee is linearly proportional to the period $\tau - T_{j-1}$.

Finally, CDSs can also be viewed as synthetic bond equivalents where the buyer of default protection has a position equivalent to shorting a bond. The seller of protection is in effect long the bond. A seller of protection has a similar economic exposure to a bond holder (they are both long credit) as both parties are adversely affected by spread widening. The converse is true for the protection purchaser.

To conclude, let us consider a simple example. Assume that protection is purchased to the tune of \$100 m. If the contractual spread is 100 bps annually and the contract is on a quarterly basis then the protection purchaser pays to the protection seller a premium of \$100 m $\times$ 100 bps paid in quarterly instalments (equal to \$0.25 m per quarter assuming a day-count basis of 90/360). If we assume a recovery rate of 40% and cash settlement, in the event of a default the protection purchaser receives a payment of \$100 m $\times$ (100% – 40%) equal to \$60 m. If the settlement were physical the protection seller would pay the protection purchaser \$100 m and receive from the protection purchaser debt obligations of the reference entity with a face value of \$100 m, but current market value of \$60 m. If the default occurred at the mid-point between two coupon dates then the protection seller would receive a payment of \$0.25/2 m from the protection purchaser to compensate for the provision of protection from the last coupon date to the default date (the accrued fee payment).

3.3 Valuation of vanilla CDSs

Before writing down valuation formulas (see O'Kane and Turnbull [2003], and Douglas [2007] for more details of the models to be presented and some valuation examples) we will briefly introduce a piece of terminology that will be used frequently. Throughout the book we will come across many products which have two streams of cashflows: the protection payments and the contingent payments. We will refer to these generically as the *fee payments* and the *contingent payments* respectively. In some contexts the protection payments are also referred to as the annuity payments. The general principle is to distinguish between periodic payments (the fee payments) and payments which are contingent upon the occurrence of specific trigger events (typically default).

To price a vanilla CDS within the context of the standard hazard rate based valuation model introduced in Chapter 2 the following model inputs are required.

- The term structure of discount factors $Z(t, T)$ for discounting future cashflows.
- The term structure of recovery rates of the obligor $\delta(t, T)$ for determining the magnitude of the contingent payment. The recovery rate can be estimated by analysis of historical default data. Typically we assume that $\delta(t, T) \equiv \delta$, i.e. the recovery rate is constant for all maturities. This is because reliable, representative default data are scarce;[2] estimating multiple recovery rates across different maturities in the event of default is prone to significant uncertainty.
- The term structure of survival probabilities $S(t, T)$ for weighting future cashflows with the probability that the cashflow will actually occur.

In addition to the model inputs listed above, a CDS contract typically requires the following contractual specifications.

- Identification of the reference entity debt obligations which the protection refers to (e.g. the 30 year bond of Argentina).
- The method of settlement (cash or physical).
- The reference obligations which qualify as the deliverable obligation for physical settlement. Suitable obligations will typically be *pari passu* with the reference obligation, e.g. equivalent bonds.
- The notional amount on which protection is purchased.
- The maturity of the protection term (and also the tenor of the fee payments).
- A specification of the events which constitute a credit event (default, failure to pay, restructuring) and also the sources of information used to monitor for the occurrence of a credit event (usually publicly available information that all obligors have access to).
- The contractually agreed spread applied to the fee payments.
- Whether or not accrued fee payments are to be paid.

Finally there are also the usual specifications of day-count conventions, holiday calendars, calculation agent etc.

3.3.1 Valuing the fee leg

The fee payments are given by the (risk-neutral) calculation of the expected cashflows which are

$$V_{\text{Fee}}(t) = \mathbf{E}\left[\sum_{j=1}^{N} N(T_j)\Delta_j Z(t, T_j)\mathbf{1}_{T_j<\tau}\right]$$

(where the expectation is calculated under an appropriate probability measure). That is, the total fee payment is a sum of discrete coupon cashflows. Each cashflow $N(T_j)$

[2] The majority of the historical data that are available have typically been collated for the North American market. This introduces additional difficulties when trying to determine recovery rates for non-North American entities.

at the set of coupon dates $\{T_j : j = 1, \ldots, N\}$ is discounted by the appropriate discount factor $Z(t, T_j)$ and multiplied by the day-count fraction. More importantly, the cashflow is weighted by the indicator function $\mathbf{1}_{T_j<\tau}$ which represents the probability of the cashflow at time T_j occurring (or conversely that the default of the obligor occurs beyond this point).

To calculate this expectation we make the assumption that interest rates are deterministic and that they are uncoupled from the default arrival process (allowing us to separate out the calculation of the expected values). Under these assumptions the fee payments are easily calculated to be

$$V_{\text{Fee}}(t) = \sum_{j=1}^{N} N(T_j)\Delta_i Z(t, T_j)S(t, T_j)$$

where we have written the survival probability as $S(t, T) = \mathbf{E}[1_{\tau>T}]$ which is the probability as seen at time t of receiving the cashflow due at time T. Clearly, determining this survival probability is crucially important to valuing the fee leg. We saw in Chapter 2 that in the simplest possible case of a constant hazard rate, the survival probability is related to the hazard rate by $S(t, T) = \mathrm{e}^{-h(T-t)}$, i.e. a homogeneous Poisson distribution.

In the limit when the premium payments are made continuously (and not at discrete times) the sum over fee payments becomes an integral

$$V_{\text{Fee}}(t) = \int_t^T N(t')Z(t, t')S(t, t')\mathrm{d}t'.$$

In practice the fee payment cashflows are paid discretely. However, for some simple analysis it can be very useful to assume that the cashflows are continuous, particularly when the interest rates and hazard rates are assumed to be deterministic and their term structures flat. In this case we can write $Z(t, T) = \mathrm{e}^{-r(T-t)}$ and $S(t, T) = \mathrm{e}^{-h(T-t)}$. The integral is (assuming unit notional)

$$V_{\text{Fee}}(t, T) = \int_t^T \mathrm{e}^{-r(s-t)}\mathrm{e}^{-h(s-t)}\mathrm{d}s$$

and is easily evaluated to give

$$V_{\text{Fee}}(t, T) = \frac{1}{r+h}\left[1 - \mathrm{e}^{-(r+h)(T-t)}\right].$$

The fee payment is also referred to as the risky dV01. This is defined as the change in value of the swap for a 1 bp shift in the credit curve. For small rates the risky dV01 is approximated as $T - t$.

If the default occurs at a time $\tau \in [T_{j-1}, T_j)$ then the accrued fee payment is computed from

$$V_{\text{Accrued}}(t) = \sum_{j=1}^{N} N(T_j)\Delta_j \mathbf{E}\left[Z(t,\tau)\frac{\tau - T_{j-1}}{T_j - T_{j-1}} \mathbf{1}_{T_{j-1}\leq\tau<T_j} \right]$$

which, evaluating the expectation operator, becomes

$$V_{\text{Accrued}}(t) = \sum_{j=1}^{N} N(T_j)\Delta_j \int_{T_{j-1}}^{T_j} Z(t,u)\frac{u - T_{j-1}}{T_j - T_{j-1}} h(u)\mathrm{e}^{-\int_0^u h(s)\mathrm{d}s}\mathrm{d}u.$$

This integral can be solved by an appropriate numerical scheme to determine the accrued fee payment. An approximate value for the accrued payment is given by

$$V_{\text{Accrued}}(t) = \frac{1}{2}\sum_{j=1}^{N} Z(0,T_j)\Delta_j\, [S(0,T_{i-1}) - S(0,T_i)].$$

Note that we have not multiplied the fee or accrued fee payments by the spread, i.e. the actual amount of dollars physically transferred is given by $s[V_{\text{Fee}}(t) + V_{\text{Accrued}}(t)]$. The reason for writing the fee payment in this manner is simply clarity (and habit!). If the spread is absorbed into the fee payment calculation it can sometimes get lost in calculations. Since the risky duration $\sum_j Z(0,T_j)\Delta_j S(0,T_j)$ is the most important quantity to capture it makes sense to leave the spread out of the fee leg calculations.

3.3.2 Valuing the contingent leg

To calculate the value of the contingent payment we have to account for the fact that the reference entity can default at any time during the lifetime of the contract $[0, T]$. The survival probability for the obligor from time 0 to time t is $S(0,t) = \mathrm{e}^{-ht}$. In a small interval Δt the probability of default is given by $h\Delta t$, i.e. the expected default rate per unit time multiplied by the time interval. The probability of defaulting in the interval $[t, t+\Delta t]$, conditional on survival up until time t is then given by

$$p_{\text{Def}}(t, t+\Delta t|t) = S(0,t)h(t)\Delta t$$

and the contingent payment is $Z(t, t+\mathrm{d}t)(1-\delta)N p_{\text{Def}}(t, t+\Delta t|t)\mathrm{d}t$ (assuming a constant hazard rate over the period $[t, t+\mathrm{d}t]$). Over the whole period $[0, T]$ the total contingent payment is calculated as the integral (summation) of these infinitesimal payments over that period. That is

$$V_{\mathrm{Cont}}(t) = (1-\delta)N \int_0^T Z(t,t')S(t,t')h(t')\mathrm{d}t'.$$

The contingent leg can be calculated via an appropriate numerical procedure. It can also be approximated as (compare this with the expression in Chapter 2)

$$V_{\mathrm{Cont}}(t) = (1-\delta)N \sum_{j=1}^{N} Z(t,T_j)[S(t,T_{j-1}) - S(t,T_j)].$$

This expression for the contingent leg is probably the most useful representation to use on a day-to-day basis.

If we assume that $h(t) = h$ and that the interest rate is also deterministic we can compute the integral above to be

$$V_{\mathrm{Cont}}(0) = (1-\delta)Nh\frac{1}{r+h}\left[1 - \mathrm{e}^{-(r+h)T}\right] = (1-\delta)hV_{\mathrm{Fee}}(0).$$

In these expressions we have again assumed that interest rates and hazard rates are independent. In addition to this we have made the assumption that the recovery rates are deterministic and independent of the other factors (interest and hazard rates).

3.3.3 Calculating the CDS PV

The CDS PV is computed according to (neglecting for the moment the accrued payment)

$$V(t) = \phi\left[-sV_{\mathrm{Fee}}(t) + V_{\mathrm{Cont}}(t)\right]$$

where s is the contractual spread of the contract. The long/short flag $\phi = \pm 1$ indicates whether we are purchasing or providing protection. If $\phi = +1$ then the PV is calculated from the perspective of the protection purchaser. That is, the PV of the CDS is equal to the fee payments that must be made $-sV_{\mathrm{Fee}}(t)$ (cash outflows) plus the contingent payments that are received $V_{\mathrm{Cont}}(t)$ (cash inflows). If $\phi = -1$ then the direction of the cashflows is reversed and the PV is computed from the perspective of the protection seller. Selling protection is referred to as being long risk.

3.3.4 Calculating the CDS par spread

The concept of the CDS par spread is analogous to that for vanilla interest rate swaps. Interest rate swap contracts are typically entered into such that the economic benefit to both parties at contract inception is equal. If this were not the case there would be no incentive for counterparties to take opposing views. Similarly CDSs are entered into on the same basis.

The par CDS spread is that spread which gives the contract a fair value of zero at inception. It is the spread to apply to the fee payments to ensure that the expected fee payments equate the expected contingent payments. It is given by setting $V(0) = 0$ and is

$$s_{\text{Par}} = \frac{V_{\text{Cont}}}{V_{\text{Fee}} + V_{\text{Accrued}}}.$$

From the expressions for the fee and contingent payments in the limit of continuous coupon payments, as $h \to \infty$ then $V_{\text{Cont}} \to \infty$ and $s \to \infty$. That is, as the obligor becomes riskier the premium demanded by the seller of protection increases (as expected).

3.3.5 Calculating the CDS mark-to-market value

At contract inception the CDS has a par value of zero. This is calculated using the market observed survival curve at that point in time. As time passes, however, the survival curve will evolve (more on this later) since the observed par CDS spreads move on a day-to-day basis depending upon the market's perception of the default risk of the reference obligor. This implies that the values of the fee and contingent payments will move away from their initial values. The overall value of the CDS will therefore change (becoming either positive or negative depending on how the market moves and from whose perspective we are calculating the PV).

Specifically the value of a CDS position at time t is the difference between the market implied value of the protection and the cost of the premium payments that have been set contractually at time zero (multiplied by the duration of the contract). Let the market spread of the CDS at time 0 be given by $s(0)$ (this is the par spread of the contract). If the current market credit spread is given by $s(t)$ then the MtM value of the CDS is given by

$$\text{MtM}(t) = V(t) - V(0) \approx [(s(t) - s(0)]V_{\text{Fee}}(0)$$

(where we have assumed that the risky duration is constant through time – which is not a bad approximation in practice).

The notation used here is somewhat non-standard (to maintain consistency with the rest of the text). It is more common to see this expression written as $\text{MtM}(t) = [(s(t) - s(0)]\text{dV01}$. If the credit spread does not vary with time then

$\text{MtM}(t) = 0$ for all time, i.e. if the market's perception of the likelihood of default of the obligor does not change over time then there is no change in the market value of the protection, as we would intuitively expect.

3.3.6 The credit triangle

Under certain simplifying assumptions there is an extremely useful rule-of-thumb relation between the credit spread, hazard rate and recovery rate of an obligor. To calculate the par spread we equate $sV_{\text{Fee}} = V_{\text{Cont}}$. In the case where the fee payments are assumed to be paid continuously we have previously calculated expressions for the fee and contingent payments. This leads to the 'credit triangle'

$$s = h(1 - \delta).$$

For example if $s = 60$ bp and $\delta = 40\%$ then $h = 1\%$. The one-year survival probability is given by $S_1 \sim 99\%$, the two-year survival probability by $S_2 \sim 98\%$ and so on. This is a simple but relatively accurate model which works well when the interest rate and credit curves are flat.

3.3.7 Valuation via Monte Carlo simulation of obligor default times

Although a closed form analytical pricing formula exists for a CDS it is useful also to value a CDS using Monte Carlo simulation of the default time of the single reference entity. This method is useful because it makes very transparent what is going on. As we will see in later chapters where we consider more complex portfolio products, thinking of the behaviour of the product in terms of Monte Carlo simulation can add considerable insight to our understanding of the product. This method is a brute force approach to valuation and would not be appropriate for rapid revaluation of a large portfolio of CDSs. However, it can serve as a good independent test of the implementation of a model.

The idea is simple (see Chaplin [2005] for more discussion of this method). We need to generate a number at random in the range $[0, \infty)$ for the obligor's default time. This is achieved by drawing a random number from virtually any distribution and applying an appropriate transformation to rescale it into the required range. For example, we can simulate a random number between 0 and 1 drawn from a uniform distribution. Given this number we calculate what time input would give a survival probability equal to this number; this is the default time. Since the default time is known all of the cashflows (and their timings) in the CDS are known and the par spread and PV etc. can be calculated. This procedure is repeated a large number of times and the average of these quantities calculated. The algorithm is as follows:

for $\alpha = 1$ to N_{Sims}

sample $x^\alpha \sim U(0, 1)$ // sample a uniformly distributed random variable

calculate $\tau^\alpha = -\frac{1}{h} \ln x^\alpha$ // default time

calculate $V^\alpha_{\text{Fee}} = N \int_0^{\min(\tau^\alpha, T)} e^{-rt} t$

calculate $V^\alpha_{\text{Cont}} = N(1 - \delta) e^{-r\tau^\alpha} \mathbf{1}_{\tau^\alpha < T}$

end loop over α

compute $V_{\text{Fee}} = \frac{1}{N_{\text{Sims}}} \sum_{\alpha=1}^{N_{\text{Sims}}} V^\alpha_{\text{Fee}}$

compute $V_{\text{Cont}} = \frac{1}{N_{\text{Sims}}} \sum_{\alpha=1}^{N_{\text{Sims}}} V^\alpha_{\text{Cont}}$

calculate $s_{\text{Par}} = V_{\text{Cont}} / V_{\text{Fee}}$.

In this example algorithm we are generating the default times by sampling from a uniform distribution. We could also sample from any other distribution. For example if $x^\alpha \sim N(0, 1)$ then the default time is computed from

$$\tau^\alpha = -\frac{1}{h} \ln N(x^\alpha)$$

(it is obvious that this is a number in the range $[0, \infty)$ as required). In order to improve the convergence of the Monte Carlo estimator we can use a variance reduction technique (see Glasserman [2003] for a good discussion of variance reduction techniques). The simplest variance reduction technique is to use antithetic variables. For the two example distributions used in the example above the antithetic variable pairs are $(x^\alpha, 1 - x^\alpha)$ for the uniform distribution and $(x^\alpha, -x^\alpha)$ for the normal distribution. Within the simulation loop the CDS can be revalued twice using each of the antithetic variates.

3.4 Calibration of the survival curve to market observed data

In previous sections we have constructed a model for valuing the fee and contingent legs of a CDS and thereby calculating the par CDS spread. In this section we outline how to calibrate the model to market observed CDS par spreads (see also Galiani [2003] for more details and some numerical examples). The corporate and sovereign CDS market is highly liquid. Like corporate and sovereign debt, CDS exposures have a time-varying value and also a maturity dependence. That is to say it is possible to purchase or sell default protection for different time horizons, for example, 1 year, 5 years etc. The liquidity in the CDS market is such that efficient price discovery in the market for a wide range of corporate and sovereign exposures is possible (for different maturities). It is important therefore that any model for valuing CDS exposures must calibrate to these market observed prices. This is particularly obvious if one considers that the market observed CDS spreads are a reflection of the aggregate market view of the default likelihood of an obligor.

The fee and contingent payments for a CDS are given by

$$V_{\text{Fee}}(0) = \sum_{j=1}^{N} Z(0, T_j)\Delta_j e^{-\int_0^{T_j} h(t)(T-t)\mathrm{d}t},$$

$$V_{\text{Cont}}(0) = \sum_{j=1}^{N} Z(0, T_j)(1-\delta)\left[e^{-\int_0^{T_{j-1}} h(t)(T-t)\mathrm{d}t} - e^{-\int_0^{T_j} h(t)(T-t)\mathrm{d}t}\right],$$

respectively (in what follows we will ignore the accrued fee payments for simplicity). In these expressions the hazard rate is no longer assumed to be a continuous function of time (or a constant value), but is written as piecewise constant. The reason for this will become clear shortly. We assume that there are a set of market observed par CDS spreads for maturities $\{s_{\text{Market}}(T_k) : k = 1, \ldots, M\}$. The spreads are typically observed for maturities 3 months, 6 months, 1 year, . . . etc. (and there are assumed to be M of them).

In the model the recovery rate is assumed to be a constant. The value can vary but is typically in the range 20–40% (the lower bound applies to sub-ordinated debt, the upper bound to senior debt and can vary depending upon the particular trading desk). In order to calibrate the model such that it replicates market observed par CDS spreads (for the M maturities) the only free parameter in the model that we can vary is the hazard rate $h(t)$. In order to proceed further it is necessary to make some assumptions about the functional form of the hazard rate. Assuming the hazard rate is constant, i.e. $h(t) = h$, enables a single point on the term structure to be calibrated to (e.g. the 5 year point). To calibrate to the entire set of par CDS spreads at times 3 months, 6 months, 1 year, . . . it is necessary to introduce more structure into the hazard rate function.

The most common (and simplest) assumption is to assume that the hazard rate is a piecewise constant (increasing) function of maturity. That is to say the hazard rate is assumed to be of the form

$$\begin{aligned} h(t) &= h_0 \qquad 0 < t < T_1 \\ h(t) &= h_k \qquad T_{k-1} < t < T_k \qquad k = 1, \ldots, M \\ h(t) &= h_M \qquad T_M < t < \infty. \end{aligned}$$

The hazard rate is constant for times $t \in [T_{k-1}, T_k]$, and jumps discontinuously at these points. The calibration problem is therefore to use the M values of h_k to fit the M market observed par CDS spreads. Since we are considering a piecewise continuous function we approximate the integrals appearing in the fee and contingent legs above as

$$\int_0^{T_j} h(t)\mathrm{d}t \approx \sum_{k=0}^{j} h_k \Delta t_k$$

where the day-count fraction is $\Delta t_k = T_k - T_{k-1}$.

The procedure for going from the market observed par credit spreads to the calibrated marginal survival curve is illustrated schematically in Figure 3.3. The calibration procedure proceeds as follows. Starting with the shortest maturity T_1 we calculate

$$V_{\text{Fee}}(T_1) = Z(0, T_1)\Delta_1 \mathrm{e}^{-h_1(T - T_1)}$$
$$V_{\text{Cont}}(T_1) = Z(0, T_1)(1 - \delta)\left[\mathrm{e}^{-h_1 T} - \mathrm{e}^{-h_1(T - T_1)}\right].$$

We then solve $s_{\text{Market}}(T_1) = s_{\text{Model}}(T_1) = V_{\text{Cont}}(T_1)/V_{\text{Fee}}(T_1)$ for the hazard rate h_1. The process is iterated for the next maturity T_2 to solve for h_2 (utilising the previously determined value for h_1). In this manner the hazard rate term structure is bootstrapped from one maturity to the next. Once the hazard rates are known, the survival curve can be computed.

From a practical point of view the calibration procedure needs to be performed using an appropriate numerical scheme such as a root finding algorithm. Although the calibration procedure is simple in concept, care must be taken in the detail of the numerical implementation to ensure that the survival probability has the correct properties (i.e. $0 \leq S(0, t) \leq 1$).

An alternative, and simpler, calibration procedure [Cherubini *et al.* 2004] is the following. For a CDS of maturity T_N the par swap rate s_N is defined from

$$s_N \sum_{j=1}^{N} Z(0, T_j)\Delta_j S(0, T_j) = (1 - \delta) \sum_{j=1}^{N} Z(0, T_j)\left[S(0, T_{j-1}) - S(0, T_j)\right].$$

For the first maturity T_1 we observe the market spread and we can solve for the survival probability up to this time according to

$$S(0, T_1) = \frac{(1 - \delta)}{s_1 \Delta_1 + (1 - \delta)}.$$

For subsequent times $M \geq 2$ we can solve for the survival probability from

$$S(0, T_M) = \frac{s_{M-1} - s_M}{Z(0, T_M)\left[s_M + (1 - \delta)\right]} \sum_{j=1}^{M-1} Z(0, T_j) S(0, T_j)$$
$$+ S(0, T_{M-1}) \frac{(1 - \delta)}{s_M + (1 - \delta)}.$$

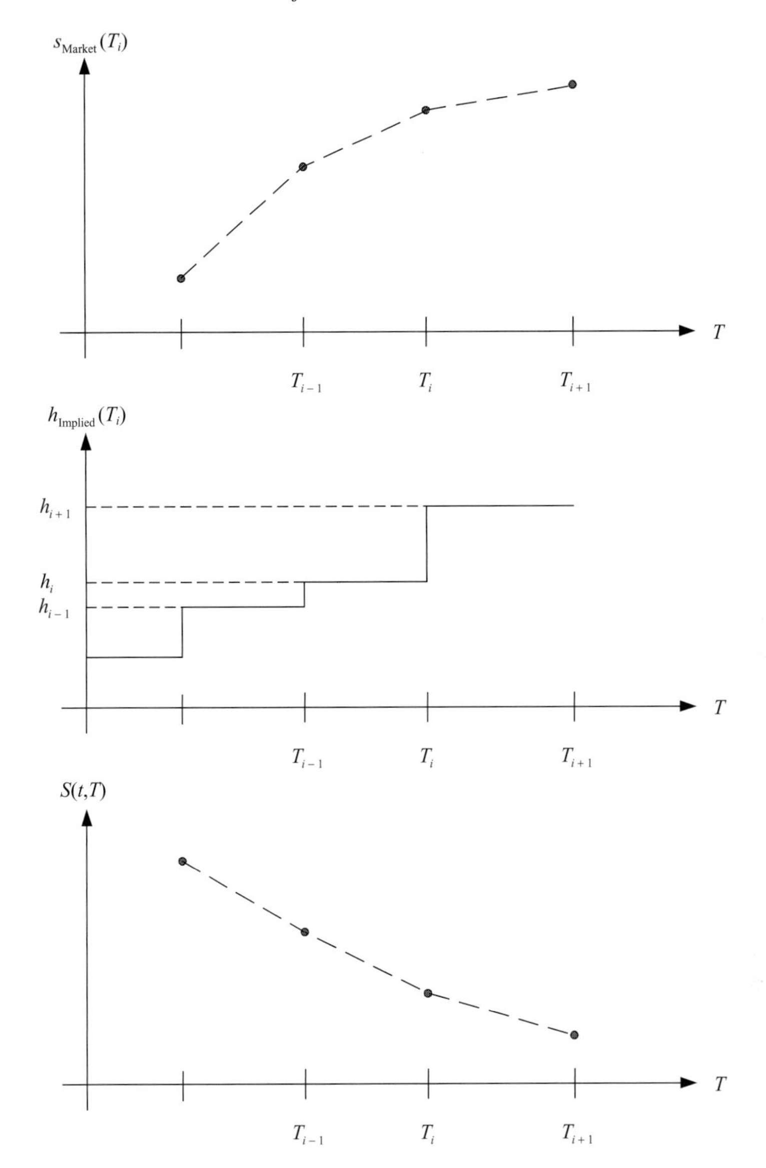

Figure 3.3 Schematic diagram illustrating the procedure for calibrating the marginal obligor survival curve to market observed par CDS spreads. The top figure shows the source market data, the middle figure the assumed parametric form for the hazard rate function and the bottom figure the bootstrapped survival curve.

Once again the survival curve can be bootstrapped from the market observed spreads. The second method is easier to implement than the first since it does not require any numerical root finding procedure.

3.5 Risk sensitivities of vanilla CDSs

In this section we describe the valuation sensitivities of a CDS (following some of the definitions to be found in Douglas [2007]). CDSs are typically subject to a number of different types of risk. Broadly speaking these can be classified as either market or credit risk. However, because a CDS is a contract with three counterparties there are additional practical risks that must also be considered.

3.5.1 Counterparty risk

Counterparty risk refers to the fact that the protection purchaser, for example, is contractually expecting to receive a contingent payment from the protection seller in the event of a default event of the reference entity. However, it is conceivable that the protection sellers may themselves default since they are after all just another reference entity (it may be inadvisable to purchase protection on a sub-prime mortgage lender from an investment bank with a large exposure to the sub-prime sector!). If the protection seller defaults before the reference entity the protection purchaser is left without any protection whatsoever. They will be forced to go back out into the marketplace and purchase protection on the reference entity again from another protection seller at an additional cost (and at the current market spread level). Conversely, the protection purchaser themselves may also default before the reference entity. In this case the protection seller's obligations to make a contingent payment cease, but on the other hand the positive impact of the fee payments is also lost.

A significant risk factor with respect to CDSs is therefore not just the default behaviour of the reference entity, but also the default behaviour of the actual contract counterparties. The correlation between the default behaviour of the three counterparties in the contract is of importance (but perversely this is not a factor that can be easily incorporated into the standard valuation model).

3.5.2 Recovery risk

In the event of a default of the reference entity (but not the other two counterparties) the protection seller is obligated to make a compensating payment to the protection purchaser. This can be in one of two forms: physical or cash delivery as explained earlier. The problem here is that throughout the life of the CDS up until this point it has been valued using an assumed amount for what would be recovered in the event of a default of the reference entity. In the actual event of a default the realised recovery rate can be very different from the assumed rate.

In particular there may be a complex interrelationship between the default of an obligor, the liquidity of its (distressed) bonds trading in the market and the value of its traded debt. For example, in the aftermath of a default of a particularly popular reference entity there may be sustained pressure on the bonds of that obligor. This is because the total notional amount of protection written on that obligor may actually exceed the amount of debt that the obligor has issued. Counterparties which are obligated to deliver these bonds in return for their face value may find they are competing with many other counterparties to purchase the required amounts of the distressed debt. This in turn can lead to upward pressure on the (distressed!) debt.

3.5.3 Quantifying market risk – CS01s

Market risks are those risks that affect the value of a financial instrument on a day-to-day basis. These include interest rate and foreign exchange rate risk and volatility risk (for derivative contracts). These are the traditional types of risk that one associates with equity, interest rate and FX instruments. Credit derivatives are also exposed to these risk factors. However, much more importantly from a single-name credit derivative point of view is the risk due to spread movements (interest rate and FX risks are typically second-order effects).

The values of the fee and contingent legs of a CDS are dependent upon the survival curve. The survival curve in turn is derived from the market observed par CDS spreads for the reference entity. If the CDS spreads widen (implying that the market perception of the credit quality of the reference entity is decreasing – meaning the entity is more likely to default according to the market), this has the impact of reducing the survival probabilities, in turn impacting the value of the fee and contingent payments. There is a clear linkage between spread movements and the value of the CDS. To quantify this risk we can calculate the *credit spread sensitivity* to a (one basis point) shift in the par CDS spreads, known as the CS01. Formally this is calculated as (using a forward difference approximation to the differential)

$$\Delta = \frac{V_{\text{CDS}}(s + \delta s) - V_{\text{CDS}}(s)}{\delta s} \approx \frac{1 - \mathrm{e}^{-(r+h)T}}{r + h}$$

(in the case where the coupons are paid continuously). That is, the sensitivity to the credit spreads is approximately equal to the risky duration $\Delta = \text{dV01}$. This can be converted into the sensitivity to the hazard rate by using the credit triangle. In the limit of low default intensity we have that $\Delta \approx T$.

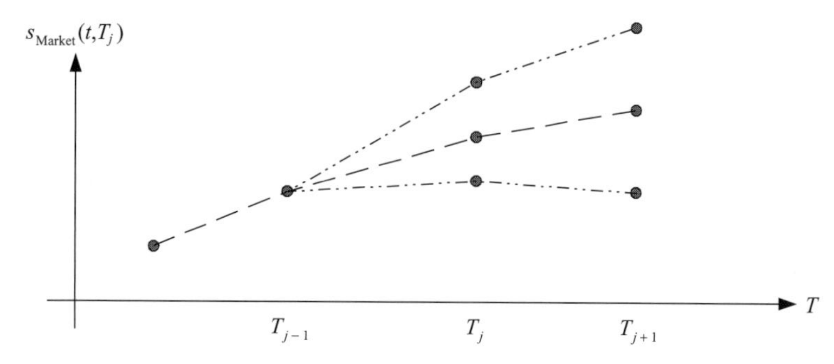

Figure 3.4 Spread curve rotation. The par spreads at maturities T_j and T_{j+1} can move relative to the spread at T_{j-1}.

In practice it is advisable to compute sensitivities using central differences rather than one-sided differences. This avoids any potential inaccuracies introduced by convexity.

In this simple example we have again assumed that the term structure of hazard rates is flat and that a single maturity par CDS spread is used to bootstrap the survival curve. In this case there is only a single spread to bump to calculate the sensitivity. In practice of course there is a spectrum of spread maturities used to compute the survival curve. This gives rise to the possibility of spread curve rotation, illustrated schematically in Figure 3.4. The fact that the spread curve can rotate, twist, kink etc. will be reflected in the CDS valuation which is sensitive not only to the level but also to the shape of the spread curve. To quantify the impact of non-parallel movements of different points on the spread curve we can calculate 'bucketed' CS01s.

To calculate bucketed CS01s we proceed as follows. For the set of market observed par CDS spreads $\{s_{\text{Market}}(T_k) : k = 1, \cdots, M\}$. Each of the spreads is bumped systematically (starting from the shortest maturity up to the longest) according to $s_{\text{Market}}(T_k) \to s_{\text{Market}}(T_k) \pm 1\text{bp}$. The new default curve is then bootstrapped from the shifted spreads. Finally the instrument is revalued and the CS01 calculated. Note that the points are bumped individually (hence the curve is not shifted in parallel). This can lead to problems with the default curve (negative probabilities) particularly if neighbouring rates are very close. If a point of maturity T_k is bumped then only hazard rates at maturities greater than this are affected.

Operationally the calculation of CS01 is the sensitivity of the CDS value to a 1 bp move in the credit spread. However, in reality credit spread movements are never bounded within this range. Because the value of a CDS is a non-linear (convex) function of the credit spread (via the risky duration) it is not possible to say that the sensitivity of an x bps move is equal to $x \times \Delta$. The reason for the convexity is the eventual onset of default which bounds losses. Simple CS01 on its own can therefore be a misleading risk measure particularly when the market is experiencing a period of significant spread volatility.

3.5.4 Quantifying default risk – value-on-default (VoD)

Credit derivatives of course also have an additional dimension of risk: default risk. The default risk is borne by the protection seller. To quantify the default risk we can introduce the *value-on-default*. This measures the change in the MtM value of the trade due to the event of a sudden default. In the event of a default two things happen.

- The protection seller is obligated to compensate the protection purchaser with the contingent payment. For example, in the case of cash settlement the protection seller makes a payment of $N(1-\delta)$ to the protection purchaser.
- The CDS terminates and the current MtM value (which can be positive or negative) of the CDS to the protection seller or purchaser is paid.

The VoD is therefore given by the sum of these two contributing factors:

$$\mathrm{VoD}(t, T) = \phi\left[-N(1-\delta) - \mathrm{MtM}(t, T)\right].$$

3.6 Chapter review

In this chapter we have introduced credit default swaps, which are the simplest and most vanilla of credit derivatives. We have described the contract mechanics, valuation within the context of the standard (intensity based) market model and their risk sensitivities. As we will see in subsequent chapters the modest CDS (and variations on the basic concept) has become a fundamental building block for the whole portfolio credit derivative industry, playing a vital role in terms of providing liquidity as well as being a mechanism whereby an investor can take synthetic exposure to a particular counterparty without actually physically owning the debt obligations of the counterparty. The importance of CDSs cannot be overestimated. In the next chapter we will introduce and analyse the first example of a portfolio product, namely credit indices, which extend the concepts of CDSs to a portfolio of reference entities. As we will see, these instruments also play an absolutely fundamental role in the functioning and ongoing development of the credit markets (they are in fact the most liquid credit derivatives currently in existence).

4
Credit indices

4.1 Introduction

In this chapter we introduce the credit indices. A credit index is a natural generalisation of an individual CDS to a portfolio of CDSs. Since their introduction around 2004 (in their current incarnation as iTraxx and CDX), they have become the most liquid of all credit derivative instruments.

As the single-name CDS market became more liquid and standardised (particularly in terms of documentation setting out standard terms and conditions) it became apparent that there was investor demand for increasingly sophisticated credit products which facilitated the trading of complex market views. The introduction of standardised indices in the same spirit as the indices which exist for equity and other markets, with transparent and stable construction and trading rules, provided an enormous impetus to the credit derivative market. Credit index products enable investors rapidly to take a macro view of the overall credit market and to trade accordingly (directional and relative value trades, leveraging credit investments and going long and short the market) – although this also introduces volatility (which can also be traded via index options). They have introduced a whole new dimension of trading and investment strategies for investors. Like single-name CDSs, index trades are traded in the privately negotiated OTC market. Liquidity was further enhanced by the introduction of tranched exposures to specific portions of the index's portfolio loss distribution (the index single-tranche synthetic CDO market). Add to this the fact that index positions can be used (with some success) to hedge tranche positions as well as broad credit market movements, and the overall effect upon the market has been significant. Furthermore, the credit indices have come to be seen as a barometer of many different facets of the financial system: a forward indicator of likelihood of default (both idiosyncratic and systemic), an indicator of leveraged buy-out (LBO) activity, a leading indicator of equity market movements. They are watched not just by credit flow traders but also by many other investors, particularly in light of the events in the credit markets in the summer of 2007.

The structure of this chapter is as follows. We firstly introduce the different credit indices and describe their current composition and construction in Section 4.2. Following this the trading mechanics of the indices will be described in Section 4.3. In terms of valuation, a portfolio of CDSs differs from a single CDS in a number of important respects which will be described in Section 4.4. The indices have been trading liquidly for a number of years meaning that there is at least some historical data to analyse to understand the dynamics of the credit market. We undertake some simple analysis of historical index data in Section 4.5 in order to understand better the properties and characteristics of the credit markets. In Section 4.6 we introduce the standardised tranches defined on the index reference pools. Finally in Section 4.7 we summarise the important points that have been introduced in the current chapter.

4.2 Description of the credit indices

We now provide a brief description of the credit indices. Full details of the composition and statistics (industrial concentrations etc.) of the current indices can be found on the index administrator's website [MarkIt]. One of the key factors in the popularity of credit indices is the fact that the rules for their composition and construction are transparent. Mostly the indices are aggregated in the Dow Jones CDX/iTraxx credit derivatives index families. Index trading is brokered by a group of investment banks who act as market makers[1] and who provide daily market quotes. Markit Group are the administrator, calculation and marketing agent for the CDX and iTraxx indices respectively (and also act for the ABX and TABX indices discussed in Chapter 11). They conduct the voting for choosing credits to be included in a new index, collect current spread levels and publish historical spreads.

Indices are constructed from equally weighted portfolios of liquid names in the CDS market. Broadly speaking there are two families of credit index:

- the iTraxx family of indices covering European and Asian credit exposures;
- the CDX family of indices covering North American credit exposures.

Within each index there are further subdivisions of obligors. For the iTraxx family we have the following.

[1] Crudely speaking, market makers buy and sell securities with the objective of providing liquidity to the market. The securities are available to clients to purchase or sell. In exchange for providing this service the market maker charges a spread. The market maker has an order book where they match buy and sell orders to arrive at a market price for an asset.

- The iTraxx main index: composed of 125 equally weighted European obligors with investment grade ratings. Maturities of 3, 5, 7 and 10 years are the most common.
- The iTraxx crossover index: this is currently composed of 50 equally weighted obligors with non-investment grade ratings. Maturities of 5 and 10 years are the most common.
- The iTraxx HiVol index: composed of 30 obligors with the widest spreads chosen from a subset of the main index.
- iTraxx Asia ex Japan: composed of 50 obligors.
- iTraxx Australia: composed of 25 obligors.
- The iTraxx Japan index: composed of 50 high-grade obligors. Typical maturities are 3, 5 and 10 years.

For the CDX family we have the following.

- The CDX investment grade index: composed of 125 equally weighted North American obligors. To be included in the index each obligor must have an investment grade rating from both Moodys and S&P. Typical maturities are 3, 5, 7 and 10 years.
- The CDX HiVol index: composed of 30 obligors, generally the credits in the investment grade index with the widest spreads.
- The CDX high yield index: composed of 100 obligors. For inclusion in this index an obligor cannot have an investment grade rating from both Moodys and S&P.
- The CDX crossover index: originally composed of 35 equally weighted obligors with a combination of B ratings from Moodys and S&P.

Low spreads and low volatility of investment grade assets lead to the creation of the sub-indices such as the crossover indices described above. These are typically composed of higher yield (riskier) assets hence providing more movement and volatility for investors to express views on.

The composition of the credit indices is determined by a poll of the various market makers. The index composition is modified (rolled) on a six monthly basis (on the 20th of March and September). At the roll the premium (spread) of the index is set. The roll process and the substitution of credits are governed by transparent rules. In most cases defaults, rating downgrades/upgrades and corporate reorganisations force credits out of the specific index. Credits can also be added/removed based on liquidity issues although this is utilised less in practice in order to keep the index composition as stable as possible over time. However, it is inevitable that the composition of the index will evolve gradually over time. Full details of the composition of the current index series can be found on the website of the index administrator.

The current version of the index is referred to as the 'on-the-run' index and is identified by its 'series' name. For example, the first CDX investment grade index was referred to as CDX.NA.IG (North American investment grade) series 1. It had a 5 year maturity of 20th March 2009. The current on-the-run index (at the time of writing) is series 8 for CDX.NA.IG and series 7 for iTraxx. The 'on-the-run' index is almost always much more liquid than off-the-run indices, although off-the-run indices can continue to trade actively. When an index rolls, an investor can roll their positions into the new index.

Indices can be traded either in funded or unfunded (swap) form – the latter being the most common form. Trade clip sizes are typically \$50–\$100 m. Maturities are standardised at 3, 5, 7 and 10 years (5 and 10 year trades are the most actively traded). However, few index trades are held to maturity as counterparties will want to ensure their positions are as liquid as possible. Counterparties will therefore 'roll' their positions into the new index every 6 months (by unwinding the old series and purchasing the new series). If an investor does not want to roll their position then the old index continues trading based on the original reference portfolio (less any defaulted counterparties).

4.3 Index trading mechanics

The mechanics of an index trade are similar to those of a plain CDS but with a number of additional features. Consider entering into a position in the iTraxx index. The two counterparties to the deal are the market maker (MM) and counterparty ABC. We assume ABC (which may be a hedge fund, for example) is selling protection, therefore the MM (which is typically an investment bank) makes the periodic coupon payments to ABC (who compensates the MM for any defaults which occur in the index). The spread payable on an index is fixed at the inception of the life of the index for the total life of the index (usually 6 months until the next roll).

Assume that there are n obligors whose CDSs make up the index. Let the notional amount of protection purchased be given by N. Let the maturity of the protection purchased be T and let the coupon payments be due on the dates $\{T_j : j = 1, \ldots, N\}$. Typical maturities are 3, 5, 7 and 10 years and the coupon tenor is usually quarterly. We assume that the index premium for series X (of maturity T) is $s_X(T_{\text{Roll}})$. This premium is set at time T_{Roll} (the index roll-time is usually 20th March and September of each year). For example, the spreads for the iTraxx main series 6 index are shown in Table 4.1. As the maturity of the contract increases the spread premium for providing protection also increases (as expected). At some later time $T_{\text{Roll}} < t$ the spread of the index is $s_X(t) \neq s_X(T_{\text{Roll}})$. The recovery rate of the index constituents is given by δ_i, the day-count fraction is $\Delta_j = T_j - T_{j-1}$. In the absence of any credit events which would trigger a contingent payment, the protection purchaser will make coupon payments to the protection seller at each coupon payment date T_j of $s_X(T_{\text{Roll}})\Delta_j N$. These payments continue until the maturity of the contract. Upon the occurrence of a credit event

Table 4.1 *iTraxx main index series 6 traded spreads*

Maturity	Premium (bps)
20 December 2009	20
20 December 2011	30
20 December 2013	40
20 December 2016	50

the protection seller must compensate the protection purchaser for the loss. In a plain vanilla CDS which references a single obligor, the occurrence of a single credit event triggers the complete unwind of the contract. In an index trade, however, the first default does not unwind the whole trade. Instead the protection seller makes good the loss and the trade continues. However, subsequent coupon payments received by the protection purchaser are made based on a notional amount which is reduced by the amount lost. If we represent the default times of the obligors in the index as $\{\tau_i : i = 1, \ldots, n\}$ then the coupon payment at date T_j is given by

$$s_X(T_{\text{Roll}})\Delta N \left[1 - \sum_{i=1}^{n} \delta_i \mathbf{1}_{\tau_i < T_j}\right].$$

This payoff mechanism continues until there are no further obligors left in the underlying pool which have not defaulted. Of course long before an entire index's portfolio has defaulted the world will have ended.

It will be noted that each of the coupon payments is based on a spread $s_X(T_{\text{Roll}})$ which is fixed for the lifetime of the particular series. As time passes, the spreads of the individual constituents evolve and hence the contemporaneous index spread evolves. However, for an investor purchasing an index mid-life (between the rolls) the same initial spread is paid. In order to compensate investors for the disparity, an upfront payment is exchanged. For example, consider selling protection on the iTraxx main index (for a particular series and a particular maturity). Assume that the spread of this iTraxx series is 30 bps. This spread is fixed at the roll of the index on either March 20th or September 20th. It does not change for the duration of this series of the index. Assume that we are entering into the contract on 20th October. At this date the spread of the index will not be equal to 30 bps, but could have moved out to, for example, 60 bps. We are selling protection on iTraxx for which the contractual spread that we will receive is 30 bps. However, the current market spread (risk) of the index is greater at 60 bps. As the seller of protection we need to be compensated for the fact that the contract spread is only 30 bps, but the market risk is 60 bps. This is facilitated by an upfront payment. In this case we, as protection seller, receive an upfront payment of (60 − 30 bps) × dV01. If the market spread was instead 20 bps and we were selling protection, we would be receiving compensation of 30 bps for providing protection in a market which is pricing risk at 20 bps. Therefore in this case we would have to make an upfront payment of (30 − 20 bps) × dV01. Returning to the case where we are selling protection, we enter into the trade on 20th October. However, we receive a full coupon for providing protection over the period 20th September–20th December. Clearly we are not providing protection for the full period. We therefore have to make a compensating payment for the accrued coupon payment (approximately one-third of the coupon payment received in this case) to the protection purchaser. Typically these upfront payments are settled at $T + 3$ days.

Index trading is highly liquid with low dealing costs and very tight bid-ask spreads. Because of this liquidity, the index spread often trades at a different value to its theoretical value (the weighted average of the spreads of the individual constituents described in the next section). That is to say, market forces of demand and supply drive the value of index contracts. The dynamics of the index may be different to the dynamics of the individual CDS constituents giving rise to a difference in spreads. This is known as the basis. A positive basis to theoretical implies that $s_{\text{Index}} > s_{\text{Theoretical}}$ and vice versa. In general a basis (positive or negative) implies that $s_{\text{Index}} \neq \sum_i w_i s_i / n$ (traded index spread is not equal to the weighted average spread) and $s_{\text{Index}} \neq s_{\text{Theoretical}}$ (traded index spread is not equal to the theoretical index spread – computed later).

The basis exists because indices, owing to their ease of execution, can react more quickly to changing market sentiment than the (125) individual CDSs and also due to differences in liquidity between the index and single-name markets. Arbitraging the basis is difficult, however, because of the costs etc. involved with trading large numbers of single-name CDSs. It is often observed that in volatile markets the magnitude of the basis is larger than in less volatile times. The basis can act as a barometer of market sentiment. A positive basis typically occurs when most participants want to buy protection on the market as a macro hedge reflecting a bearish sentiment. A negative basis typically occurs when most participants want to go long the market, pushing index spreads below their theoretical value due to high demand for credit risk reflecting a bullish sentiment.

Hedge funds are big users of index trades because of their low execution costs and liquidity (allowing positions to be entered and liquidated quickly with tight bid-offer spreads). A typical example of usage of an index trade is to express a macro view of the market, i.e. it can go long or short the index depending on the particular view of systemic spread movements. Other examples of index trading strategies can be found in Rajan *et al.* [2007].

Table 4.2 *Payoffs of the index and portfolio of CDSs*

	Index payoff (bps)	Portfolio payoff (bps)
No defaults	$(s_A + s_B)/2$	$(s_A + s_B)/2$
Obligor A defaults	$((s_A + s_B)/2)/2$	$s_B/2$
Obligor B defaults	$((s_A + s_B)/2)/2$	$s_A/2$
Obligors A and B default	0	0

4.4 Valuation of credit indices

A key difference between a CDS index product and an individual CDS is in the number of possible contingent payments. For an individual CDS there is only a single contingent payment upon default and the contract terminates. For an index product there is a contingent payment every time there is a default of one of the constituents.

A credit index is simply a portfolio of (equally weighted) individual CDSs. Its present value can therefore be computed by a simple addition of the present values of the individual components. This is simple. However, what is the fair par spread of the index product? As it turns out the answer to this is intuitive, but not quite so simple.

Consider constructing an index from an equally weighted portfolio of two individual CDSs referencing two obligors denoted by A and B. Let the par CDS spreads of the two obligors be given by s_A and s_B respectively. Let the value of the two CDSs be given by $V_A = -s_A V_{\text{Fee}}^A + V_{\text{Cont}}^A$ with a similar expression for CDS B. The value of an equally weighted portfolio composed of the two CDSs (paying their respective spreads) is given by $\Pi = (V_A + V_B)/2$. The value of an index, with an associated index spread s_{Index}, is given by $\Pi_{\text{Index}} = (s_{\text{Index}} V_{\text{Fee}}^A + V_{\text{Cont}}^A + s_{\text{Index}} V_{\text{Fee}}^B + V_{\text{Cont}}^B)/2$.

If we choose the index spread as an arithmetic average of the spreads of the individual obligors

$$s_{\text{Index}} = \frac{1}{n} \sum_{i=1}^{n} s_i,$$

then the payoff of the index under various default scenarios is not the same as the payoff of the individual trades as shown in Table 4.2. Clearly the payoffs are not the same. Choosing the index spread to be the arithmetic average of the individual spreads is not the correct choice.

Consider now that we have a portfolio of n obligors making up the index. Each obligor i has a CDS PV of $V_i = -s_i V_{\text{Fee}}^i + V_{\text{Cont}}^i$. We calculate the index par spread according to

$$s_{\text{Index}} = \sum_{i=1}^{n} w_i s_i$$

where $s_i = V^i_{\text{Cont}}/V^i_{\text{Fee}}$ is the par spread of the CDS of obligor i and

$$w_i = \frac{\text{dV01}_i}{\sum_{i=1}^{n} \text{dV01}_i}$$

and $\text{dV01}_i = \sum_{j=1}^{N} Z(t, T_j)\Delta_j S_i(t, T_j)$ is the risky dV01 of obligor i. Clearly as $s_i \to \infty$, $S_i(t, T_j) \to 0$ and $\text{dV01}_i \to 0$, i.e. the risky duration approaches zero.

The index spread is therefore the risky duration weighted average of the spreads in the constituent portfolio. If all the obligors are identical then $s_{\text{Index}} = \sum_{i=1}^{n} s_i/n$. The risky duration weighted average accounts for the dispersion of obligor risk in the portfolio.

4.5 Time series analysis of credit indices

However beautiful the strategy, you should occasionally look at the results.

(Winston Churchill)

The various series of the main indices have now been trading liquidly for over 3 years. This means there is a time series of data available to analyse in order to try and uncover the drivers and dynamics of the credit markets. In this section we present some simple statistical analysis of the iTraxx main 5 year index. Figure 4.1 shows the evolution of the iTraxx index over the period September 2004 to November 2007. Also shown in Figure 4.2 for comparison is the FTSE 100 index over the same period. In general when the market's perception of the credit quality of an individual obligor declines this has the effect of driving up the CDS par spread of protection sold on that obligor. This is because if an obligor's ability to repay their debt or their general survival prospects decrease, the premium demanded by a counterparty to sell protection on that obligor will increase. Conversely, when the perceived credit quality increases the CDS spreads decrease. Similar considerations apply when scaling this argument up to an entire index of trades. However, there is also the impact of systemic and idiosyncratic behaviour. An individual obligor subject to an LBO rumour may not move the index significantly (even though the obligor's credit spread may move by several hundred per cent over a very short period of time), but a sustained period of uncertainty regarding, for example, exposure of the financial sector to sub-prime debt can move the index significantly. The credit indices can therefore act as a barometer for the health of the overall market.

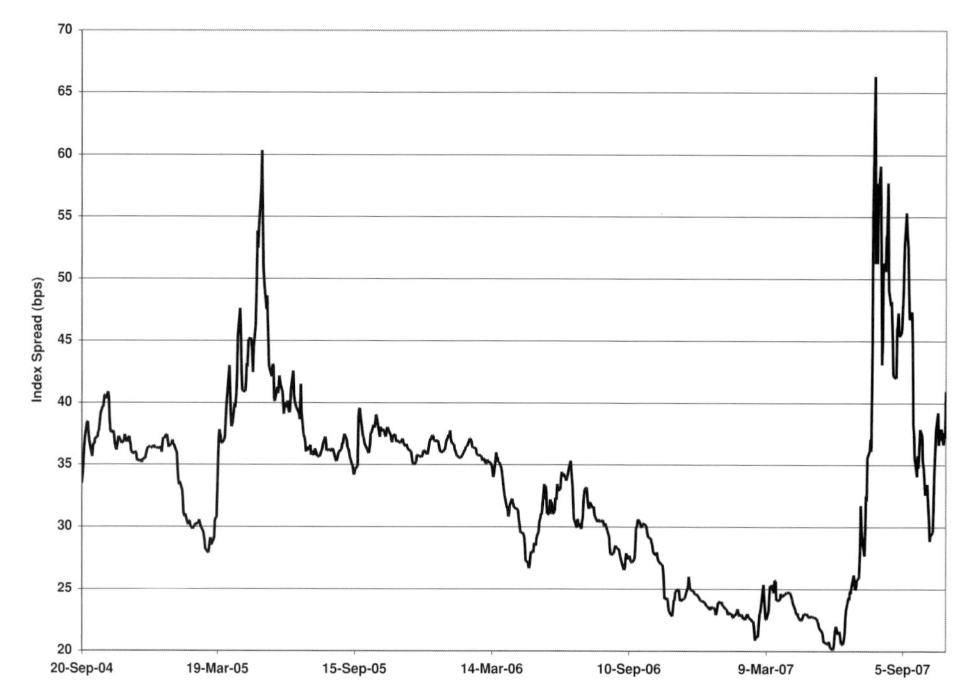

Figure 4.1 Plot of the iTraxx 5 years main index over the period September 2004 to November 2007. (Source: Bloomberg Finance L. P.)

In terms of notable events there are a number of these that are of interest. The spike in spreads around May 2005 to about 60 bps is the 'correlation crisis'. This was followed by a long period where the perceived default risk of the index's constituents was gradually decreasing towards a minimum of 20 bps until the summer of 2007. Then the credit crunch happened. This is reflected in the extremely rapid rise of the index to almost 70 bps.

It is worth noting the correlation between the behaviour of the credit index and the (in this example FTSE) stock market index. Credit and equity markets tend to move together since many investors equate the two asset classes. But credit is not like equity: it has a term structure and is much more like a bond. In particular, assuming no default, a CDS will always mature and continue to pay out on its coupons. In general, as the credit index rises, indicating a worsening in perceived credit quality, stock market indices will fall precisely because the default risk of the constituents has increased (thereby posing a threat to corporate earnings). Over the long term, credit spreads have been strongly correlated with equity market volatility. As the equity market rises, volatility of the index declines. Rising equity values imply a decrease in the default probability of the equity index constituents

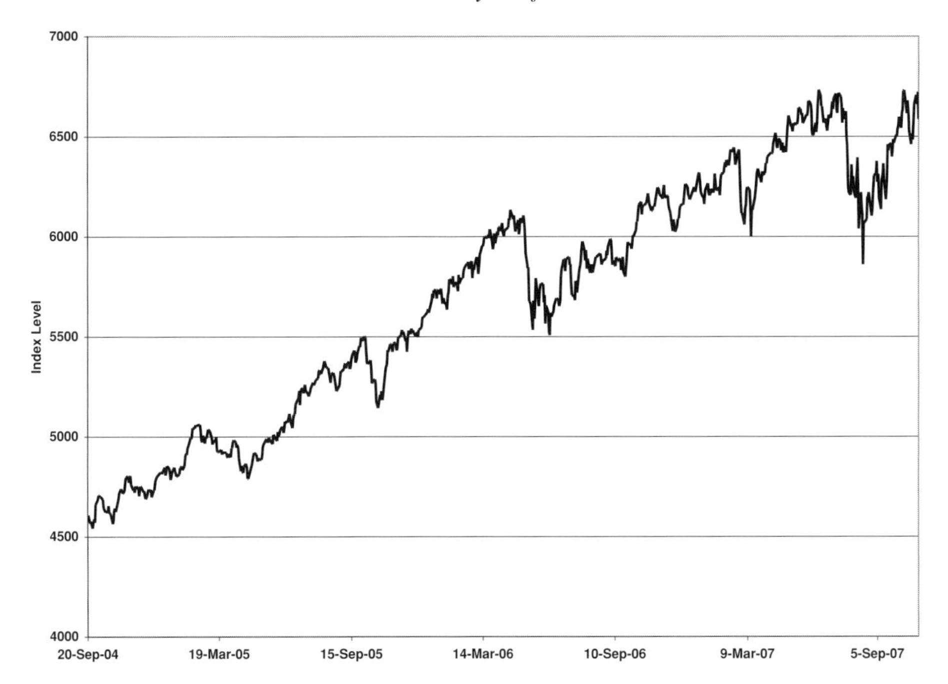

Figure 4.2 Plot of the FTSE 100 main index over the period September 2004 to November 2007. (Source: Bloomberg Finance L. P.)

leading to a tightening of par CDS spreads and hence a lowering of the credit indices. Historically the correlation between the VIX (volatility) index (also known as Wall Street's 'fear gauge') and credit indices has also been high. If the VIX is climbing, this may be a precursor to the credit indices coming off their lows for this particular credit cycle.

Figure 4.3 shows the normalised log returns of the iTraxx index over this period. The (non-normalised) return at time t is defined as $R_t = \ln s_t - \ln s_{t-1}$ where s_t is the value of the index (in bps) at time t. This variable is normalised by rescaling it according to $\hat{R}_t = (R_t - \mu)/\sigma$ where μ and σ are the mean and volatility of the data respectively (hence the normalised time series will have a mean of 0 and volatility of 1). Normalising the returns enables us to compare directly the statistical properties of the index returns and a normal distribution. Also shown on this figure is the normalised FTSE returns and a time series generated from $X_t = \mu + \sigma\varepsilon_t$, where $\varepsilon_t \sim N(0, 1)$ is a standard normal deviate and $\mu = 0$ and $\sigma = 1$. The two systemic credit events (the correlation crisis and the credit crunch) are clearly visible. The credit crunch in particular represents the most severe fluctuations in the short life of the index. It is interesting to note that during the correlation crisis the FTSE did not fluctuate by much. However, during the credit crunch the stock indices were also

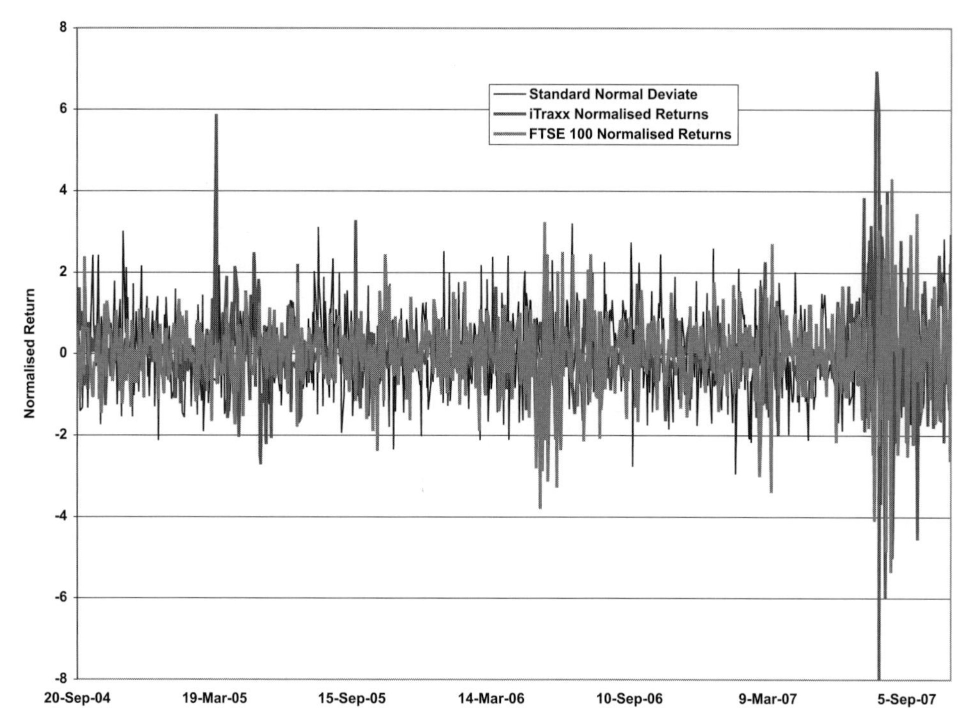

Figure 4.3 Normalised log-returns of the iTraxx 5 year main index over the period September 2004 to November 2007. Also shown on this figure is the normalised FTSE return series and a time series of random $N(0, 1)$ deviates.

significantly impacted. This is indicative of the much more systemic (throughout the entire financial system) nature of the credit crunch compared to the (relatively) idiosyncratic nature of the correlation crisis.

A histogram of the normalised log-returns is shown in Figure 4.4 on the left-hand side and a quantile-quantile plot is shown on the right-hand side. Clearly, from these plots the return distribution of the iTraxx main index over this period is not normally distributed. In fact the skewness is approximately 0.3 and the excess kurtosis is 15 (although this is due largely to the two extreme events identified – if these are removed from the data the returns become far more normal). The fact that the index returns are not normally distributed has a serious consequence for the valuation of instruments where the payoff of the instrument depends on the index. For example, a European credit default swaption on the index will have a payoff of the form $V_T(s_T) = (s_T - X)^+$ where X is the option strike. To calculate the value of the trade we have to calculate

$$\frac{V_0(s_0)}{Z_0} = \mathbf{E}_Q \left[\frac{V_T(s_T)}{Z_T} | \Im_0 \right].$$

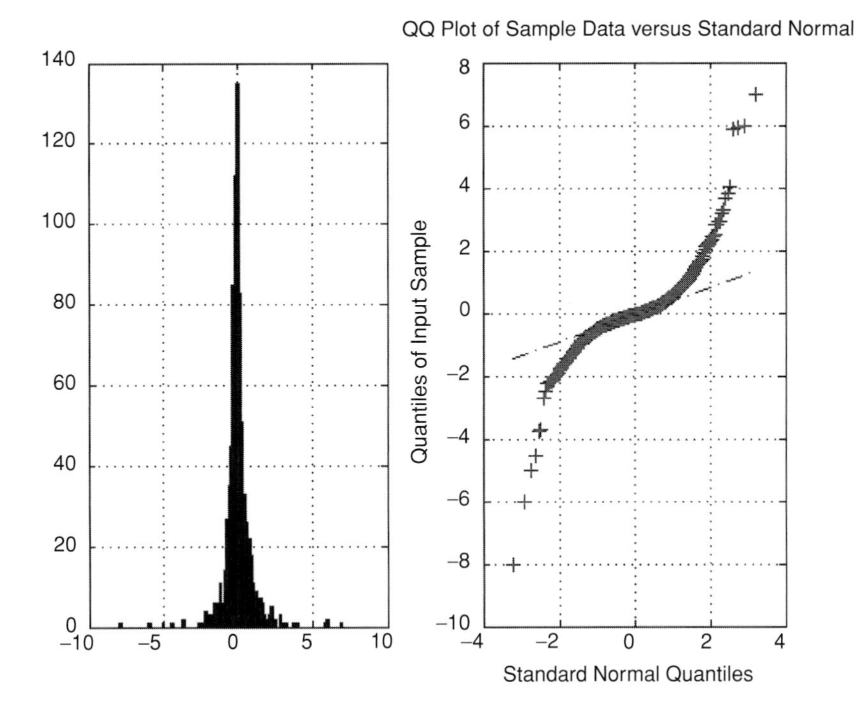

Figure 4.4 Two different plots of the normalised log-returns of the iTraxx index. On the left-hand side is a histogram, on the right-hand side is a quantile-quantile plot. The quantile-quantile plot compares the cumulative probability density distribution of the index data against the cdf of a standardised normal deviate. A Kolmogorov–Smirnov goodness-of-fit test also rejects the hypothesis that the index returns are normally distributed.

The calculation of the expectation value (under the probability measure Q) requires a modelling assumption to be made about the distributional properties of the underlying variable. Assuming a normal (Gaussian) distribution for the return of a financial variable makes the calculation of expected values trivial due to the analytic tractability of Gaussian integrals. If we assume that the dynamics of the variable s_t are log-normal then it is a simple matter to come up with analytical expressions for the expectation value. This naturally leads to expressions for the value analogous to those for vanilla calls and puts (Black–Scholes like). There is a strong desire therefore for financial variables to be normally distributed. Unfortunately the reality does not always match the desire.

Finally, Figure 4.5 plots the auto-correlation function. The auto-correlation function measures correlations between measurements separated by k time steps and is defined by

$$C(k) = \langle \hat{R}_t \hat{R}_{t+k} \rangle = \frac{1}{N-k} \sum_{t=1}^{N-k} \hat{R}_t \hat{R}_{t+k}.$$

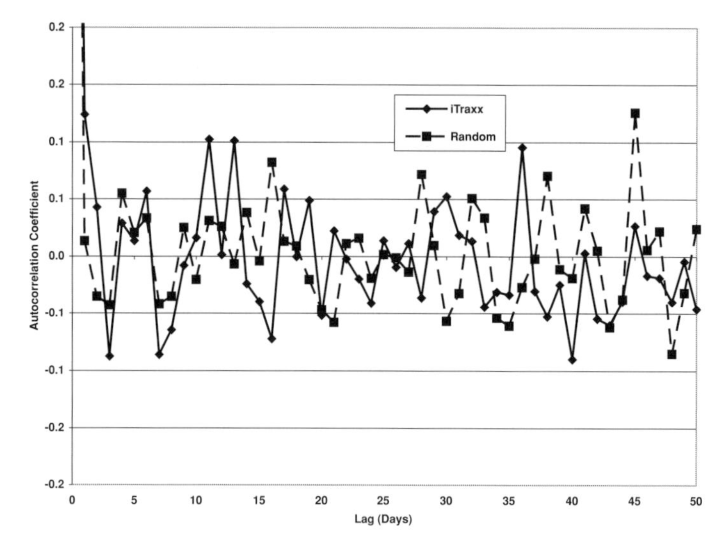

Figure 4.5 Auto-correlation function of the normalised iTraxx returns.

If the measurements are uncorrelated then we would expect $C(k) = 0$ for $k > 0$. If the measurements have short-range correlations then we would expect $C(k) \sim e^{-k/k_X}$ where k_X is the characteristic temporal range of the correlations. If the measurements display long-range correlations then we would expect a power-law decay in the correlation function of the form $C(k) \sim k^{-\gamma}$ where $0 < \gamma < 1$ is an exponent characterising the rate of decay of the correlations.

Also plotted in Figure 4.5 is the auto-correlation function of the randomly generated time series. There is some evidence that for lags of less than about 20 days there is some memory in the iTraxx returns. However, the evidence of memory in the time series of iTraxx returns is by no means clear-cut (the auto-correlation coefficients quickly decay to below the threshold of random noise).

4.6 Tranched credit index exposures

The indices described thus far reference a portfolio of CDSs each referencing individual obligors. It is of course a portfolio product since the payoff of the instrument depends upon the behaviour not of a single but of multiple obligors. In index form a seller of protection is liable to provide compensation to a protection purchaser for each and every obligor in the portfolio that defaults.

The application of tranche technology to a standardised index enables the portfolio losses to be sliced up and sold to those investors who are only interested in that particular slice of the risk. This allows the tranche investor to take a leveraged (compared to the index) position since the protection sold or purchased is based on a single slice of the overall portfolio exposure. For example, selling protection

Table 4.3 *Standardised tranches defined on the iTraxx main index*

Tranche	Attachment point (%)	Detachment point (%)	Attachment point ($ m)	Detachment point ($ m)	Tranche notional ($ m)
Equity	0	3	0	37.5	37.5
Junior mezz	3	6	37.5	75	37.5
Senior mezz	6	9	75	112.5	37.5
Senior	9	12	112.5	150	37.5
Super-senior	12	22	150	275	125

Table 4.4 *Standardised tranches defined on the CDX investment grade main index*

Tranche	Attachment point (%)	Detachment point (%)	Attachment point ($ m)	Detachment point ($ m)	Tranche notional ($ m)
Equity	0	3	0	37.5	37.5
Junior mezz	3	7	37.5	87.5	50
Senior-mezz	7	10	87.5	125	37.5
Senior	10	15	125	187.5	62.5
Super-senior	15	30	187.5	375	187.5

Table 4.5 *Standardised tranches defined on the CDX high yield index*

Tranche	Attachment point (%)	Detachment point (%)	Attachment point ($ m)	Detachment point ($ m)	Tranche notional ($ m)
Equity	0	10	0	12.5	12.5
Equity	10	15	12.5	18.75	6.25
Mezz	15	25	18.75	31.25	12.5
Senior	25	35	31.25	43.75	12.5
Super-senior	35	100	43.75	125	81.25

on the full index might yield a positive carry of 40 bps. Selling protection on the 0–3% first-loss slice of the portfolio yields a running spread of 500 bps as well as an upfront payment which is a large fraction of the initial tranche notional.

Just as a standardised portfolio of obligors provided impetus to the trading of index positions, defining exposures to standardised tranches of the standardised portfolio encouraged the rapid adoption of trading in index tranches and provided the impetus for the development of the single-tranche synthetic CDO market (STCDO). Standardised index tranches can also act as a partial hedge for bespoke tranche positions, encouraging further liquidity in the STCDO market. The standardised slices of the index portfolios are defined in Tables 4.3, 4.4 and 4.5 for the iTraxx and CDX investment grade and CDX high yield indices respectively Each of these tranches has liquid maturities of 3, 5, 7 and 10 years.

4.7 Chapter review

This chapter has introduced the liquidly traded credit indices iTraxx and CDX. The introduction of these standardised portfolio products provided an enormous impetus to the development of the synthetic credit market, based on trading of both standardised indices and bespoke products. The introduction of tranched exposures to the index portfolios also led to the emergence of correlation as an actively traded financial variable.

In the next chapter we will introduce and analyse default baskets. These are a natural stepping stone on to a discussion of synthetic CDOs.

5

Valuation of default baskets

5.1 Introduction

In this chapter we introduce and analyse default baskets. As we will see in subsequent chapters there are many similarities between the modelling of default baskets and synthetic CDOs, and a lot of the technology applied to default baskets carries over to CDOs. It is these modelling similarities that motivate starting with default baskets; much of what we learn about them will prove to be useful when looking at synthetic CDOs.

The chapter begins in Section 5.2 with a brief description of what default baskets are, their economic rationale (for both buyers and sellers of protection) and the mechanics of default basket trades. Section 5.3 describes the general approach to valuing default baskets. We have already met in Chapter 2 the concepts of modelling individual obligor default as a Poisson process and saw how default correlation could be captured by the use of copula functions. In the limits of no correlation between default events or perfect correlation between default events, a default basket can be valued analytically within the standard hazard rate model. Section 5.4 describes the details of these calculations. Although these analytic limits correspond to unrealistic limiting cases, they can provide useful insight into the overall behaviour of default baskets and are therefore worthy of study.

Section 5.5 considers the general case where $0 \leq \rho_{\text{Default}} \leq 1$. The simplest method for valuing default baskets is Monte Carlo simulation of the correlated default times. The behaviour of the model as a function of its input parameters will be quantified and explained in Section 5.6. Within the Monte Carlo framework it is a straightforward matter to calculate sensitivities to key market parameters such as CDS par spreads by perturbing the model inputs (often simply referred to as 'bumping'). However, when calculated by means of Monte Carlo simulation, estimates of sensitivities suffer from the well-documented problems of simulation noise. Credit derivatives have the additional complication that the contingent payoff is non-zero only for a default event that occurs before the maturity of the product; bumping the input parameters can introduce discontinuities into the Monte Carlo estimator which are difficult to overcome thereby increasing the problem of simulation noise.

The use of variance reduction techniques [Joshi and Kainth 2003] can mitigate some of the problems of calculating sensitivities for individual trades, however in practice other solutions must be sought.[1] This is because, from a risk management perspective, managing and understanding the risks within a complex structured credit trading book, involves daily revaluation of not one but many trades (possibly many thousands when one includes all of the additional trades such as vanilla CDSs used within a book to hedge different exposures). Even an efficient Monte Carlo method to revalue all of these trades may not be sufficient to produce the daily risk statistics required by senior management. To obviate the difficulties associated with Monte Carlo simulation semi-analytic valuation methodologies were developed. These will be described briefly in Section 5.7 in order to introduce the concept of factor-model approximations to the full copula dependence structure (a key concept in the modelling of synthetic CDOs). Finally in Section 5.8 we review the important concepts introduced in the chapter and set the scene for the next part of the book on valuing synthetic CDOs.

5.2 Brief overview of default baskets

A plain vanilla CDS involves one of the counterparties purchasing protection against the occurrence of certain credit events (such as outright default, debt restructuring etc.) of a reference obligor. This protection is purchased from a protection seller who is another counterparty willing to take an opposite view of the prospects of the reference obligor. Should a credit event occur, the protection seller must compensate the protection purchaser with a one-off contingent payment for the loss incurred by the distress to the reference obligor. In return for providing this protection (and taking the risk), the protection purchaser compensates the protection seller with a stream of periodic coupon payments for the specified lifetime of the trade. However, once a contingent payment is made, there are no further periodic fee payments and the protection ceases and the contract terminates.

A default basket is a trivial extension of this concept of purchasing protection on a single obligor to a basket (portfolio) of obligors. The contingent payment is now triggered on the occurrence of, say, the second default event that occurs within the basket of obligors. Although the conceptual leap from a CDS to a default basket is simple, the modelling becomes significantly more complicated. This is due solely

[1] It is also the case that a particular variance reduction technique that works well for one type of trade may not work as well for other trades.

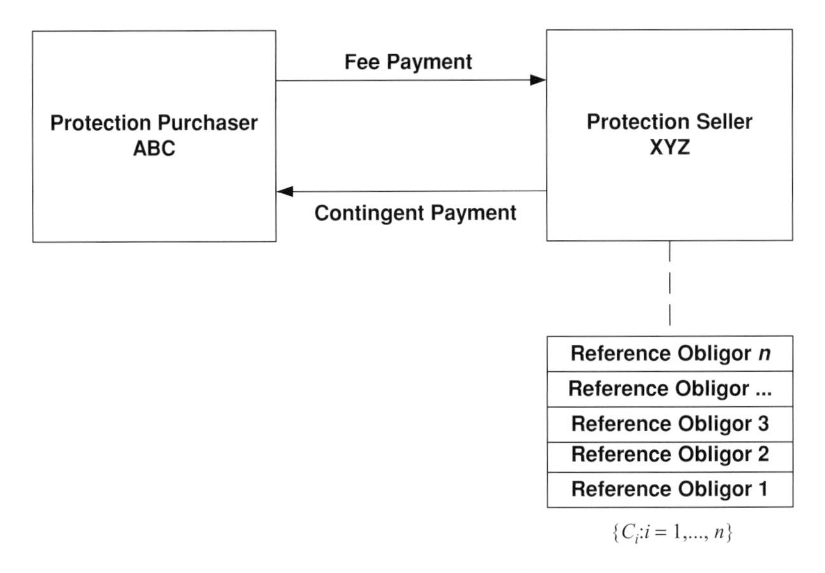

Figure 5.1 Schematic diagram showing the relationship between the different counterparties in a default basket contract. In this figure counterparty ABC is purchasing protection from counterparty XYZ. The protection references the basket $\{C_i : i = 1, \ldots, n\}$ of obligors.

to the need to characterise the mechanism whereby the credit quality of one obligor influences the other obligors in the basket.

Figure 5.1 shows the relationships between the different counterparties in a default basket. Default baskets are typically structured to provide protection on a particular sector or group of obligors that the protection purchaser has exposure to. For example, a computer manufacturer may purchase default protection on a basket of their component suppliers to protect themselves in the event of a default of a critical component maker. More recently default baskets can be used to provide protection against certain exposures in synthetic CDOs (particularly when delta hedging at an individual obligor level is impractical). Further details of the economic rationale and investment strategies involving default baskets may be found elsewhere [Chaplin 2005], and modelling details found in d-fine [2004], Iscoe and Kreinin, and Penaud and Selfe [2003].

The cashflow mechanics of a default basket are shown in Figure 5.2 (the notation in this figure will be defined more precisely later). We define the critical default time, τ_c, as the time of the default event that triggers a contingent payment from the protection seller to the protection purchaser. In this example it is the time of the second default of an obligor in the reference basket. The periodic fee payments from the protection purchaser to the protection seller are made up until the last coupon date before this time. At time τ_c the contingent payment to the protection

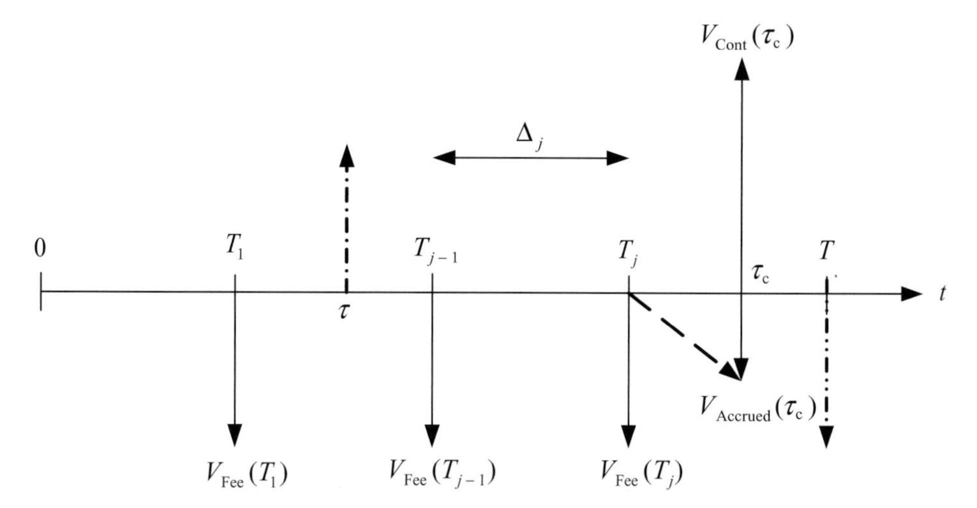

Figure 5.2 Schematic diagram showing the cashflows of a default basket. The full coupon payments are made up until the last coupon date before the critical default time τ_c. At this time the contingent payment is made. No further coupon payments are made (except for an accrued fee payment for the period $[T_j, \tau_c)$ if this is specified in the contract termsheet).

purchaser is made by the protection seller. There may also be an accrued payment that is made by the protection purchaser to the protection seller to compensate the protection seller for providing protection for the period from the last coupon date to the critical default time. After the critical default time no further fee payments are made.

5.3 General valuation principles for default baskets

Consider a default basket with $i = 1, \ldots, n$ obligors in the underlying basket. The maturity of the basket is given by $T_N = T$. We assume that the periodic protection (coupon) payments are made (in arrears) at the times $\{T_j : j = 1, \ldots, N\}$ and that $\Delta_j = T_j - T_{j-1}$ is the day-count fraction (measured in years) between adjacent coupon payment dates. Furthermore, assume that the contractually agreed coupon payment at coupon date T_j is given by $s(T_j)$ (multiplied by the notional). The notional on which the coupon is paid at this date is denoted by $N(T_j)$. In what follows we will almost always assume that the contractual spread and the notional amounts are the same for all coupon payment dates (as is usually the case in practice). This assumption will have little material impact upon the qualitative modelling conclusions to be drawn. Let the current (valuation) time be t. Typically for valuation purposes we will always consider $t = 0$. We let the recovery rate of the obligors be δ_i where $0 \leq \delta_i \leq 1$. Typically we assume that all of the obligors are homogeneous (it is usually the case for vanilla default baskets that the notional amounts and recovery rates of the obligors are assumed to be homogeneous).

The (deterministic) discount factors discounting a payment from t' to t (where $t < t'$) are given by $Z(t, t')$. If we assume that the term structure is constructed from a single (spot) interest rate r then $Z(t, t') = e^{-r(t'-t)}$. In practice, of course, the term structure of discount factors is constructed from a whole range of money market rates, swap rates and futures rates. However, for our purposes making this single factor approximation is adequate since the interest rate behaviour of portfolio credit derivatives is not the primary focus (credit spread and default correlation fluctuations are much more important).

The value of the nth-to-default (NtD) basket from the perspective of the protection purchaser at time t is given by

$$V^{\text{NtD}}(t) = -sV^{\text{NtD}}_{\text{Fee}}(t) + V^{\text{NtD}}_{\text{Cont}}(t),$$

where

$$V^{\text{NtD}}_{\text{Fee}}(t) = \sum_{j=1}^{N} \mathbf{E}\left[Z(t, T_j)\Delta_j \mathbf{1}_{t<T_j<\tau_c}\right]$$

and

$$V^{\text{NtD}}_{\text{Cont}}(t) = \mathbf{E}\left[(1-\delta)Z(t, \tau_c)\mathbf{1}_{t<\tau_c<T}\right]$$

are the expected periodic protection (fee) payments and contingent payment respectively. We are explicitly assuming that the protection payments are made discretely (as is usually the case in practice, of course). The trade set-up and cashflows are shown in Figure 5.2 for the example of a second-to-default basket. The first default at time τ does not trigger the contingent payment (and hence the fee payment stream continues past this default event). However the second default event at τ_c triggers the contingent payment (and the accrued fee payment). Subsequent fee payment cashflows then cease.

Having calculated the fee and contingent legs of the default basket (swap), it is a simple matter to determine the par spread which gives the default basket a value of zero at trade inception. This is given by

$$s_{\text{Par}} = \frac{V^{\text{NtD}}_{\text{Cont}}(0)}{V^{\text{NtD}}_{\text{Fee}}(0)},$$

explicitly writing $t = 0$ to show that the values of the two legs of the trade are the values at trade inception where the remaining time to maturity is T.

The indicator function in the expectation values for the fee and contingent legs is such that $\mathbf{1}_{\tau<T}$ is equal to 1 if $\tau < T$ and zero otherwise. The actual indicator functions in the expectation values refer to the critical default time τ_c. This is defined as the time of the default event that triggers a contingent payment (and hence termination of subsequent fee payments). For example, if the default times of the obligors are $\{\tau_i : i = 1, \ldots, n\}$ and the basket is of the first-to-default variety, the critical default time is given by $\tau_c = \min(\{\tau_i\})$. Similar considerations also lead to the critical default time for second and higher order default baskets.

The simplest method for computing the relevant expectations is Monte Carlo simulation. Monte Carlo, or statistical sampling, methods have existed for a long time. Excellent reference works on the application of Monte Carlo methods to financial problems are Jackel [2002] and Glasserman [2003] . An overview of the application of these methods to problems in statistical physics (one of the original applications of modern Monte Carlo techniques was the Metropolis algorithm developed in the 1950s) is described in Newman and Barkema [1999].

5.4 Analytic valuation of default baskets in simple limiting cases

It is instructive to consider first how to value a default basket in some limiting cases. These calculations will prove a useful check of the implementation of more sophisticated models (in particular providing pricing bounds – a nice visual means of interpreting these results can be found in Schonbucher [2003]). In addition to this, the analytic models will be of use in understanding hedging simulations to be introduced in a later chapter. The material in this section is based on the description given in Chaplin [2005].

5.4.1 Valuation of first-to-default baskets (completely independent defaults)

Consider a two-name basket where the obligors in the basket are identified as A and B. Assume interest rates and hazard rates are constant. Therefore the discount factors are given by $Z(t, T) = \mathrm{e}^{-r(T-t)}$ and the survival curves by $S_A(t, T) = \mathrm{e}^{-h_A(T-t)}$ for obligor A (with a similar expression for obligor B). Assume that the fee payment (annuity stream) is paid continuously (hence the summation in the expression for the fee payment leg is replaced by an integral). Counterparty risk is ignored. In the absence of correlation the joint survival probability over the period $[0, t]$ is $S_A(0, t)S_B(0, t)$. The probability density that A is first-to-default at time t is $S_A(0, t)S_B(0, t)h_A$ (conditional on both names surviving until time t), and conversely for obligor B.

At time t the fee payment is $s_{\mathrm{FtD}} Z(0, t)S_A(0, t)S_B(0, t)\mathrm{d}t$ where s_{FtD} is the contractual spread on the protection leg payments. Over the interval $[0, T]$ the total fee payment is given by

$$V_{\mathrm{Fee}}(0) = s_{\mathrm{FtD}} \int_0^T Z(0,t) S_{\mathrm{A}}(0,t) S_{\mathrm{B}}(0,t) \mathrm{d}t.$$

This integral is easily evaluated when interest and hazard rates are constant. If both A and B survive until t and then either A or B defaults, a contingent payment of

$$V_{\mathrm{Cont}}(0) = (1-\delta_{\mathrm{A}}) \int_0^T h_{\mathrm{A}} S_{\mathrm{A}}(0,t) S_{\mathrm{B}}(0,t) Z(0,t) \mathrm{d}t + (1-\delta_{\mathrm{B}}) \int_0^T h_{\mathrm{B}} S_{\mathrm{A}}(0,t) S_{\mathrm{B}}(0,t) Z(0,t) \mathrm{d}t$$

is made. This is easily evaluated to give

$$V_{\mathrm{Cont}}(0) = [(1-\delta_{\mathrm{A}})h_{\mathrm{A}} + (1-\delta_{\mathrm{B}})h_{\mathrm{B}}] \frac{1 - \mathrm{e}^{-(h_{\mathrm{A}}+h_{\mathrm{B}}+r)T}}{h_{\mathrm{A}}+h_{\mathrm{B}}+r}.$$

Since we are assuming that the premium payments are being made continuously we have that $s_i = (1-\delta_i)h_i$ and since $V_{\mathrm{FtD}}(0) = -s_{\mathrm{FtD}} V_{\mathrm{Fee}}(0) + V_{\mathrm{Cont}}(0)$ the value of the FtD is given by

$$V_{\mathrm{FtD}}(0) = [s_{\mathrm{A}} + s_{\mathrm{B}} - s_{\mathrm{FtD}}] \frac{1 - \mathrm{e}^{-(h_{\mathrm{A}}+h_{\mathrm{B}}+r)T}}{h_{\mathrm{A}}+h_{\mathrm{B}}+r}.$$

These expressions generalise in a simple manner to the case where we have n obligors in the basket and where obligor i has a constant hazard rate h_i, to

$$V_{\mathrm{Fee}}(0) = s_{\mathrm{FtD}} \frac{1 - \mathrm{e}^{-(r+\sum_{i=1}^n h_i)T}}{r + \sum_{i=1}^n h_i}$$

$$V_{\mathrm{Cont}}(0) = \sum_{i=1}^n s_i \frac{1 - \mathrm{e}^{-(r+\sum_{i=1}^n h_i)T}}{r + \sum_{i=1}^n h_i}.$$

At trade inception we require $V_{\mathrm{FtD}}(0) \equiv 0$ which implies that the par spread in this limiting case is given by

$$s_{\mathrm{FtD}} = \sum_{i=1}^n s_i.$$

That is, in the limit of zero correlation, the par spread of an FtD is simply equal to the sum of the par CDS spreads of the underlying obligors. This is slightly counter-intuitive; the same premium is paid for an FtD which terminates on the first default as would be paid for purchasing individual CDSs (when an obligor defaults only that CDS terminates, the others continue paying out). This is due to the assumption that the hazard rates are constant; for obligor i its par spread is given by $s_i = (1 - \delta_i)h_i$ which is independent of time. Therefore when an obligor defaults, the FtD unwinds, but it is possible to purchase CDS protection on the remaining names in the basket for the same premium as at time $t = 0$.

Obviously, since we have analytic expressions for the fee and contingent payments of the basket, it is a simple matter to obtain analytic expressions for the sensitivity of the basket's PV to fluctuations in for example, obligor par CDS spreads.

These calculations can be extended in a straightforward (although algebraically tedious) manner to second and higher order default baskets. The interested reader is referred to Chaplin [2005] for further details.

5.4.2 Completely dependent defaults

Consider again a basket of two obligors A and B. Obligor A has a hazard rate $h_{\text{A}} = 2\%$ and obligor B has $h_{\text{B}} = 1\%$. This implies that if obligor A defaults at τ, B will default at time 2τ. In terms of simulation, if for simulation α we generate a random number x^{α} (from an arbitrary distribution) then the default times for all the obligors $i = 1, \ldots, n$ are calculated from

$$\tau_i^{\alpha} = -\frac{1}{h_i} \ln[\Phi(x^{\alpha})]$$

(using the same random number since the default behaviour of the two obligors is completely correlated) where $\Phi(x)$ is the cdf of the random variable x (not necessarily the normal distribution). For the basket of two obligors if $h_{\text{A}} > h_{\text{B}}$ this implies $\tau_{\text{A}} < \tau_{\text{B}}$ then the probability that A and B survive until time t is equal to the probability that A survives until t.

For an FtD, the contract pays out the contingent payment only if a CDS referencing obligor A pays out. A second-to-default pays out only if a CDS on B pays out. Therefore the value of the FtD is simply s_{A}, the par CDS spread on obligor A. A position in an FtD can be fully hedged by an offsetting position in a CDS referencing obligor A. The par spread of the second-to-default is simply the par spread of the CDS referencing obligor B. These arguments generalise to larger baskets.

In the more realistic case where the default correlation lies in the range $0\% \leq \rho \leq 100\%$ these analytic calculations are no longer applicable. However, from a validation point of view it is always useful to have some limiting cases where the results are known a priori against which more sophisticated models can be benchmarked.

5.5 Monte Carlo valuation of default baskets

In the context of portfolio credit derivative modelling, the normal (Gaussian) copula model has emerged as the market standard methodology used for valuing instruments and communicating the views of market participants on default correlation. We will now describe the application of this model to the valuation of default baskets. In general the valuation is a two-step process:

- firstly generate a default scenario;
- secondly value the contractual cashflows given the particular default scenario.

These steps are repeated many times in order to obtain estimates for the various contract cashflows. One of the great advantages of the Monte Carlo method is its flexibility. Virtually any specification of cashflows and cashflow triggers, exemptions, redirections etc. can be coded up. It is very easy, for example, to go from valuing a first-to default basket to valuing a second-to-default basket simply by changing a single line of code.

5.5.1 Generating default time scenarios

Within the normal copula framework it is a simple matter to calculate the correlated default times of the obligors. We assume the following inputs are available:

- the default correlation matrix $\mathbf{C}$ (which is an $n \times n$ matrix);
- the marginal survival curve $S_i(t, T)$ for each obligor i (stripped from their par CDS curves).

The default correlation matrix may be a fully specified matrix with $n(n-1)/2$ distinct off-diagonal entries or a simplified matrix where all the off-diagonal entries are given by a single number ρ (or some block diagonal form in-between the two extremes possibly representing inter- and intra-sector correlations). The entries in this matrix represent the joint, pairwise, propensity of obligors to default together. Given these inputs the procedure to generate correlated default times for a particular simulation α is as follows:

1. calculate $\mathbf{A}$ from $\mathbf{C} = \mathbf{A}\mathbf{A}^{\mathrm{T}}$ Cholesky decomposition of correlation matrix
2. sample $\vec{\varepsilon}^{\alpha} \sim N(\mathbf{0}, \mathbf{1})$ vector of standard normal deviates
3. calculate $\vec{\phi}^{\alpha} = \mathbf{A}\vec{\varepsilon}^{\alpha}$ vector of correlated standard normal deviates
4. calculate $\tau_i^{\alpha}(\phi_i^{\alpha})$ correlated default times for $i = 1, \ldots, n$.

We have included the index α to indicate that the default times generated depend upon the particular simulation scenario ($\alpha = 1, \ldots, N_{\text{Sims}}$) (but the Cholesky decomposition of the input correlation matrix need only be done once).

In implementation terms all of the results reported in this chapter (and indeed book) have utilised the (pseudo) random number generators provided in the

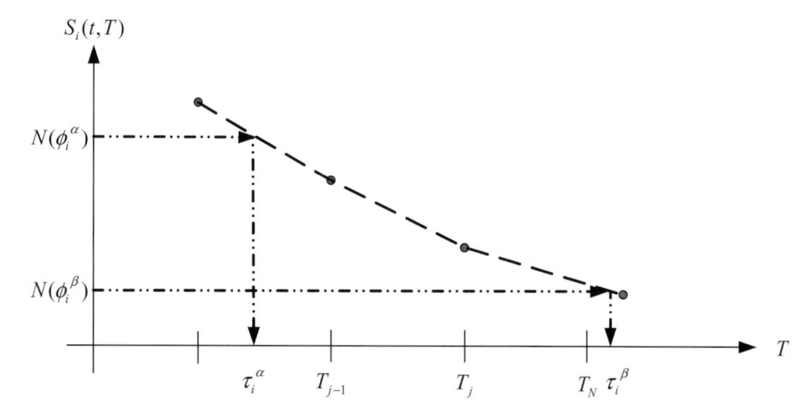

Figure 5.3 Schematic diagram representing the procedure for calculating correlated default times. In this example the random number generated for obligor i for simulation α is mapped to a default time $\tau_i^\alpha < T_N$. For a different simulation β the default time is $\tau_i^\beta > T_N$.

standard 'Numerical Recipes in C++' text. There are other more advanced generators around with essentially infinite periodicity, such as the Mersenne twister, and these can also be utilised. It is in point of fact a sensible precaution to test random number generators very thoroughly and to undertake pricing comparisons using different generators[2] to mitigate model risk. There are batteries of statistical tests that random number generators can be subjected to in order to test their properties. Monte Carlo simulation is wholly reliant on good quality random number sequences; if your generator is faulty, your results will be too. Time invested in reliable and well understood generators is time very well spent. An additional note of caution is to be very careful when attempting any modifications to existing random number generator code. Often pseudo-random number generator algorithms exploit subtle numerical and computational features of how numbers are represented on a machine. Modifying these algorithms can lead to disastrous consequences and should only be attempted by those who are absolutely sure they know what they are doing.

The correlated obligor default times are computed from the survival curve by inverting the relationship $S_i(0, \tau_i^\alpha) = N(\phi_i^\alpha)$ (where $N(x)$ is the cumulative normal distribution function), i.e. $\tau_i^\alpha = S_i^{-1}[N(\phi_i^\alpha)]$ for $i = 1, \ldots, n$. Typically the inversion of the survival curve to calculate the default time must be undertaken numerically. A simple bi-section or Newton–Raphson type technique is adequate for this task. This process is illustrated schematically in Figure 5.3. A clear dependency in this process is the calculation of the Cholesky decomposition of the input correlation matrix. Cholesky decomposition will only work for a matrix which

[2] Including a radioactive source to generate truly random events if you have one to hand on the trading desk!

is positive semi-definitive meaning $\mathbf{x}^{\mathrm{T}}\mathbf{C}\mathbf{x} \geq 0$ for some vector $\mathbf{x}$. Noting that the eigenvalues λ_i and eigenvectors $\mathbf{v}_i$ of the correlation matrix $\mathbf{C}$ are defined from $\mathbf{C}\mathbf{v}_i = \lambda_i \mathbf{v}_i$, then positive semi-definite implies that $\lambda_i \geq 0$. Therefore if the correlation matrix is such that any of its eigenvalues are less than zero, the decomposition will fail. In these circumstances there are some techniques that can be applied to rescue the situation (the simplest being to set the offending eigenvalues to zero, and hope that the corresponding eigenvectors do not contribute much to the overall matrix!). Some techniques for salvaging a correlation matrix are described in Jackel [2002], Higham, and Higham [2006]. Problems with positive semi-definiteness typically arise not from the actual market data used to estimate the correlation matrix (unless there are not enough market data for the true nature of the correlation to be apparent), but from risk management requirements to assess particular scenarios, for example setting all the correlations to unity.

In practice, for a homogeneous correlation matrix (where all the off-diagonal elements are identical) Cholesky decomposition will only fail when $\rho \to 1$. Since such a high correlation is almost never observed in practice the problem is not a major one. However, when dealing with general inhomogeneous correlation matrices it is advisable to check carefully the decomposition part of the process;[3] if it fails or (worse) succeeds and gives nonsense values then the default times will also be nonsensical. Keeping track of the statistical properties of the simulated default times and comparing them with their expected values is often a sensible precaution.

In the simplified case where the survival curve is constructed from a single CDS par spread (and hence a single hazard rate) the survival curve is given by $S(t, t') = \mathrm{e}^{-h(t'-t)}$. In this case the default times can be determined for each simulation α analytically according to

$$\tau_{\mathrm{i}}^{\alpha} = -\frac{1}{h_i} \ln N\left(\phi_i^{\alpha}\right).$$

This expression will be of great use throughout the subsequent analysis to be presented.

From the expressions for the default times we can see that calculating sensitivities of a default basket's PV to fluctuations in the input parameters eventually leads to fluctuations in the default times. Bumping the CDS par spread affects the hazard rate (and hence the default time), bumping the default correlation affects the Cholesky decomposition of the correlation matrix and hence the random numbers generated (and therefore the default time). Much insight into the behaviour of default baskets and also synthetic CDOs can be obtained by thinking about the impact on the default times of parameter fluctuations.

[3] For example, adding a check to verify that $\mathbf{C} - \mathbf{A}\mathbf{A}^{\mathrm{T}}$ is within numerical precision of $\mathbf{0}$ can spot gross errors. This adds a computational cost, but for small baskets the matrix operations are not particularly onerous (and the Cholesky decomposition is only performed once for each basket valuation).

Finally we note that it is a relatively simple matter to replace the normal copula characterisation of the obligor dependence structure with another copula. The only difficulty with this lies in developing an appropriate algorithm for sampling from the copula. See Cherubini *et al.* [2004] for details of simulating from other copulas.

5.5.2 Valuing cashflows

Consider one of the (many) default time simulations α. Assume that we have successfully generated the set of default times for this scenario and also determined the critical default time τ_c^α (for example, the time of the first default). For a plain vanilla default basket the contingent leg is calculated according to

$$V_{\text{Cont}}^\alpha\left(t, \tau_c^\alpha\right) = N(1-\delta)Z\left(t, \tau_c^\alpha\right)\mathbf{1}_{\tau_c^\alpha < T}.$$

That is, if the critical default occurs before the maturity of the basket there is a contingent payment; if it occurs after maturity there is no contingent payment. Similarly the fee payments are given by

$$V_{\text{Fee}}^\alpha\left(t, \tau_c^\alpha\right) = sN\sum_{j=1}^{N} Z(t, T_j)\Delta_j \mathbf{1}_{\tau_c^\alpha < T_j},$$

i.e. a coupon payment is only made if the critical default time occurs after that coupon payment. If the critical default occurs before maturity and between two coupon payments $[T_{m-1}, T_m]$ the coupon payments are made in full up until and including the payment at time T_{m-1}. There is also an additional accrued coupon payment resulting in a total fee payment of

$$V_{\text{Fee}}^\alpha\left(t, \tau_c^\alpha\right) = sN\sum_{j=1}^{m} Z(t, T_j)\Delta_j + V_{\text{Accrued}}\left(t, \tau_c^\alpha\right)$$

for $T_{m-1} < \tau_c^\alpha < T_m < T$ where the accrued coupon payment made for the period $[T_{m-1}, \tau_c^\alpha]$ is given by

$$V_{\text{Accrued}}\left(t, \tau_c^\alpha\right) = sNZ\left(t, \tau_c^\alpha\right)\frac{\tau_c^\alpha - T_{m-1}}{T_m - T_{m-1}}$$

for $T_{m-1} < \tau_c^\alpha < T_m$ (assuming linear accrual of coupon between coupon payment dates). The Monte Carlo estimators for the fee and contingent legs can now be calculated from

$$V_{\text{Fee}}(t) = \frac{1}{N_{\text{Sims}}}\sum_{\alpha=1}^{N_{\text{Sims}}} V_{\text{Fee}}^\alpha\left(t, \tau_c^\alpha\right),$$

$$V_{\text{Cont}}(t) = \frac{1}{N_{\text{Sims}}}\sum_{\alpha=1}^{N_{\text{Sims}}} V_{\text{Cont}}^\alpha\left(t, \tau_c^\alpha\right),$$

and the par spread is calculated from $s_{\text{Par}} = V_{\text{Cont}}/V_{\text{Fee}}$. Note that it is not advisable to calculate the par spread estimator on a simulation-by- simulation basis. If $\tau_c^\alpha < T_1$ (the first coupon payment date), and the accrued fee payments are being ignored this results in $V_{\text{Fee}}^\alpha = 0$ and $s_{\text{Par}}^\alpha \to \infty$ for this simulation which is clearly incorrect.

One of the major advantages of the Monte Carlo simulation methodology is evident from these formulas – its flexibility. To incorporate more complex payoffs (e.g. a ratcheting up of the coupon payment over time) one simply has to modify the formulas for the fee and contingent legs accordingly. A small additional implementation complexity is that when the obligor loss given default amounts $N_i(1 - \delta_i)$ are inhomogeneous the contingent payment depends upon the identity of the defaulting obligor. If the obligors are all homogeneous this is not a problem, but in the more general case where the obligors are inhomogeneous it is necessary to keep track within the valuation algorithm of the relevant defaulting obligor.

To summarise, the central problem of pricing default baskets using Monte Carlo simulation is calculating the default times for each of the individual obligors in the basket. Once the default times are known the appropriate time of the critical default (e.g. the second default time for a second-to-default basket) is then used in the pricing of the payment streams.

5.6 Phenomenology of default baskets

Having defined what a default basket is and developed a valuation methodology we will now examine the properties of the model.

5.6.1 Baseline case to be analysed

The following simplifying assumptions are made.

- The term structure of discount factors is constructed from a single spot rate hence the discount factors are simply given by $Z(t, t') = e^{-r(t'-t)}$.
- Accrued fee payments are ignored (this has little material impact upon the conclusions to be drawn).
- The obligors are assumed to be homogeneous (in terms of their notional amounts and recovery rates – hence the loss given default is identical for each obligor). Again this assumption has little material impact upon the conclusions that we will draw.
- The par CDS spread term structure is approximated with a single spread value. This implies that a single hazard rate is sufficient to characterise the cumulative survival curve where survival probabilities are given by $S(t, t') = e^{-h(t'-t)}$. It is an easy matter to move between par spreads and hazard rates via the credit triangle relationship $s = h(1 - \delta)$.

This final assumption simplifies the modelling significantly since we do not have to go through the difficulties of bootstrapping a survival curve from market observed par CDS spreads for each obligor in the basket. It also means that we can calculate analytically the default times for each simulation without having to invert the survival curve numerically. This is a strong simplification to make in terms of the impact upon the basket's PV, but qualitatively the consequences are not severe as the phenomenology of the models remains unchanged.

In addition to these simplifying modelling assumptions the following contract and simulation parameters are assumed.

Number of obligors in basket	5 (typical traded baskets have 4–20 obligors)
Basket maturity	5 years
Basket contractual fee spread	150 bps
Fee payment tenor	quarterly (in arrears)
Day-count basis	30/360
Homogeneous obligor notional	\$10 m
Homogeneous obligor recovery rate	40%
Homogeneous obligor par CDS spread	180 bps
Homogeneous obligor spread bump	10 bps
Homogeneous obligor hazard rate	3% (calculated from $s = h(1 - \delta)$)
Default basket order (first, second etc.)	varies
Risk-free rate	5%
Number of default time simulations	varies

These parameters are chosen as being fairly typical of the sorts of default basket actively traded in the marketplace.

5.6.2 Convergence of PV estimators

The first, and possibly most important, characteristic of the model to assess is its performance as a function of the number of simulations used to generate the fee and contingent leg estimators. If the PV estimator only becomes stable after a huge number of simulations it is important to know this as it will impact significantly on how the model is used in practice in a live environment.

Figure 5.4, shows the estimated basket PV (assuming a contractual coupon spread of 150 bps) and also the simulation error as the number of simulations is varied in the range 100–100 000 000 simulations. The Monte Carlo simulation error $\hat{\varepsilon}$ is given by the standard formula

$$\hat{\varepsilon} = \frac{\hat{\sigma}}{\sqrt{N_{\text{Sims}}}}$$

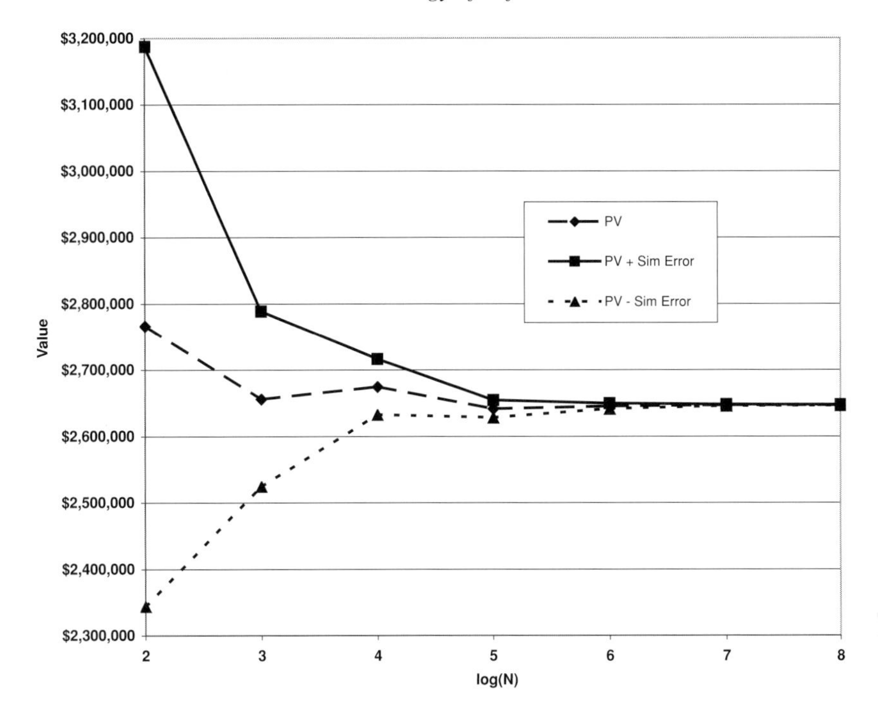

Figure 5.4 Convergence of Monte Carlo PV estimator as a function of the number of default scenario simulations. The horizontal axis is the (base 10) logarithm of the number of simulations.

where $\hat{\sigma}^2 = \langle V^2 \rangle - \langle V \rangle^2$ is the sample standard deviation (in this case of the estimated basket PV V) and

$$\langle V \rangle = \frac{1}{N_{\text{Sims}}} \sum_{\alpha=1}^{N_{\text{Sims}}} V^\alpha$$

$$\langle V^2 \rangle = \frac{1}{N} \sum_{\alpha=1}^{N} (V^\alpha)^2.$$

Similar results are obtained for second-, third- and fourth-to-default baskets although it is worth remembering that as we go from first- to second-to-default baskets and so on, more simulations are necessary to achieve good convergence.

Clearly, as the number of simulations increases the estimator converges on a long-run value and the simulation error becomes reduced in magnitude. It is a useful verification of the Monte Carlo implementation to demonstrate that $\hat{\varepsilon} \sim N^{-1/2}$. Figure 5.5 shows the convergence of the logarithm of the simulation error $\log \hat{\varepsilon}$ against $\log N$ for the first- to fourth-to-default baskets as the number

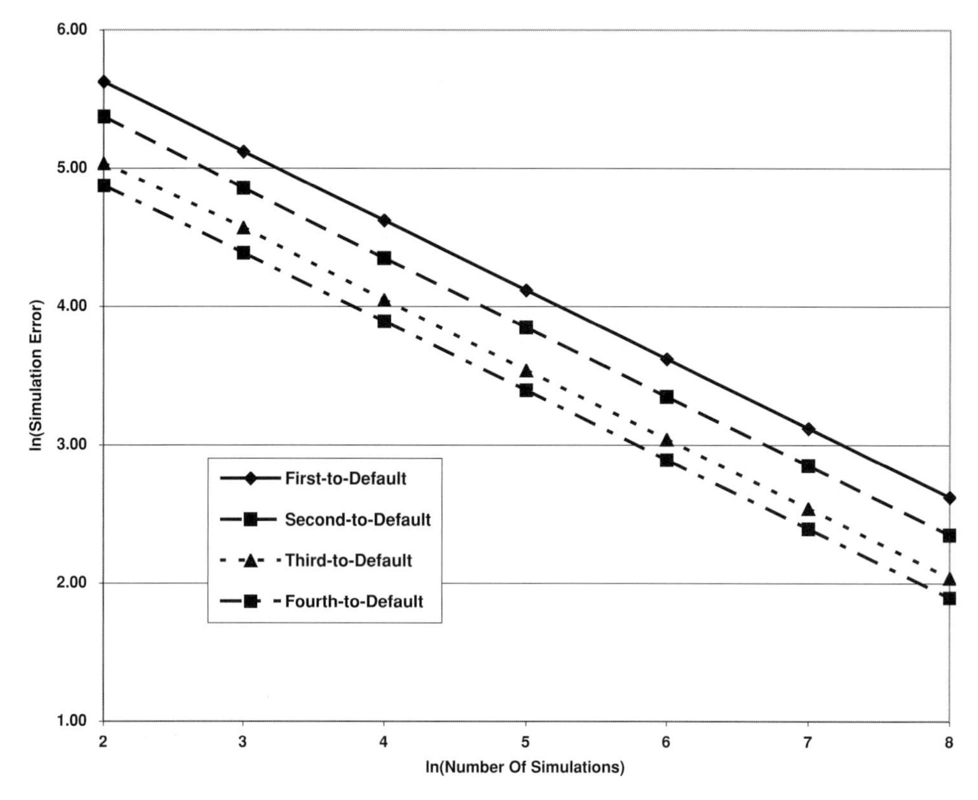

Figure 5.5 Plot of $\log \hat{\varepsilon}$ against $\log N$ for first- through fourth-to-default baskets. Fits of linear models to these plots yield gradients very close to the expected value of -0.5 for Monte Carlo simulation.

of simulations increases. Fitting linear models to these results yields exponents of -0.5, -0.503, -0.502 and -0.497 for the gradients for the first- to fourth-to-default baskets respectively. If we were to increase the number of simulations the results for the higher order default baskets would converge even more closely to their theoretically expected value of -0.5. It is worth noting that these results have been obtained using only a single trial simulation seed (to seed the random number generator). To characterise the performance of the model fully it would also be beneficial to rerun the convergence analysis using multiple trials for the simulation seed. This ensures that a different random number sequence is used in each trial to generate the correlated obligor default times leading to differing cashflow streams. For large enough numbers of simulations the different estimators should all converge to within the simulation error and it is important to know what this number is (since it removes 'seed risk' from the valuation). It is left as an exercise for the reader to verify this.

The analysis in this section has highlighted some of the problems encountered when valuing default baskets using Monte Carlo simulation, notably the relatively slow convergence of the estimators. These problems also exist, of course, when applying Monte Carlo techniques to other asset classes, such as interest rate products. There is a large body of work that has been developed in order to improve the performance of Monte Carlo simulation (see Glasserman [2003] for a detailed description of these methods). Some of these techniques have been adapted quite successfully to credit derivatives. The reader is referred to [Joshi [2003], [2004], Rott and Fries [2005] Chen and Glasserman [2006], for further details of these methods. In general we will not consider improvements to basic Monte Carlo performance further since, as we will see, there are perfectly adequate semi-analytic methodologies that have been developed for the valuation of default baskets and synthetic CDOs.

5.6.3 CDS par spread sensitivities

The basket default model has three key inputs: the default correlation between the obligors, the recovery rate of individual obligors and a hazard rate representing the default likelihood of an individual obligor. The first two of these inputs are typically unobservable parameters in the marketplace (for which appropriate reserving methodologies must be developed in order to mitigate parameter uncertainties). However, as we have seen in previous chapters, there is a liquid market that trades in buying and selling protection on the default risk of individual obligors for a range of maturities – the par CDS spreads. From these spreads the survival curve is bootstrapped.

If we wish to hedge a default basket position against fluctuations in these observable parameters, we require quantification of the baskets' sensitivity to fluctuations in the par CDS spreads. Figures 5.6–5.8 show the sensitivity of the default basket PV to fluctuations in the credit spreads of the underlying obligors for first- through to third-to-default baskets. Note that all of these graphs are plotted on the same scale (to highlight the magnitude of the relative sensitivities of the different baskets).

For each of these graphs the following procedure has been followed to obtain the results reported. The obligors are all assumed to be homogeneous with par CDS spreads of 180 bps and recovery rates of 40%. This corresponds to a homogeneous hazard rate of 3%. For these parameters the par spread is calculated for each of the first- to third-to-default baskets (obtained to be 875 bps, 177 bps and 25 bps respectively) and this par spread is subsequently used as the baskets' contractual spread. Hence at a spread bump of 0 bps from the original spreads (of 180 bps), the PV of each of the baskets is identically equal to zero. The par CDS spreads of each of the obligors is then simultaneously perturbed from 180 bps through the range –180

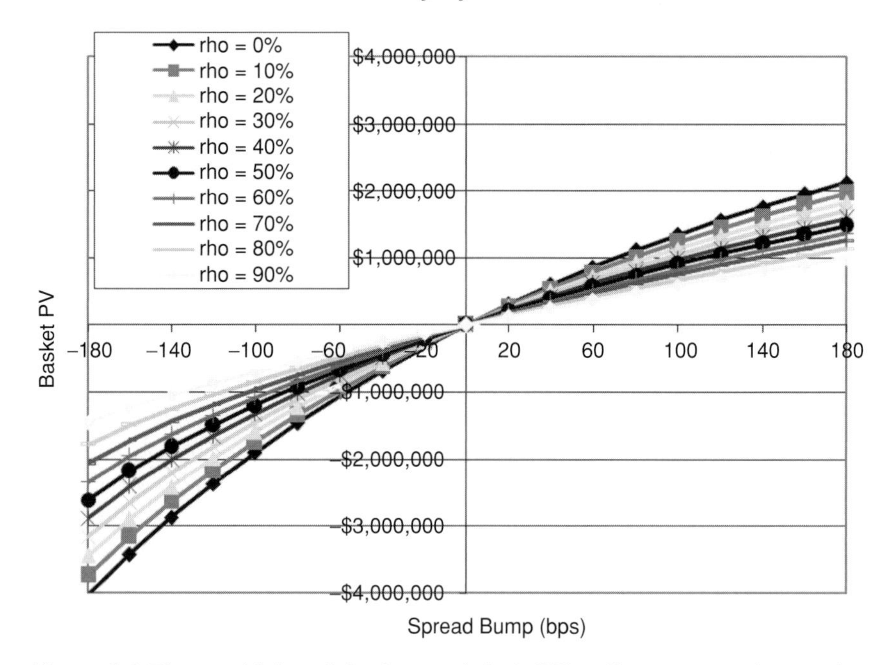

Figure 5.6 The sensitivity of the first-to-default PV to (homogeneous) perturbations in the par CDS spreads for a range of different default correlations. The important point to note is that the dependence is non-linear.

bps to +180 bps and the resultant PV calculated. For a spread bump of −180 bps the hazard rate will be 0% (and in this limit we would expect the contingent payment to become zero). This is repeated for different correlations in the range [0%, 90%].

The most important observation to make about the form of these graphs is that the PV is a non-linear function of the spread bump. This is significant since it impacts the calculation of the delta sensitivities (using a large spread bump to calculate the delta will introduce convexity into the bumped PV estimator). To better understand the qualitative form of these results it is instructive to consider the fee and contingent leg payments separately. For the first-to-default basket when $\rho = 0\%$ these are shown in Figure 5.9.

If we ignore interest rates and assume that all the fee payment coupons are paid at equally spaced intervals so that $\Delta_j \equiv \Delta$, for all j, the Monte Carlo estimators for the fee and contingent legs are given by

$$V_{\text{Fee}}^{\text{NtD}}(t) = \Delta \sum_{j=1}^{N} \mathbf{E}[\mathbf{1}_{\tau_c > T_j}],$$

$$V_{\text{Cont}}^{\text{NtD}}(t) = (1 - \delta)\mathbf{E}[\mathbf{1}_{\tau_c < T_N}].$$

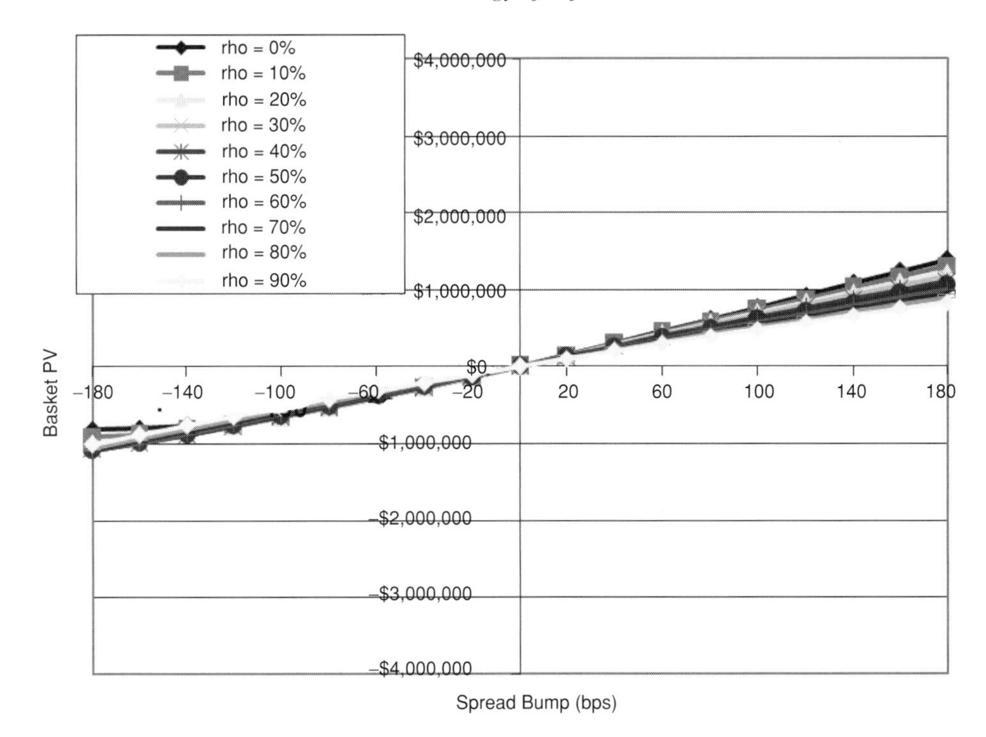

Figure 5.7 The sensitivity of the second-to-default PV to (homogeneous) perturbations in the par CDS spreads for a range of different default correlations. The important point to note is that the dependence is non-linear.

The expectation values are calculated in the Monte Carlo simulation according to

$$\mathbf{E}\left[\mathbf{1}_{\tau_c > T_j}\right] = \frac{1}{N_{\text{Sims}}} \sum_{\alpha=1}^{N_{\text{Sims}}} \Theta\left(\tau_c^\alpha - T_j\right)$$

(and a similar expression for the contingent leg estimator) where $\Theta(x)$ is the Heaviside unit step function.

As $h \to +\infty$ then from $S(t, T) = \mathrm{e}^{-h(T-t)}$ the simulated default times for each obligor for each path α will be such that the critical default time $\tau_c^\alpha \to 0$. In this case we would expect each simulated default scenario to generate a non-zero contingent payment. That is $V_{\text{Cont}}^\alpha \neq 0$ for a very large fraction of the simulated paths. In the limit as the spread bump becomes extremely large (and the default times extremely small) we expect $(1-\delta)\mathbf{E}[\mathbf{1}_{\tau_c < T_N}] \to N_{\text{Sims}}(1-\delta)/N_{\text{Sims}}$ (multiplied by the contingent leg notional). Hence as the magnitude of the spread bump becomes very large and positive the contingent leg will become saturated and tend towards a limiting value of the loss given default (which in this case is given by $(100\% - 40\%) \times \$10$ m). This is the reason for the appearance of convexity in the PV of

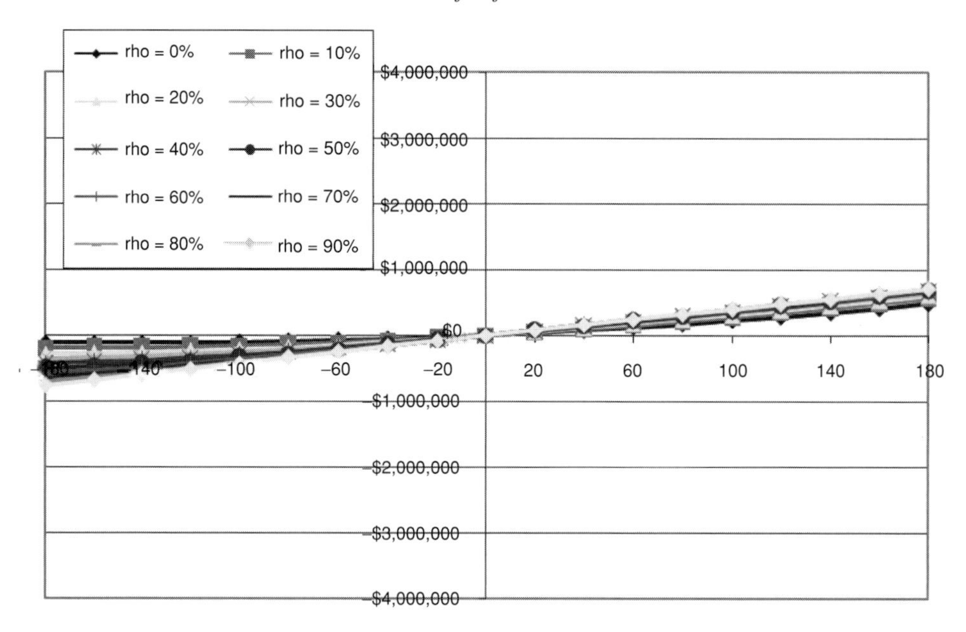

Figure 5.8 The sensitivity of the third-to-default PV to (homogeneous) perturbations in the par CDS spreads for a range of different default correlations. The important point to note is that the dependence is non-linear.

the FtD for large spread bumps. Additionally in this limit $\mathbf{E}[\mathbf{1}_{\tau_c > T_j}] \to 0$ for all coupon payments at times T_j, i.e. the critical default has such a high probability of happening there is no time to accrue any coupons and $V_{\text{Fee}} \to 0$. This implies that the convexity is due only to the saturation of the contingent payment at its limiting value.

Conversely as $h \to 0$, $\tau_c^\alpha \to +\infty$ for a large fraction of simulated paths. In this case we would expect $V_{\text{Cont}} \to 0$ (as is observed in the figure above) and $V_{\text{Fee}} \to s_{\text{FtD}} \sum_{j=1}^{N} \Delta_j$ since the critical default time occurs after all of the coupon payments. This behaviour is also intuitively obvious; as the hazard rate becomes infinitesimally small, the probability of default vanishes and the FtD comes to resemble a simple risk-free bond with periodic coupon payments.

It is a simple matter to verify numerically that as the spread bump becomes very large (e.g. 10 000 bps), the contingent payment slowly approaches its limiting value. This is left as an exercise for the reader.

5.6.4 Correlation and basket maturity sensitivities

The next important sensitivity of the model to understand is the variations of PV as a function of the (homogeneous) correlation assumption used to characterise the obligors' propensity for synchronous default. We will also examine the baskets'

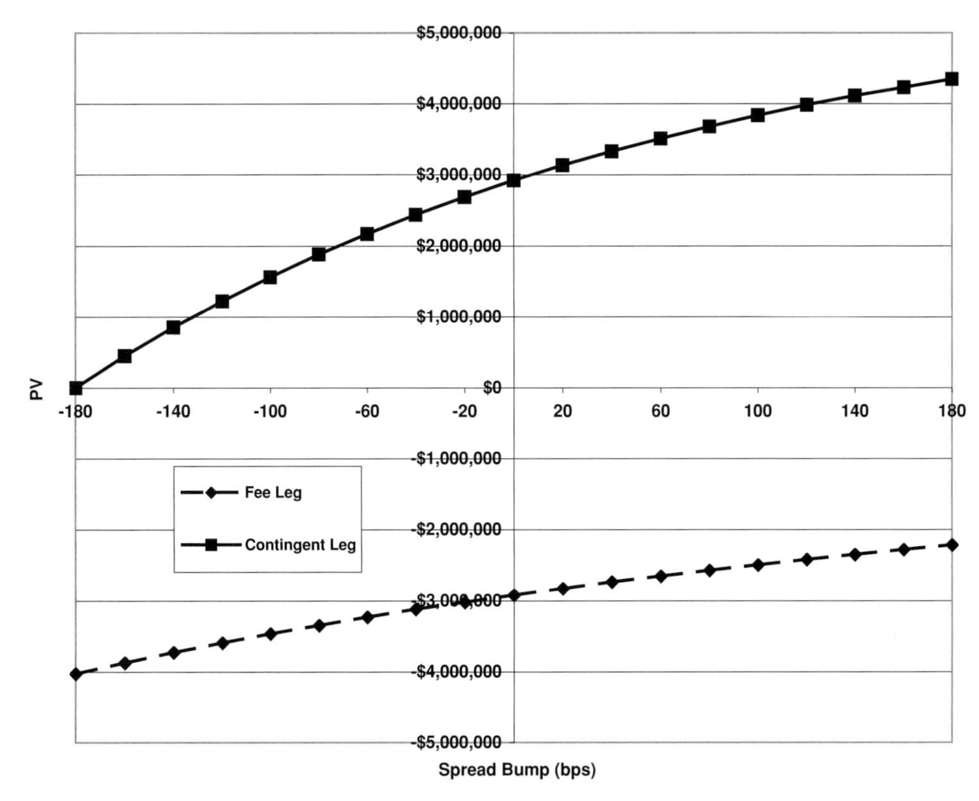

Figure 5.9 Decomposition of the PV credit spread sensitivity into separate fee and contingent leg components (for a first-to-default basket with $\rho = 0\%$).

behaviour as a function of maturity since this will prove to be very informative in terms of understanding the phenomenology of the model.

Figures 5.10–5.12 show the par spreads of first-, second- and third-to-default baskets as a function of the homogeneous correlation assumption. Each plot also shows the par spread for different basket maturities ranging from 1 to 20 years (typical maturities traded in the market are 3, 5, 7 and 10 years; the 15 and 20 year plots serve to illustrate certain features of the behaviour of the model). As expected as $\rho \to 0\%$ the par spread of the FtD approaches 900 bps (the value obtained using the analytical formulas). Additionally as $\rho \to 100\%$ all the default baskets' par spreads approach the same limiting value.

The following observations are made. For the first-to-default basket the par spread is a monotonically decreasing function of the correlation; this behaviour does not change significantly as the maturity of the basket is increased. For the second-to-default basket at short maturities the par spread is monotonically increasing and as the maturity increases the shape of the graph evolves to become firstly non-monotonic and then monotonically decreasing at very long maturities. Finally, for

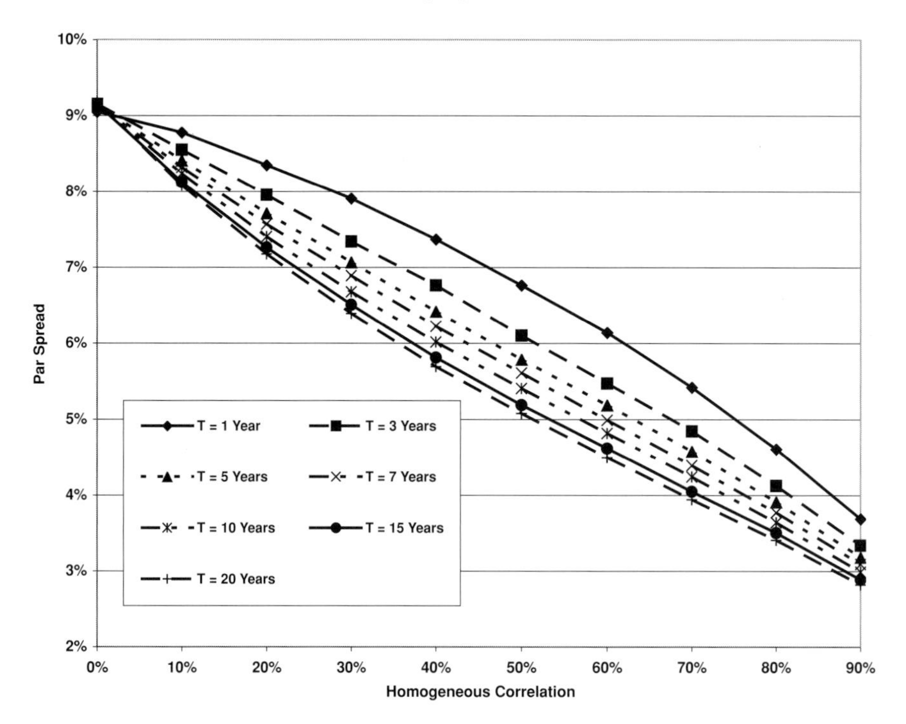

Figure 5.10 The par spread as a function of the homogeneous correlation for a range of maturities for the first-to-default basket.

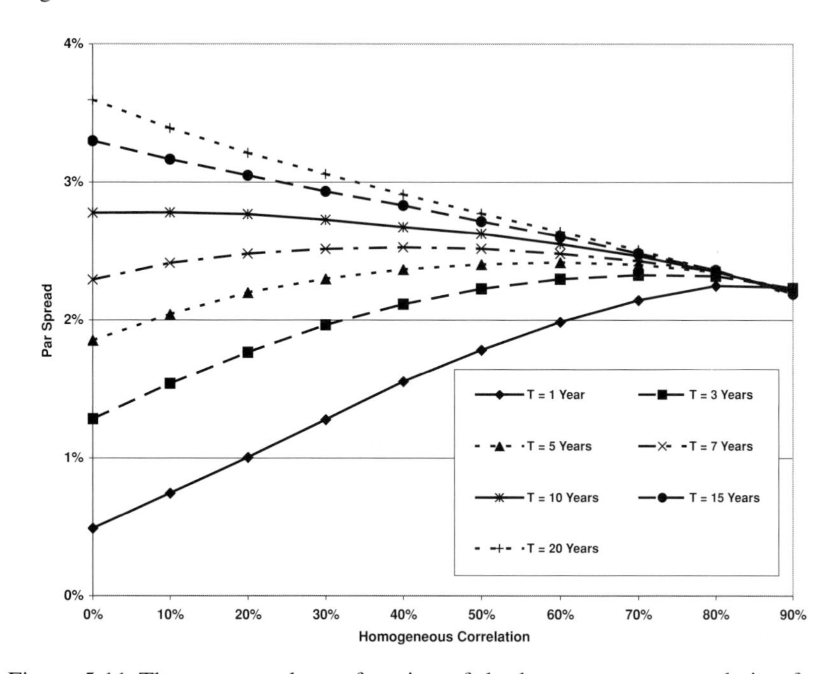

Figure 5.11 The par spread as a function of the homogeneous correlation for a range of maturities for the second-to-default basket.

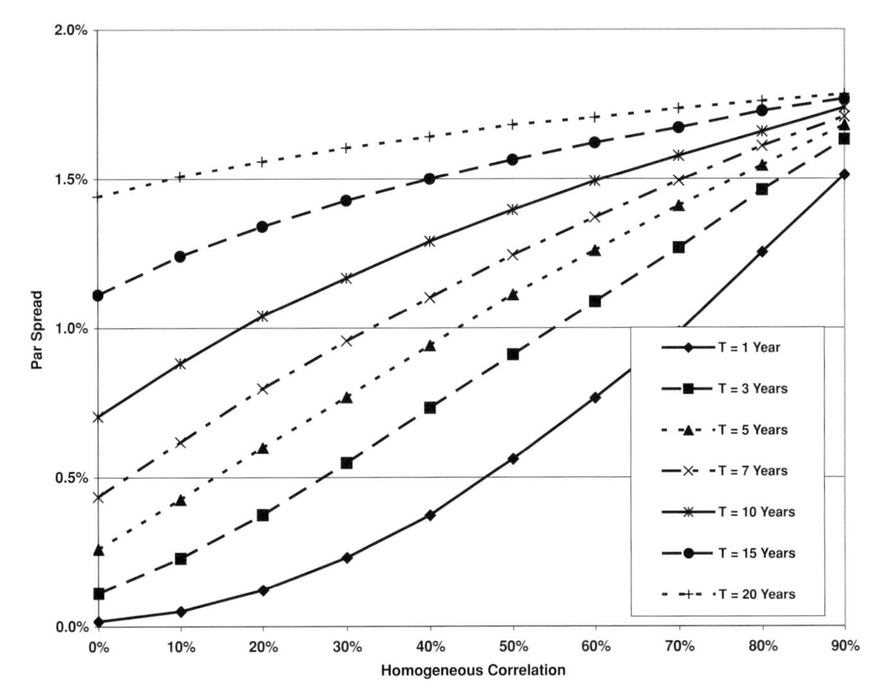

Figure 5.12 The par spread as a function of the homogeneous correlation for a range of maturities for the third-to-default basket.

the third-to-default basket the par spread is monotonically increasing for all of the maturities analysed.

Clearly there is a lot going on here. In Chapter 7 analogous plots for synthetic CDO par spreads will also yield similar behaviour. In that context it will be shown that the behaviour can be explained in terms of the portfolio loss distribution. We will therefore postpone further analysis of this phenomenology for default baskets until Chapter 7.

5.6.5 Impact of spread bump magnitude upon delta estimators

Having explained the reason for the convexity of the basket's PV as a function of the spread bump, it is important now to quantify the impact of this convexity. The reason for this is simple. We are interested in calculating delta values for the basket with respect to fluctuations in the market observed CDS par spreads. The calculation of delta values via Monte Carlo simulation is noise influenced (it is tempting to say noise dominated). Calculating delta by a spread bump of for example 1 bp will produce only a small fluctuation in the basket's PV and the subsequently calculated delta may be swamped by noise. Conversely, calculating the delta with a very

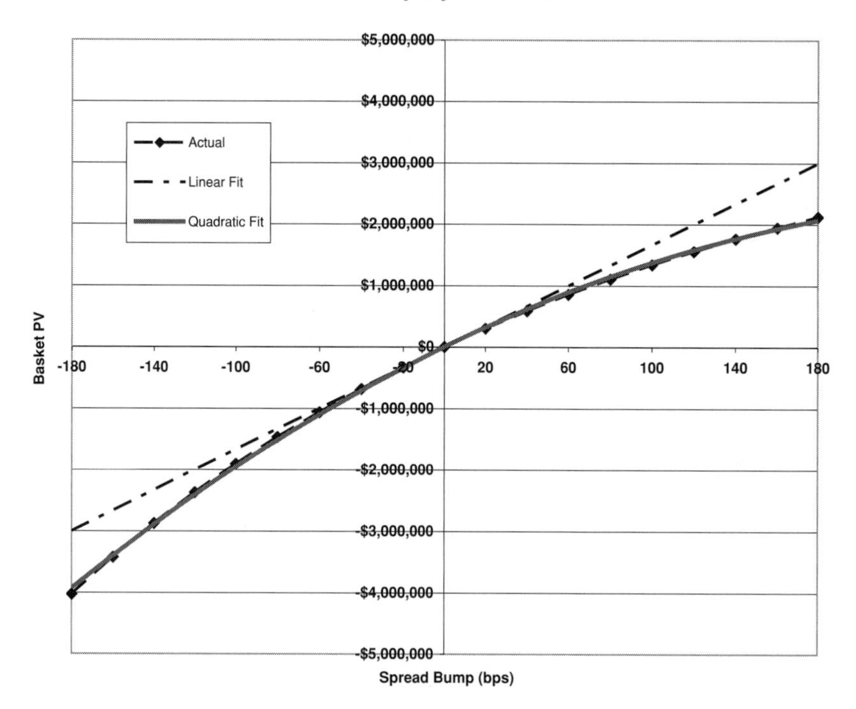

Figure 5.13 The PV of a first-to-default basket as a function of the spread bump for $\rho = 0\%$. Overlaid on this are linear and quadratic fits to the data. The linear fit to the data is only valid for a small range (approximately ± 20 bps) about the unperturbed case.

large bump removes some of the noise but at the expense of introducing non-linear convexity effects. Somewhere in-between is the optimal magnitude of spread bump to use to calculate the best delta estimator. The following table summarises these observations.

Spread bump magnitude	Delta estimator
Small	noise dominated
Medium	optimal
Large	convexity influenced

Figure 5.13 re-plots the first-to-default basket PV as a function of the par spread bump for the case where $\rho = 0\%$. Overlaid on this plot are linear and quadratic fits to the actual data. It is easy to observe that the quadratic model provides a very good fit to the data, and the linear model is less good for extreme bumps. From this graph we can conclude that linear sensitivities can be calculated using spread bumps up to a magnitude of approximately 20 bps. For spread bumps larger than this, non-linear effects begin to manifest themselves.

5.7 Semi-analytic valuation of default baskets

The majority of this chapter has valued default baskets using full Monte Carlo simulation. This approach is simple to understand and implement but suffers from the general problem of simulation-induced noise in estimators, particularly so when calculating sensitivities. These problems are made worse by the discontinuous nature of the default basket payoff. In addition to this, in the Monte Carlo framework a fully populated obligor default correlation matrix is required. This can be difficult to obtain or estimate, particularly for bespoke baskets where some of the obligors may be relatively obscure (and hence not have liquidly traded debt). To mitigate these problems semi-analytic approaches to valuation have been developed. We will now briefly describe this method and its application to default baskets. The primary motivation for introducing semi-analytic methods at this stage is to prepare the ground for the application of these methods to synthetic CDOs (which will be discussed fully in the next chapter).

5.7.1 Approximating the full dependence structure within a factor model framework

Factor models are popular in many areas of credit portfolio modelling. The basic idea is to construct a model representing an obligor's asset value/return. When the obligor's asset value falls below a certain threshold level, default is assumed to have occurred. The obligor value is decomposed into two components: a systemic component which represents a factor common to all the obligors (an intuitive way to think of this is as 'the state of the economy' or 'the state of the business cycle') and an idiosyncratic component which is specific to each individual obligor. In some credit applications (e.g. CreditMetrics) more than a single systemic factor is used to characterise the collective behaviour of the obligor asset values. These factors may correspond to different macroeconomic variables (e.g. GDP, interest rates etc.). The strength of the dependence of an obligor's asset value on a particular factor is characterised by a weighting factor. Although these models are intuitively appealing it is often very difficult in practice to parameterise them. For this reason it is highly unusual to see anything other than single-factor models used within a credit derivative valuation context. The most popular single-factor model is as follows.

5.7.2 A single-factor model

Let the value of the assets of obligor i be X_i where $X(T) \sim N(0, 1)$. It is assumed that default of this obligor occurs if $X_i < B_i$ where B_i is the default barrier level of obligor i, i.e. default occurs if the value of the obligors assets fall below some specified threshold. This approach is clearly similar to structural models of default – in fact it is a multivariate version of the Merton model. A drawback of these models is that to characterise the default barrier requires detailed knowledge of an obligor's balance sheet. In many cases this information is simply not publicly available. In the credit markets however there is information liquidly available which (in principle) reflects the markets' aggregate opinion as to the likelihood of default of an obligor: their par CDS spreads. The default barrier (representing the probability of default) is therefore calibrated to the marginal default probability q_i of obligor i by $q_i(t) = N(B_i(t))$. Hence $B_i(t) = N^{-1}(q_i(t))$ (remembering that the marginal default probabilities are exogenously implied from the market observed par CDS spreads of the individual obligors).

Now assume that the 'state of the economy' is reflected by a single market risk factor V. The correlation between obligor i and the economy is given by ρ_i (hence obligors are coupled indirectly through the state of the economy). If we assume a single-factor model then the returns on the assets of obligor i are given by

$$X_i = \rho_i V + \sqrt{1 - \rho_i^2}\varepsilon_i$$

where $V \sim N(0, 1)$ and $\varepsilon_i \sim N(0, 1)$ are assumed to be standard normal deviates. The random variable X_i is also normally distributed since the normal distribution is stable under convolution. If an individual obligor's behaviour is completely independent of the overall macroeconomic environment then $\rho_i = 0$; conversely if the obligor's behaviour is driven exclusively by macro factors then $\rho_i = 1$. The most important point to note is that conditional on the realisation of the common factor V, the variables X_i and X_j, and hence the default behaviour of the obligors are independent (since the idiosyncratic factors are independent).

The default condition $X_i < B_i$ implies that

$$\rho_i V + \sqrt{1 - \rho_i^2}\varepsilon_i < B_i.$$

Conditional on V, the default probability for obligor i is therefore

$$q_i(t|V) = N\left(\frac{B_i(t) - \rho_i V}{\sqrt{1 - \rho_i^2}}\right)$$

where $N(x)$ is the cumulative normal distribution. This is the conditional probability that $\tau_i < t$ or equivalently $X_i < B_i$. The conditional survival probability for obligor i is given by

$$S(t < \tau_i|V) = 1 - q_i(t|V) = N\left(\frac{\rho_i V - B_i(t)}{\sqrt{1 - \rho_i^2}}\right).$$

The conditional independence assumption is that conditional on the state of the economy, the asset returns are independent, i.e. $\{X_i|V : i = 1, \ldots, n\}$ are independent. The conditional default probability for obligors $i = 1, \ldots, n$ is then given by the product of the marginal default probabilities

$$Q(t|V) = \prod_{i=1}^{n} q_i(t|V)$$

and the conditional joint survival probability is given by

$$S(t|V) = \prod_{i=1}^{n} [1 - q_i(t|V)].$$

This expression represents the conditional survival probability of the basket. The joint unconditional default probability is obtained by averaging out the common factor and is obtained as

$$Q(t) = \int_{-\infty}^{+\infty} Q(t|V)P(V)\mathrm{d}V$$

where

$$P(V) = \frac{1}{\sqrt{2\pi}}\mathrm{e}^{-V^2/2}$$

is the pdf of the common factor (assuming it is normally distributed). The joint survival probability is defined as $S(t < \tau_1, \ldots, t < \tau_n)$ and is given by $S(t < \tau_1, \ldots, t < \tau_n) = 1 - Q(t < \tau_1, \ldots, t < \tau_n)$.

The basket unconditional survival probability is

$$S_{\text{Basket}}(t) = \int_{-\infty}^{+\infty} S(t|V)P(V)\mathrm{d}V = \mathrm{e}^{-h_{\text{Basket}}t}.$$

The integral in the expression for the unconditional default and survival probabilities can be evaluated numerically much more rapidly than by outright simulation. The conditional independence framework can also be extended to include more systemic risk factors, as well as to incorporate other copulas (to capture different dependency structures). This is an important point that we will return to in Chapter 10 when we consider extending the standard synthetic CDO model to facilitate calibration to market observed spreads.

Now that we have obtained an expression for the basket survival probability (at each coupon date) the fee and contingent legs can be valued according to

$$V_{\text{Fee}}(0) = \sum_{j=1}^{N} Z(0, T_j)\Delta_j S_{\text{Basket}}(0, T_j)$$

$$V_{\text{Cont}}(0) = \sum_{j=1}^{N} Z(0, T_j)\left[S_{\text{Basket}}(0, T_j) - S_{\text{Basket}}(0, T_{j-1})\right].$$

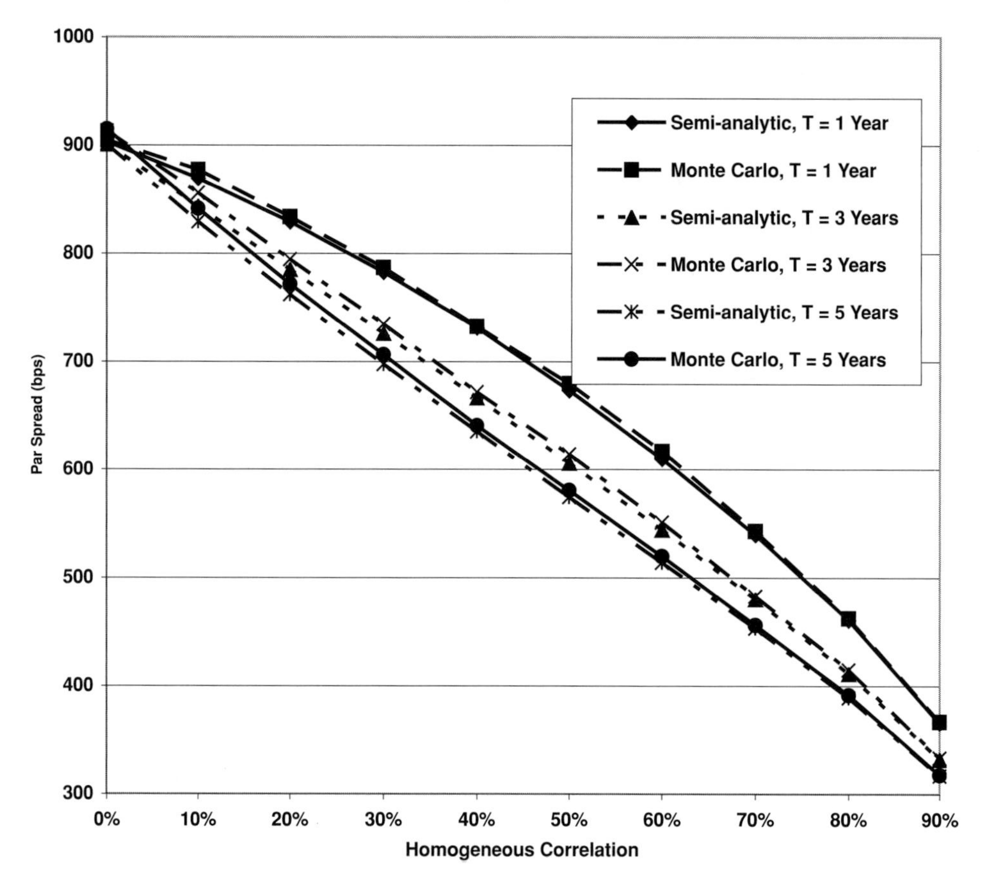

Figure 5.14 The par spread as a function of homogeneous correlation (for varying maturities) for an FtD basket. Shown are the results for the Monte Carlo and semi-analytic models. There is good agreement between the two different valuation methodologies.

Figure 5.14 is a comparison between the results of the Monte Carlo FtD basket model presented earlier and the results of the semi-analytic factor model for varying correlation and maturities. It will be observed that there is good agreement between the two models. The semi-analytic models described can also be extended (with some difficulty) to value more complex baskets such as second-to-default, m-out-of-n defaulting etc. The reader is referred to the relevant research papers for more details of the models [Laurent and Gregory 2003].

5.8 Chapter review

This chapter has introduced and analysed default baskets. The market standard model for valuation of default baskets is to use the normal copula model as the mechanism for generating correlated default times. A large part of this chapter has been devoted to the application of Monte Carlo techniques to value the resulting cashflows. The limitations of this technique (noise) when applied in other areas of quantitative finance are well known. Credit derivatives also introduce additional difficulties. However, as a technique it is simple to understand conceptually and also to implement. For non-time-critical activities (such as independent validation of the models used for reporting firm-wide PV and risk numbers) where running more simulations is acceptable it is a very useful modelling approach. The real benefit of Monte Carlo simulation in the context of modelling portfolio credit derivatives however is that it facilitates a clear understanding of why the phenomenology of an instrument is the way it is. Cannot understand the correlation sensitivity of the PV? Not a problem, look at the simulated default times or the distribution of simulated losses as the correlation is varied.

To mitigate some of the problems of Monte Carlo simulation methods, semi-analytic methods have been developed. These approximate the full dependence structure between the obligors with a low-dimensional factor based approach. Semi-analytic methods are widely used for the valuation of vanilla trades (e.g. standardised baskets based on the liquid credit indices such as iTraxx). However, for bespoke deals the contract specifications may include some non-standard features (e.g. non-standard cashflow waterfalls which redirect cashflows under certain circumstances). In these cases Monte Carlo simulation is often the only appropriate method, although if there is not a huge volume of such bespoke trades the computational demands of Monte Carlo simulation may be tolerable.

In the next chapter we will begin our analysis of synthetic CDOs. Much of what has been learnt in this chapter in the context of default baskets will prove to be extremely useful.

6

Valuation of synthetic CDOs

6.1 Introduction

In this chapter we construct and analyse a number of different models for the valuation of synthetic CDO tranches. Subsequent chapters will extend the models and analysis introduced so it is a very important chapter to the overall scope of the text. As we will see, much of the preceding discussion on valuation of default baskets will prove to be of use here.

This chapter begins in Section 6.2 with a description of the cashflow mechanics of synthetic CDOs, highlighting a number of important features that a suitable model must capture. The basic principles of synthetic CDO pricing are introduced in Section 6.3. As we will see in this section, the key quantity to determine when valuing a synthetic CDO is the portfolio loss distribution. The remainder of the chapter is devoted to describing different methods for the construction of the portfolio loss distribution. Section 6.4 describes the Monte Carlo method for implementing the standard market model. As was found for default baskets in the previous chapter, the Monte Carlo method is a flexible and intuitive methodology, but the problems of simulation noise are ever present. This coupled with the typically large size of the underlying pool in a tranche make the Monte Carlo method unsuitable for many time-critical applications within a bank (such as revaluation of a tranche portfolio for the purposes of calculating value-at-risk (VaR) numbers). In addition to this, a senior tranche high up in the capital structure may require many defaults to occur before the tranche is impacted. If the underlying obligors are of high credit quality (with a small probability of defaulting) a prohibitively large number of simulations may be required in order to obtain statistically meaningful estimates of the fee and contingent payments. For these reasons the development of fast semi-analytic valuation methodologies was a crucial milestone in the evolution of the CDO market. Because traders were able to value and risk manage positions quickly, the whole trading process became easier, leading to greater liquidity. Semi-analytic methods form the bulk of the chapter and are covered in Section 6.5. Section 6.6 gives a brief overview of structural models. Finally in Section 6.7 we review what has been introduced in the chapter and motivate the direction in which we next go.

6.2 Synthetic CDO cashflow mechanics

In Chapter 1 we introduced synthetic CDOs. As discussed there the important contractual features of a synthetic CDO are the following.

- The composition of the underlying pool of CDSs.
- The capital structure of the tranches. This is characterised by the attachment points of the tranches (the amount of $ losses in the pool that can be absorbed before a tranche is impacted) and the detachment points of the tranches (the subsequent losses in the pool that impact the tranche; losses greater than the detachment point no longer have any effect on the tranche since its notional is exhausted).
- The maturity of the tranche protection.
- The tranche par spread (the coupon rate applied to the periodic fee leg payments).

There are of course other tranche characteristics that are of importance which we will introduce as necessary. In this section we describe the ordering and timing of cashflows for a plain vanilla synthetic CDO, at the same time introducing some of the notational conventions to be used throughout the analysis.

A synthetic CDO is essentially a bilateral swap contract. The two counterparties to the contract agree to swap a contractually agreed stream of cashflows. We will refer to the two counterparties generically as the *protection seller* and the *protection purchaser*. The protection purchaser makes periodic fee payments to the protection seller in return for the protection provided by the protection seller. The magnitude and timing of the fee and contingent payments are dependent upon the default behaviour of the obligors in the reference pool. Figure 6.1 shows a number of possible cashflow scenarios. We assume in each of these figures that each default impacts the tranche, i.e. it is an equity tranche with a sufficiently large tranche width to absorb all the losses. In Figure 6.1(a) there are no defaults of the obligors before the maturity T of the tranche. In this case the protection purchaser makes all the fee payments at the coupon dates $\{T_j : j = 1, \ldots, N\}$ based on the original notional of the tranche. In Figure 6.1(b) we introduce a number of defaults at times $\tau < T$. As each default occurs the cumulative loss to the underlying pool increases. If the cumulative losses are sufficient to eat up all of the tranche's subordination, then that tranche begins to suffer losses. This reduces the notional amount on which the spread for the fee payments is paid. If the losses are such that the total pool loss amount exceeds the tranche notional (and subordination) then the fee and contingent payments are zero.

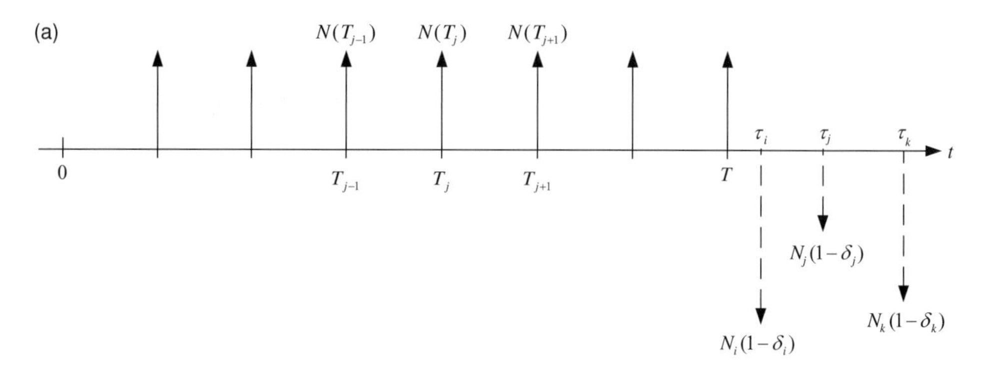

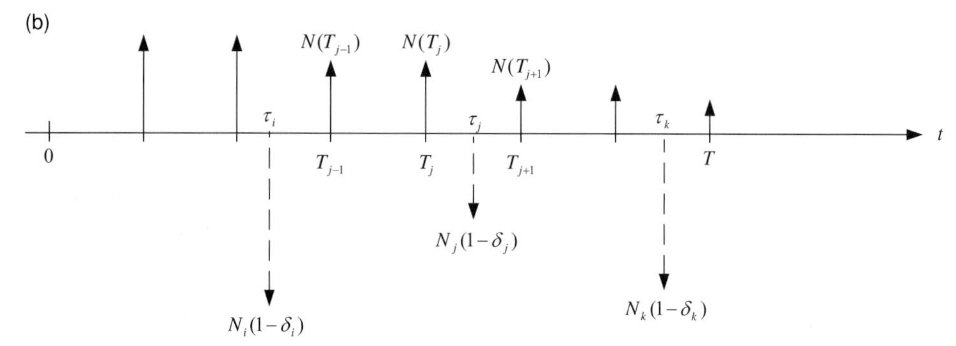

Figure 6.1 (a) A synthetic CDO cashflow stream when there are no defaults occurring before the maturity T of the tranche. In this case the coupon payments made by the protection purchaser are based on the full notional amount and are constant through time. (b) A synthetic CDO cashflow stream when there are a number of defaults occurring before the maturity T of the tranche. In this case the coupon payments made by the protection purchaser (and received by the protection seller) reduce as the defaults impact the tranche notional. (c) Impact of front-loaded defaults upon synthetic CDO cashflows. In this case the defaults quickly wipe out the notional amount of the tranche. The payments received by the protection seller are therefore smaller than if the defaults were back-loaded (see (d)). (d) Impact of back-loaded defaults upon synthetic CDO cashflows.

It is clear from these figures that the timing of defaults has a major impact upon the magnitude of the cashflows. If all the defaults occur post tranche maturity the tranche is not impacted; if all the defaults are front loaded at short times (Figure 6.3(c)), tranches are quickly impacted. The more rapidly the tranche notional is impacted the smaller the fee payments to the protection purchaser are since the coupon is based on the outstanding tranche notional. If all the defaults occur near the maturity of the tranche (Figure 6.4(d)) there is a different stream of fee cashflows. These figures demonstrate that the timing and clustering of defaults has a significant impact upon the valuation of a synthetic CDO (and from a modelling point of view the expected losses at each coupon date will be of importance).

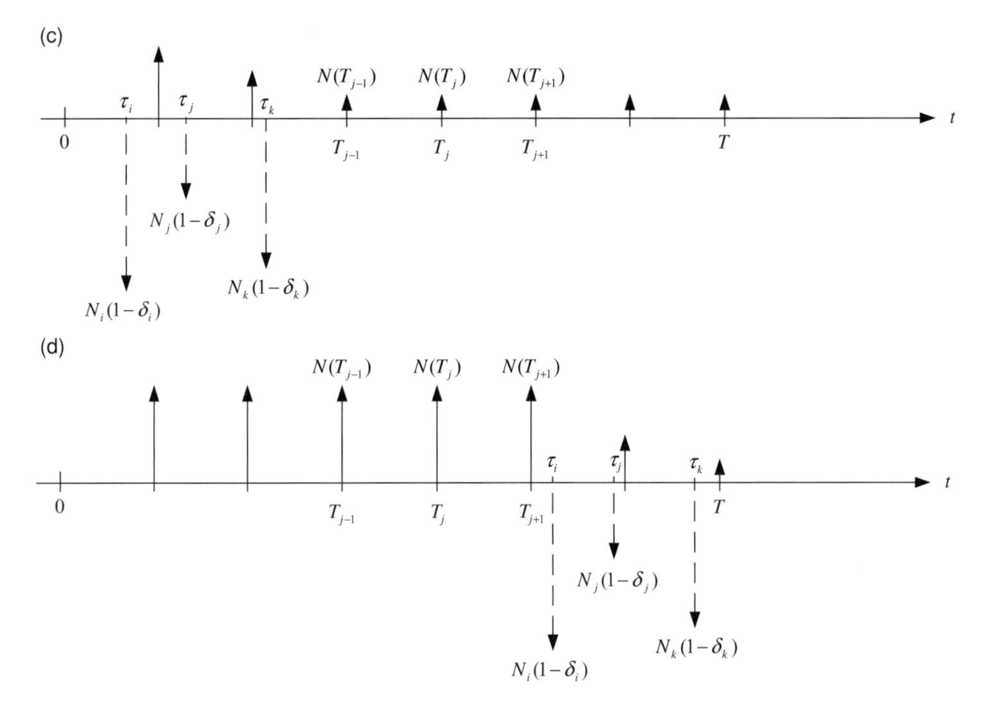

Figure 6.1 (*cont.*)

In all cases when a default occurs the protection seller compensates the protection purchaser for the losses incurred if the cumulative pool losses exceed the tranche attachment point (until the total pool losses exceed the tranche detachment point).

6.2.1 Tranche upfront payments and running spreads

Tranches are usually traded in unfunded form meaning that at contract inception the value of the spread applied to the fee leg is chosen to balance the expected value of the contingent leg (and neither counterparty has to pay any amount upfront). For some tranches however (in particular the equity tranche of STCDOs) there is also an upfront payment that is exchanged at the contract inception at $t = 0$. The purpose of the upfront payment is to compensate the protection seller for the risk they are taking in providing protection. The upfront payment is usually quoted as a percentage of the tranche notional amount (it is usually a substantial fraction of the tranche notional, for example 30% of the \$37.5 m iTraxx equity tranche notional).

It is the single-tranche market convention to pair an upfront payment with a fixed running spread. For example, when trading 0–3% equity tranches referencing the standardised iTraxx portfolio, the market quote consists of a value for the upfront payment as well as a value for the running spread. The running spread is the (constant) spread applied to the fee payment leg to determine the total value of the fee payments. Market convention is to quote a running spread of 500 bps (5%) which is paid quarterly in arrears. This value is fixed for the lifetime of the contract (although of course the notional amount which the spread is applied to will decrease over time as defaults eat into the underlying pool). The upfront payment however is a function of the day-to-day fluctuations in the market variables.

6.2.2 Amortisation of the super-senior tranche

Default events affect the attachment points of each tranche as the subordination of the tranches is reduced. For example, consider a synthetic CDO with 125 equally weighted names. If one of the obligors defaults, with a recovery rate of 40%, then the loss to the portfolio is 60%/125 = 0.48% of the total portfolio notional. The loss to a 0–3% tranche is 16% of the tranche notional. Subsequent coupons paid on the 0–3% tranche would be on a notional of 84% of the initial amount. After the credit event the 0–3% tranche attachment and detachment points become 0 and 2.52% respectively. For the mezz tranches both the attachment and detachment points are reduced by 0.48%, but the tranche notional and width remain the same (since the default has not yet impacted the mezz tranches). For the super-senior tranche (30–100% for the iTraxx index) the attachment point is reduced by 0.48%. In addition, the notional amount of this tranche is reduced since the default of an entity means the amount recovered on the defaulting obligor can no longer be lost.

More formally, as defaults occur the defaulting name, of notional amount N_i, is removed from the pool (reducing the notional of the pool by the loss amount $N_i(1-\delta_i)$ where δ_i is the recovery rate) and the notional of the tranches which are impacted (who have had their subordination eaten away) is written down by the loss amount. In addition to this the super-senior ($x-100\%$) tranche's notional is also written down by the recovered amount $N_i\delta_i$. This mechanism ensures that the pool notional and the total possible tranche protection notional are matched.

6.3 Basic principles of synthetic CDO pricing

In this section we outline the general approach to valuing a synthetic CDO. We assume that there are $i=1,\ldots,n$ obligors in the underlying pool. Let the notional of obligor i be N_i and their recovery rate be δ_i. The total pool notional is given by

$$N_{\text{Pool}}=\sum_{i=1}^{n}N_i.$$

If the default times of the obligors are denoted by $\{\tau_i : i = 1, \ldots, n\}$, the total (cumulative) pool loss at time t, $L(t)$ is given by

$$L(t) = \sum_{i=1}^{n} N_i(1 - \delta_i)\mathbf{1}_{\tau_i < t}$$

(where the indicator function is defined as standard).

Let us consider a tranche γ. We assume that the tranche attachment/ detachment points are given by (K_L^γ, K_U^γ) (we assume that the attachment/ detachment points are converted to dollar amounts and that the attachment point is less than the detachment point). The loss (payoff) that this tranche experiences when the cumulative pool loss is $L(t)$ (also expressed in dollars) is given by $L^\gamma(K_L^\gamma, K_U^\gamma, t)$ and is calculated as a function of the pool loss according to

- if $L(t) < K_L^\gamma$ then $L^\gamma(K_L^\gamma, K_U^\gamma, t) = 0$,
- if $K_L^\gamma \le L(t) \le K_U^\gamma$ then $L^\gamma(K_L^\gamma, K_U^\gamma, t) = (K_U^\gamma - K_L^\gamma) - L(t)$,
- if $K_U^\gamma < L(t)$ then $L^\gamma(K_L^\gamma, K_U^\gamma, t) = (K_U^\gamma - K_L^\gamma)$.

This payoff structure can be summarised as a call-spread payoff

$$L(K_L^\gamma, K_U^\gamma, t) = \max\left\lfloor \min(L(t), K_U^\gamma) - K_L^\gamma, 0 \right\rfloor.$$

The shape of this payoff for the attachment/detachment points defining the iTraxx capital structure is shown in Figure 6.2. In this plot we have assumed a homogeneous recovery rate of 40% and an obligor notional of \$10 m, hence each successive default in the pool contributes an additional \$6 m of losses. The tranche payoffs show that losses only impact the tranche once the cumulative pool losses exceed the tranche attachment point. Because the equity tranche has no subordination, defaults immediately begin to impact it. After the seventh default the equity tranche is completely wiped out; subsequent losses contribute no further to the payoff of this tranche. For the 3–6%, the 3% subordination means that there must be seven defaults in the pool before its notional begins to be eaten away.

The effects of subordination are clear to see. The equity tranche has no subordination and therefore the first losses start contributing to the tranche's payoff (meaning the protection purchaser receives payments from the protection seller). On the other hand the 12–22% tranche is completely isolated from the effects of the first 25 defaults. The downside however is that once defaults start to eat into the tranche, the notional of the tranche is large (in comparison to the other lower lying tranches) and therefore the tranche is sensitive to the next 20 defaults (after 46 defaults the tranche is completely exhausted and the losses are saturated).

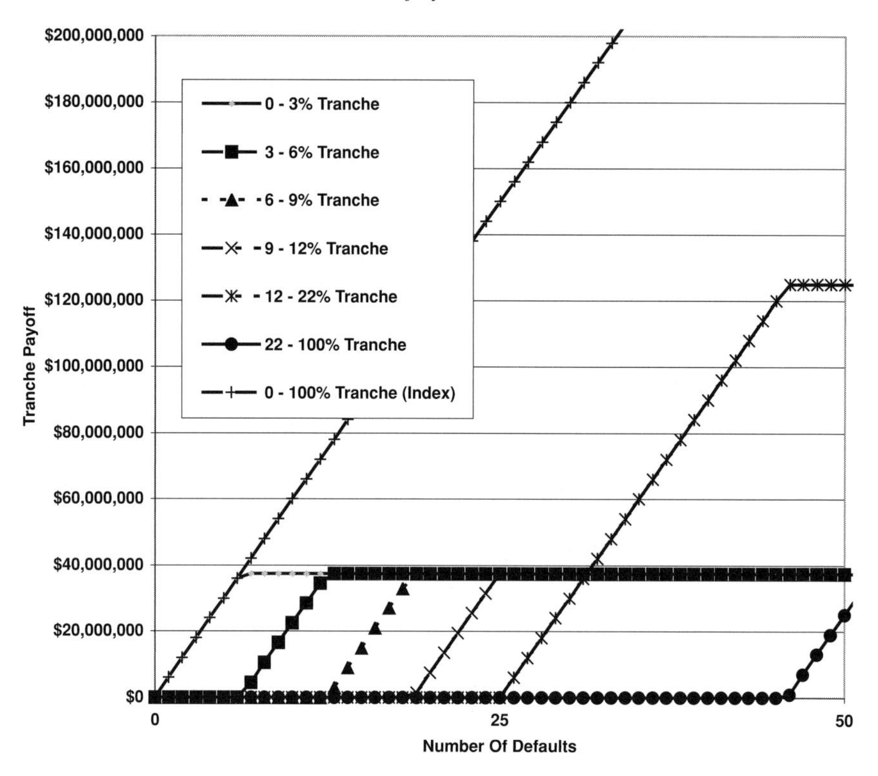

Figure 6.2 The payoff profiles for the tranches corresponding to the iTraxx capital structure (the index corresponds to the 0–100% tranche).

The value of the contingent leg in the case where the contingent payments are made continuously at any time within the period $[0, T]$ is given by

$$V^{\gamma}_{\text{Cont}}(0) = \mathbf{E}_Q\left[\int_0^T Z(0,t)\mathrm{d}L\big(K^{\gamma}_{\text{L}}, K^{\gamma}_{\text{U}}, t\big)\right],$$

i.e. the expected discounted value of the cumulative tranche loss. If we make the approximation that the contingent payments are only made at coupon dates this is approximated as

$$V^{\gamma}_{\text{Cont}}(0) = \mathbf{E}_Q\left[\sum_{j=1}^{N} Z(0,T_j)\left[L\big(K^{\gamma}_{\text{L}}, K^{\gamma}_{\text{U}}, T_j\big) - L\big(K^{\gamma}_{\text{L}}, K^{\gamma}_{\text{U}}, T_{j-1}\big)\right]\right].$$

In practice it turns out that this is a reasonable assumption to make. The expectation operator may be taken inside the summation and

$$\mathbf{E}_Q\left[L\big(K^{\gamma}_{\text{L}}, K^{\gamma}_{\text{U}}, t\big)\right] = \bar{L}\big(K^{\gamma}_{\text{L}}, K^{\gamma}_{\text{U}}, t\big) = \frac{1}{N_{\text{Sims}}}\sum_{\alpha=1}^{N_{\text{Sims}}} L^{\alpha}\big(K^{\gamma}_{\text{L}}, K^{\gamma}_{\text{U}}, t\big)$$

is the Monte Carlo estimator for the expected loss of the tranche with attachment/detachment points (K_L^γ, K_U^γ). The contingent leg payment can therefore be written as

$$V_{\text{Cont}}^\gamma(0) = \sum_{j=1}^{N} Z(0, T_j)\left[\bar{L}\left(K_L^\gamma, K_U^\gamma, T_j\right) - \bar{L}\left(K_L^\gamma, K_U^\gamma, T_{j-1}\right)\right].$$

The (discretely paid) fee payments at each coupon date T_j are given by

$$V_{\text{Fee}}^\gamma(0) = \mathbf{E}_Q\left[\sum_{j=1}^{N} Z(0, T_j)\Delta_j\left[\left(K_U^\gamma - K_L^\gamma\right) - L(K_L, K_U, T_j)\right]\right],$$

that is, the expected value of the discounted remaining tranche notional $(K_U^\gamma - K_L^\gamma) - L(K_L^\gamma, K_U^\gamma, T_j)$ at each coupon date T_j, summed over all the coupon dates. Replacing the expectation operator with the expected value this becomes

$$V_{\text{Fee}}^\gamma(0) = \sum_{j=1}^{N} Z(0, T_j)\Delta_j\left[\left(K_U^\gamma - K_L^\gamma\right) - \bar{L}\left(K_L^\gamma, K_U^\gamma, T_j\right)\right].$$

This can also be written as

$$V_{\text{Fee}}^\gamma(0) = \sum_{j=1}^{N} Z(0, T_j)\Delta_j \times\left[\left(K_U^\gamma - K_L^\gamma\right) - \frac{1}{2}\left[\bar{L}\left(K_L^\gamma, K_U^\gamma, T_{j-1}\right) + \bar{L}\left(K_L^\gamma, K_U^\gamma, T_j\right)\right]\right]$$

(approximating defaults occurring at the mid-point between coupon payment dates).

The tranche par spread is given by $s_{\text{Par}}^\gamma = V_{\text{Cont}}^\gamma / V_{\text{Fee}}^\gamma$. This is the spread to be applied to the fee leg payments which ensures the PV of the tranche at contract inception is zero. If we assume that the contractually agreed tranche spread is s^γ the PV of the tranche is given by

$$V^\gamma = \phi^\gamma\left(-s^\gamma V_{\text{Fee}}^\gamma + V_{\text{Cont}}^\gamma\right).$$

The long/short flag $\phi^\gamma = +1$ if we are calculating the tranche PV from the perspective of the protection purchaser (short correlation – see Chapter 12 for an explanation of this terminology) and $\phi^\gamma = -1$ if it is from the perspective of the protection seller (long correlation).

If $s^\gamma = s_{\text{Par}}^\gamma = s_{\text{Par}}^\gamma(0)$ is the tranche par spread at contract inception at $t = 0$ then $V^\gamma(0) = 0$. If $s_{\text{Par}}^\gamma(t)$ is the market observed tranche par spread at some later time t

then the tranche MtM value given by $\text{MtM}^\gamma(t) = V^\gamma(t) - V^\gamma(0)$ is

$$\text{MtM}^\gamma(t) = \left(s^\gamma_{\text{Par}}(t) - s^\gamma_{\text{Par}}(0)\right) V^\gamma_{\text{Fee}}(t)$$

since the tranche par spread at time t is given by $s^\gamma_{\text{par}}(t) = V^\gamma_{\text{Cont}}(t)/V^\gamma_{\text{Fee}}(t)$.

Finally the tranche upfront payment (if there is one) can be computed from $-u^\gamma + s^\gamma_{\text{Running}} V^\gamma_{\text{Fee}} = V^\gamma_{\text{Cont}}$ where $s^\gamma_{\text{Running}}$ is the contractually agreed running spread.

It is apparent from this analysis that the central problem in synthetic CDO pricing is to determine the expected tranche losses at each coupon date. Once the expected losses have been computed all other quantities of interest can be calculated. This also includes risk sensitivities obtained by perturbing the model inputs and revaluing the tranche.

6.4 Valuation in the standard market model using Monte Carlo simulation

The standard market model is most easily implemented using Monte Carlo simulation of correlated default times (see Bluhm [2003], Bluhm and Overbeck [2007] for further details of tranche valuation models). Valuation is a two-step process: firstly generate the default times and secondly value the subsequent fee and contingent cashflows. Monte Carlo simulation is a flexible and intuitive method particularly when the cashflows have any non-standard features or when trying to value multiple instruments as part of a more complex structured solution (for example linking the payoffs of several tranches together). It can be very useful for providing significant insight into the model's behaviour. For non-time-critical applications (for example, independent back-office verification of front-office valuation models) it is highly recommended as a first approach to gain insight into a particular product.

To reiterate, the central problem in pricing CDO tranches is the determination of the portfolio loss distribution. We now present an algorithm for obtaining this in the standard market model using Monte Carlo simulation. The approach to be outlined is very simple. The correlated default times of the obligors are generated (using the same algorithm as described in the previous chapter for default baskets), and the subsequent losses are assigned to an appropriate time bucket. We assume that the lattice for losses coincides with the coupon payment dates (this is only to make the pseudo code easier to follow – this assumption is not necessary in practice). The Monte Carlo procedure builds up the time bucketed loss distribution function at each coupon date simulation-by-simulation. From this the expected loss at each coupon date can be calculated. The final step of the valuation is to use the

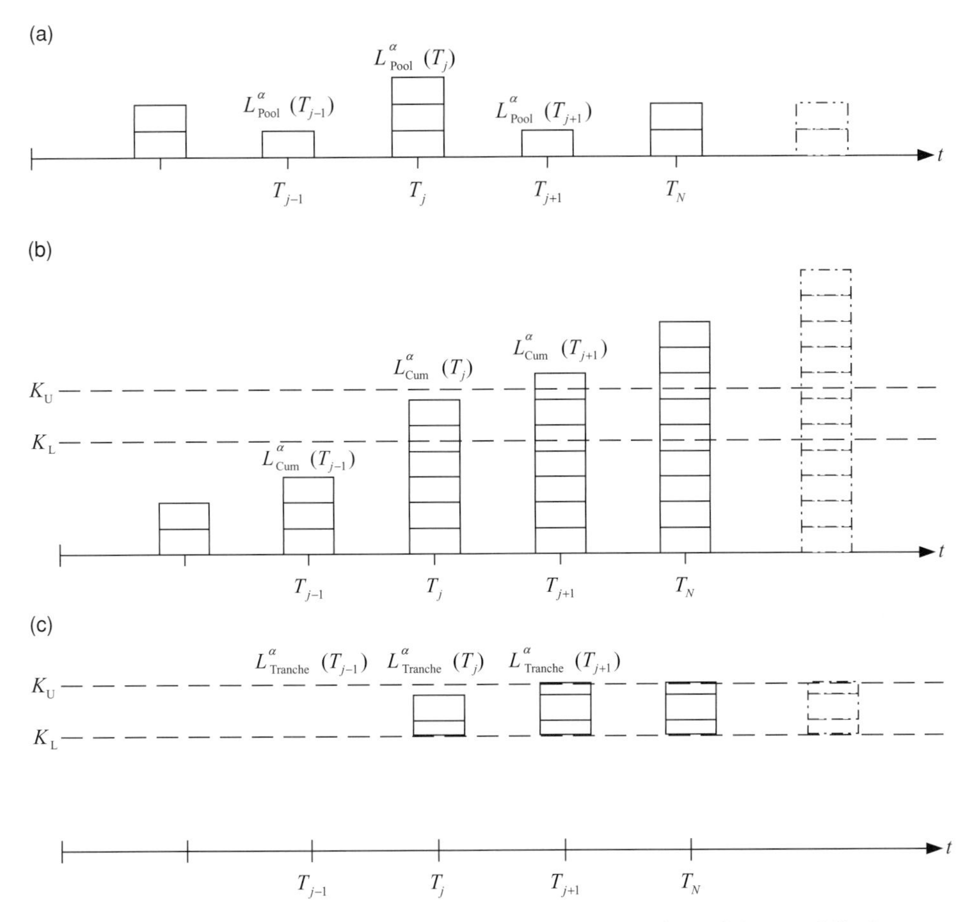

Figure 6.3 (a) Schematic diagram showing the construction of the portfolio loss distribution for an individual simulation. In this particular simulation α there is one loss in the time bucket at T_{j-1}, three at time T_j and so on (for simplicity the losses due to each individual obligor are assumed to be identical). (b) Schematic diagram showing the simulated portfolio cumulative loss at each timenode. Also shown are the levels of the tranche attachment and detachment points and where the cumulative loss lies in relation to them. (c) Schematic diagram showing the simulated tranche (not portfolio) loss at each timenode. In this particular simulation the cumulative losses only begin to eat into the tranche from time T_j onwards. However, the cumulative losses at time T_{j+1} are sufficient to have eaten up all of the tranche's notional, and subsequent losses have no impact upon the tranche.

expected losses to determine the fee and contingent legs. Figure 6.3 illustrates the algorithm schematically. As was described for the valuation of default baskets we assume that the obligor default correlation matrix $\mathbf{C}$ (and its Cholesky decomposition computed from $\mathbf{C} = \mathbf{A}\mathbf{A}^{\mathrm{T}}$) and the marginal obligor survival curves $S_i(t, T)$ are available as inputs. Obligor default times are obtained by inverting the marginal survival curve. The algorithm for constructing the portfolio loss distribution is as follows:

for simulation $\alpha = 1$ to N_{Sims}
  generate $\vec{\varepsilon}^{\alpha} \sim N(\mathbf{0}, \mathbf{I})$ // vector of uncorrelated deviates
  calculate $\vec{\phi}^{\alpha} = \mathbf{A}\vec{\varepsilon}^{\alpha}$ // vector of correlated deviates
  for obligor $i = 1$ to n
    calculate $\tau_i^{\alpha} = S_i^{-1}(N(\phi_i^{\alpha}))$
    if $\tau_i^{\alpha} < T$ // did this obligor default before the maturity of the tranche?
      determine j such that $\tau_i^{\alpha} \in [T_{j-1}, T_j)$
      $L_{\text{Pool}}^{\alpha}(T_j) += (1 - \delta_i)N_i$
    end if $\tau_i^{\alpha} < T$
  end loop over i

$$\text{calculate } L_{\text{Cum}}^{\alpha}(T_j^G) = \sum_{k=0}^{j} L_{\text{Pool}}^{\alpha}(T_k) \qquad j = 1, \ldots, N$$

$$\text{Calculate } L_{\text{Tranche}}^{\alpha}(T_j^G) = \max\left\lfloor \min\left\lfloor L_{\text{Cum}}^{\alpha}(T_j^G), K_{\text{U}}\right\rfloor - K_{\text{L}}, 0\right\rfloor \qquad j = 1, \ldots, N$$

end loop over α

$$\text{calculate } \bar{L}_{\text{Tranche}}(T_j) = \frac{1}{N_{\text{Sims}}} \sum_{\alpha=1}^{N_{\text{Sims}}} L_{\text{Tranche}}^{\alpha}(T_j) \qquad j = 1, \ldots, N$$

Once the default times are generated the simulated pool losses are assigned to the appropriate time bucket $L_{\text{Pool}}^{\alpha}(T_j)$. The cumulative pool losses at each coupon date $L_{\text{Cum}}^{\alpha}(T_j)$ can be computed and from this the tranche losses determined. These quantities are computed for each simulation. Finally the expected tranche loss at each coupon date $\bar{L}_{\text{Tranche}}(T_j)$ is computed by averaging over the ensemble of default simulations. The fee and contingent legs can now be computed from the expected losses (utilising the expressions in Section 6.3).

The above algorithm is really quite simple and elegant. However, Monte Carlo simulation of course suffers from the problem of simulation noise. This problem is particularly severe for synthetic CDOs where a large number of defaults may be necessary in order to impact a tranche (for the 12–22% tranche example discussed earlier 25 defaults are necessary before the tranche notional begins to be written down). For this reason the development of semi-analytic valuation methodologies was a key event in the evolution of the synthetic CDO market (since it allows rapid re-pricing of tranches and calculation of their risk sensitivities).

Semi-analytic methods are well adapted to synthetic CDOs but they are not so for cashflow CDOs. This is because a cashflow CDO will have a very bespoke cashflow waterfall structure. Monte Carlo simulation on the other hand is perfectly adapted to cashflow CDOs since it is a simple matter to incorporate the particular waterfall rules into the valuation algorithm. Cashflow CDOs will be touched upon briefly in Chapter 11 (doing the topic justice would require an entire book in its own right!).

6.5 Valuation in the standard market model using semi-analytic techniques

The Monte Carlo approach to valuing synthetic CDOs is intuitively appealing and simple to understand: generate a set of default times, build up a representation of the expected losses in the underlying pool as a result of this default scenario and calculate the resulting cashflows. But the usual problems remain, namely simulation noise and the convergence of estimators, particularly sensitivities. In addition to this there are other problems.

Firstly default baskets are usually small with between 5–20 obligors referenced in each basket. An efficient Monte Carlo implementation can churn through a default basket quite quickly. Synthetic CDOs on the other hand typically reference an order of magnitude more obligors (if not greater) than default baskets. Indeed, as we have seen, the standardised credit indices reference portfolios of 125 obligors. Secondly the payoff of a senior tranche requires a large number of defaults to occur (before the maturity of the tranche) in order for the tranche to be impacted (and a non-zero contingent leg payment to be made). Scenarios with large numbers of defaults will of course be less likely than scenarios with few defaults. The estimators for the contingent leg for senior tranches will be generated essentially just by a very small number of paths. This means senior tranches require more simulations than junior tranches to obtain statistically reliable estimators.

If synthetic CDOs were relatively illiquid compared to default baskets the problem at the trading book level might be manageable. We could just bite-the-bullet and run large numbers of simulations. Unfortunately the most liquid trading activity is in tranches defined on the standardised indices. Default baskets tend to be relatively illiquid bespoke instruments traded to hedge exposures to a particular group of obligors in a larger synthetic CDO. It is inescapable therefore that more sophisticated solutions must be sought. In this section we will describe the most popular of the semi-analytic methods that have been developed to value synthetic CDOs.

6.5.1 The single-factor approximation to the full dependence structure

One of the major steps forward was the application of factor models to synthetic CDO pricing [Vasicek 1987, Laurent and Gregory 2003]. Factor models simplify the dependence structure between the obligors. Factor models are similar in spirit to the classic Merton model [Merton 1974].

The purpose of a factor model is to model the value of the assets of the obligor. As in the Merton model, default is assumed to occur when the value of the assets of obligor i X_i fall below a specified barrier level B_i. The general form of a single-factor model is to model the obligor asset value as a combination of an idiosyncratic component specific to that obligor and a systemic component which is common to all obligors. In that spirit we construct the asset return X_i as

$$X_i = \beta_i V + w_i \varepsilon_i$$

where $V \sim N(0, 1)$ is the common systemic component and $\varepsilon_i \sim N(0, 1)$ is the idiosyncratic component (hence the index i which indicates that this shock is specific to this particular obligor). These two components are mixed together with weights β_i and w_i respectively. An important point to note is that because the normal distribution is stable under convolution [Grimmett and Stirzaker 2004], the sum of the systemic and idiosyncratic factors is normally distributed. Imposing the constraints that $\mathbf{E}[X_i] = 0$ and $\text{var}[X_i] = 1$ leads to the relations $w_i = \sqrt{1 - \rho_i^2}$ where $\beta_i = \rho_i$. The one-factor model is then represented as

$$X_i = \rho_i V + \sqrt{1 - \rho_i^2} \varepsilon_i.$$

The correlation between the asset values for two obligors i and j is given by $\rho_{ij} = \text{cov}(X_i, X_j)/\sqrt{\text{var}(X_i)}\sqrt{\text{var}(X_j)}$ which when evaluated is given by $\rho_{ij} = \rho_i \rho_j$. Because the ε_i are independent, the correlation between the obligors is driven purely by the systemic factor. The crucial point about the factor models is that conditional on the value of the systemic factor, V, the defaults of each obligor are independent (hence the conditional probabilities of default will also be independent). See Gou *et al.* [2007] for an extension to this basic approach.

The name single-factor refers to the fact that there is only a single macro factor V present in the model which simultaneously drives the systemic (common) behaviour of all the obligors. However, there may be cases where it is desirable to introduce greater variety into the factors which drive the asset value. The single-factor model can be generalised to an M factor model where there are M independent factors which drive the evolution of an obligor's asset value. In this case the obligor coupling to the single factor becomes $w_i \to w_{ik}$ where $k = 1, \ldots, M$ are the independent factors. The obligor asset return now becomes

$$X_i = \sum_{k=1}^{M} w_{ik} V_k + \sqrt{1 - \sum_{k=1}^{M} w_{ik}^2} \varepsilon_i$$

where $\mathbf{E}[V_k V_l] = \delta_{kl}$ (factors are independent) and all the other factors are also uncorrelated. It is a simple matter to verify that $\mathbf{E}[X_i] = 0$ and $\text{var}[X_i] = 1$ as required.

6.5.2 Constructing the portfolio loss distribution

6.5.2.1 A recursive method

The distribution of losses in the underlying portfolio is composed of a set of discrete values since an obligor either defaults or does not incurring a loss of either 0 or $L_i = N_i(1 - \delta_i)$. If all the obligors are homogeneous then the loss distribution will be a series of $n + 1$ discrete values. The following methodology is based upon that due to Andersen *et al.* [2003] (and follows the notation of [Gibson 2004]). It is very similar to another semi-analytic methodology introduced in Hull and White [2004].

The recursive method for portfolio loss construction and synthetic CDO valuation can be decomposed into four different steps as follows.

1. Calculation of the marginal conditional obligor default probabilities.
2. Calculation of the portfolio conditional loss distribution.
3. Calculation of the tranche expected loss.
4. Calculation of the tranche fee and contingent payment streams.

We now describe each step in detail.

6.5.2.2 Calculation of the marginal conditional obligor default probabilities

Within the one-factor framework it is a simple matter to calculate the conditional default probabilities of the obligors, at the same time maintaining the marginal default probabilities of the obligors. Assume that the creditworthiness/default state of an obligor can be represented by the obligor's asset value. The notional of obligor i is given by N_i and the recovery rate is δ_i. Let the (time-dependent) risk-neutral default probability of obligor i be $q_i(t)$ which is defined such that $\tau_i < t$. In a single-factor model the asset value of obligor i is given by $X_i = \rho_i V + \sqrt{1 - \rho_i^2}\varepsilon_i$ where $i = 1, \ldots, n$ and $0 \leq \rho_i \leq 1$ is the coupling between the obligor and the systemic factor representing the state of the economy.

X_i, V, ε_i are assumed to be zero mean unit variance random variables. Their probability distribution functions are $F_i(X_i)$, $G(V)$ and $H_i(\varepsilon_i)$ respectively (in general these need not be normal but in the case of the standard model the random variables are standard normal deviates). Obligor i defaults if $X_i < B_i$ where B_i is some barrier level. The condition $X_i < B_i$ implies that $F_i(B_i(t)) = q_i(t)$ and therefore $B_i(t) = F_i^{-1}(q_i(t))$.

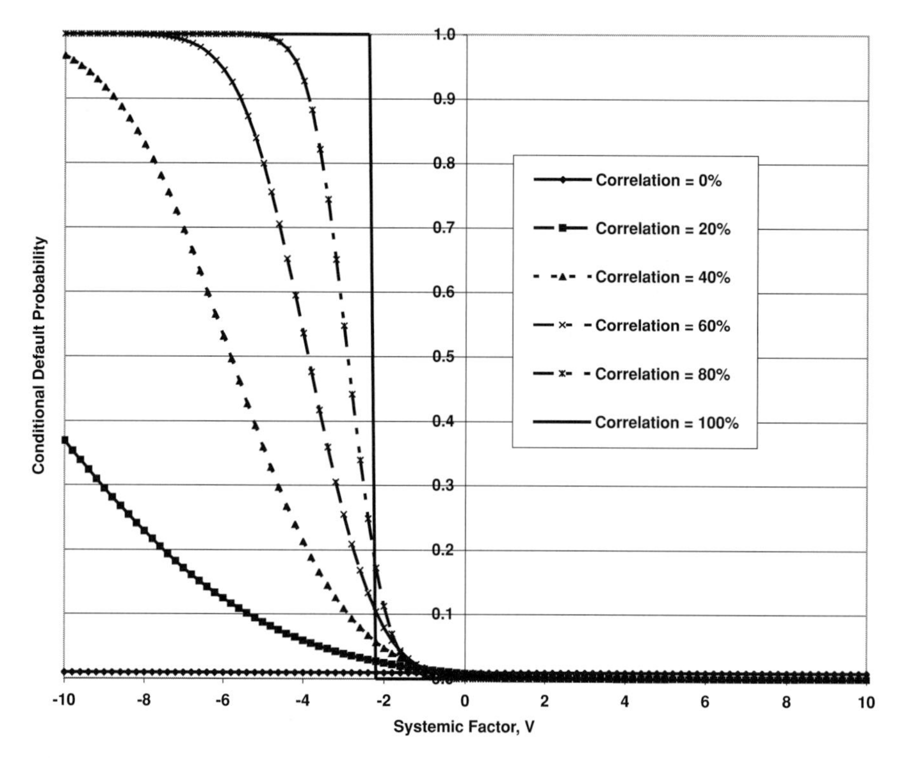

Figure 6.4 The marginal conditional default probability as a function of the systematic factor (for different levels of correlation).

The conditional (on the state of the economy V) default probability for obligor i is given by $P(X_i < B_i|V)$ and is calculated as

$$q_i(t|V) = H_i\left(\frac{B_i(t) - \rho_i V}{\sqrt{1-\rho_i^2}}\right)$$

(note the time dependence which arises because of the time dependence of the marginal default probability). The functional form of the conditional default probability (when the variables are normally distributed) is shown in Figures 6.4 and 6.5, both for the case where $h = 1\%$ and $q_i(t) = 1 - \mathrm{e}^{-h_i t}$. Figure 6.4 shows the ρ dependence for a fixed time horizon of $T = 1$ year. Figure 6.5 fixes $\rho = 50\%$ and plots the time horizon dependence. From $q_i(t) = 1 - \mathrm{e}^{-h_i t}$ as $t \to +\infty$, $q_i \to 1$ and hence $B_i(t) \to +\infty$. Therefore the condition $X_i < B_i$ will almost always be satisfied for any V, ε_i. This is the only point in the model where temporal dependence enters. Some intuition about the model can be obtained by considering a number of limiting cases. As $V \to +\infty$ (the economy doing 'well')

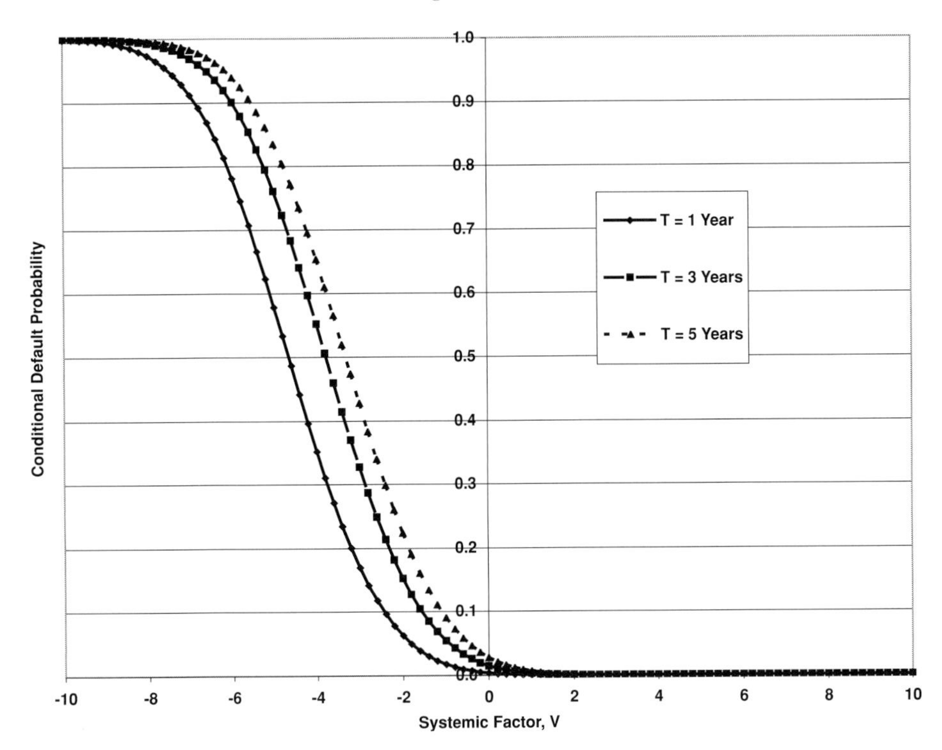

Figure 6.5 The marginal conditional default probability as a function of the systematic factor (for a correlation of $\rho = 50\%$ for different time horizons).

$q_i(t|V) \to 0$, i.e. an obligor has little chance of defaulting whatever their idiosyncratic behaviour. Conversely as $V \to -\infty$ (the economy doing 'badly') $q_i(t|V) \to 1$, i.e. an obligor has a high chance of defaulting irrespective of their idiosyncratic behaviour due to their coupling to the macro state of the economy. Finally when $\rho = 0\%$, $q_i(t|V) = q_i(t)$, i.e. with no coupling to the economy, an obligor's conditional default probability is simply their individual marginal probability.

The marginal default probability at time t (which is an input to the model) can be recovered from the conditional default probability by averaging the marginal density over the systemic factor

$$q_i(t) = \int_{-\infty}^{+\infty} q_i(t|V)P(V)\mathrm{d}V.$$

It is simple to calculate this integral numerically to high precision using an appropriate quadrature, such as Gauss–Hermite [Press *et al.* 2002]. However, this approach is only valid when $P(V)$ is normally distributed. For the standard model this is the case but when considering more general distributions for the systemic factor (for example to enable calibration of the model to market observed spreads – to be discussed in Chapter 10) other numerical integration schemes must be used. It is also important to consider carefully the range of integration. For the standard model where $P(V)$ is normally distributed, the range $V \in [-10, +10]$ is adequate. However, when $P(V)$ is non-normal, and in particular heavy tailed, it is necessary to extend the range of integration in order to avoid probability mass leakage. It is useful to monitor the sum of the portfolio losses over all losses (which should be equal to 100%) as a means of testing the model's performance and implementation. If the value deviates significantly from 100% it is a fair bet the numerical integration scheme is at fault.

A further important point to consider is that throughout this analysis the survival curve is static. That is, the credit spreads observed at $t = 0$ for each of the obligors do not evolve during the course of the calculation. The model is therefore not a dynamic credit spread model. A final important point to note is that since $q_i(t|V) \equiv q_i(\rho, t|V)$ is dependent upon the correlation assumption, the full conditional portfolio loss distribution $p(l, t|V)$ will also be a function of the correlation assumption ρ. This is of central importance to the calculation of implied and base correlations, as we will see in Chapter 9.

6.5.2.3 Calculation of the portfolio conditional loss distribution

We assume that we have available the conditional marginal default probability of each obligor denoted by $q_i(t|V)$ (conditional on the systemic factor V) calculated in the previous step. The distribution of the number of defaults conditional on the systemic factor can be determined using a simple recursion method and the fact that, conditional on the systemic factor, the default behaviour of the obligors is independent.

In the analysis to be presented it is assumed that the obligors' notional amounts and recovery rates are homogeneous meaning that the loss given default is identical for each and every obligor. This simplifies the calculation of the loss distribution since there are now only a discrete set of $n + 1$ values that the loss distribution can take. The model can be extended quite straightforwardly at the cost of only slightly greater algorithmic complexity to incorporate inhomogeneous obligors [Parcell 2006].

Consider a portfolio of k obligors. Let the probability of there being l defaults by time t conditional on the systemic factor V be $p_k(l, t|V)$ where $l = 0, 1, \ldots, k - 1, k$ (i.e. can have zero losses, one loss, up to the total portfolio loss). Now add a $(k + 1)$th obligor to the portfolio. This extra obligor has a conditional default probability given by $q_{k+1}(t|V)$. The default distribution of the new portfolio of

$k + 1$ obligors is given by (remembering that conditional on the systemic factor V the obligor defaults are independent),

$$\begin{aligned} p_{k+1}(0, t|V) &= p_k(0, t|V)(1 - q_{k+1}(t|V)), \\ p_{k+1}(l, t|V) &= p_k(l, t|V)(1 - q_{k+1}(t|V)) \\ &\quad + p_k(l-1, t|V)q_{k+1}(t|V) \qquad \text{for } l = 1, \ldots, k-1, k, \\ p_{k+1}(k+1, t|V) &= p_k(k, t|V)q_{k+1}(t|V). \end{aligned}$$

The recursion is started for $k = 0$ with $p_0(0, t|V) = 1$. From this we can recursively build up the size of the portfolio to arrive eventually at $p_n(l, t|V)$ for $l = 0, 1, \ldots, n-1, n$ to obtain the full conditional loss distribution for the entire portfolio of obligors. A nice visual way to interpret the recursive method for portfolio loss distribution construction is as a non-recombining binomial tree.

Note that if we wish to introduce an obligor into the portfolio that has already defaulted (e.g. for calculating value-on-default as described in Chapter 8) then we would set $q_{k+1}(t|V) \equiv 1$ and proceed with the calculation as normal (or alternatively set $h_i \to \infty$ which has the same impact in the single-factor model for the marginal conditional default probability).

6.5.2.4 Calculation of the tranche expected loss

Once the conditional loss distribution is obtained the unconditional loss distribution can be determined by numerical integration of

$$p_n(l, t) = \int_{-\infty}^{+\infty} p_n(l, t|V)P(V)\mathrm{d}V.$$

Hence to compute the full unconditional loss distribution requires $n + 1$ integrations (corresponding to each value of $l = 0, 1, \ldots, n-1, n$) at each time t of interest. The same considerations outlined for the previous step regarding the numerical integration also apply here. It is worth noting that the effects of probability mass leakage (due to a poor numerical integration scheme) can be monitored by calculating the par spread of the 0–100% tranche of a portfolio. This by definition should be equal to the weighted average par spread of the CDSs in the index, i.e. the index spread. Additionally this value should be independent of the correlation assumption and also independent of the copula choice (which characterises the dependence structure amongst the obligors in the pool). Monitoring this value as a function of the correlation assumption provides a very good test of the model's implementation and performance.

The expected loss of the tranche at a coupon date T_j is given by

$$\bar{L}(T_j) = \sum_{l=0}^{n} p(l, T_j) L(l)$$

where the loss to the tranche if l defaults occur in relation to the attachment and detachment points is given by the standard payoff $L(l) = \max[\min(lN(1-\delta), K_U) - K_L, 0]$ (we have explicitly assumed in this formula that all the obligors' loss amounts are homogeneous) and $p(l, T_j)$ is the probability of the losses in the portfolio being equal to l at time T_j.

Note that the expected loss of the whole portfolio is given by

$$\bar{L}(T_j) = \mathbf{E}\left[\sum_{i=1}^{n} N_i(1-\delta_i)\mathbf{1}_{\tau_i < T_j}\right] = \sum_{i=1}^{n} N_i(1-\delta_i)\left[1 - S_i(T_j)\right],$$

i.e. the expected loss of the portfolio is independent of the correlation between the obligors. However, the expected loss of a tranche, which is only exposed to a particular fraction of the total portfolio loss, is heavily correlation dependent. This is the fundamental opportunity offered by single-tranche CDO trading; the expected tranche loss (and hence the fee and contingent leg values) is dependent upon the correlation assumption. It is not independent of the correlation. The expected loss of the portfolio is also useful for assessing how valuable tranche subordination is to an investor. If a portfolio has an expected loss of 5% then a tranche with only 3% subordination will not be as attractive as a tranche with subordination of 6%.

Having now obtained an expression for the tranche expected loss at each coupon date, we can use the expressions derived previously to determine the fee and contingent leg payments (and hence other quantities such as the par spread etc.).

6.5.2.5 The Fourier transform approach to computing the portfolio loss distribution

An alternative (but closely related) method for constructing the portfolio loss distribution is to utilise the characteristic function of portfolio losses [Debuysscher and Szego 2003, Grimmett and Stirzaker 2004].

For each individual obligor i the loss-given-default is given by $L_i = N_i(1-\delta_i)$ and the total portfolio loss is $L_{\text{Total}} = \sum_{i=1}^{n} L_i$. The characteristic function of the total portfolio loss variables distribution is $\psi(t) = \mathbf{E}\lfloor e^{jtX} \rfloor$ where $t \sim 0$ and $j^2 = -1$ is an imaginary number.

Conditional on the random variable V, the conditional characteristic function describing the distribution of portfolio loss is $\psi_V(t) = \mathbf{E}\lfloor e^{jtL_{\text{Total}}} | V \rfloor$. For an individual obligor they will either default with a loss $(1-\delta_i)$ with probability $q_i(\cdot|V)$ or not default with loss 0 and probability $1 - q_i(\cdot|V)$. Therefore

$\mathbf{E}\lfloor e^{jtL_{\text{Total}}}|V\rfloor = e^{jt(1-\delta_i)}q_i(\cdot|V) + (1 - q_i(\cdot|V))$. Applying the conditional independence property we arrive at

$$\psi_V(t) = \prod_{i=1}^{n} \left[1 + q_i(\cdot|V)\left(e^{jt(1-\delta_i)} - 1\right)\right]$$

for the conditional characteristic function. The unconditional characteristic function can be obtained by integrating out the dependence on the conditional variable

$$\psi(t) = \int_{-\infty}^{+\infty} \psi_V(t)P(V)\mathrm{d}V.$$

To get the actual portfolio loss distribution the characteristic function is inverted using an appropriate Fourier transform algorithm (see Press *et al.* [2002] of course for appropriate routines). This method is very efficient for dealing with inhomogeneous portfolios.

6.5.2.6 The normal proxy method

A very nice and simple method for the computation of the tranche expected loss is the normal proxy model [Shelton 2004]. As before, the pool and tranche loss are given by

$$L_{\text{Pool}}(t) = \sum_{i=1}^{n} N_i(1 - \delta_i)\mathbf{1}_{\tau_i \le t},$$

$$L_{\text{Tranche}}(t) = \min\left[\max(L_{\text{Pool}}(t) - K_{\text{L}}, 0), K_{\text{U}} - K_{\text{L}}\right],$$

respectively. From the single-factor model $X_i = \rho V + \sqrt{1-\rho^2}\varepsilon_i$ we can calculate the conditional default probability $q_i(t|V)$ in the usual way. Conditional on the systemic factor V all defaults are independent. For the loss distribution, conditional on V, the expected loss is given by

$$\mu|V = \mathbf{E}[L(t)|V] = \sum_{i=1}^{n} N_i(1 - \delta_i)(1 - S_i(t|V))$$

where $S_i(t|V) = 1 - q_i(t|V)$ is the conditional survival probability. The covariance of the expected loss distribution is given by

$$\text{cov}(L_k(t), L_{k'}(t)|V) = \sum_{i=1}^{n} N_i^k N_i^{k'}(1 - \delta_i)^2 S_i(t|V)\left[1 - S_i(t|V)\right].$$

Now we make the crucial assumption that conditional on the systemic factor V, the distribution of losses is multivariate normal. Under this assumption $\mathbf{E}[L_{\text{Tranche}}(t)|V]$

is easily computable since the payoff of the tranche loss is simply a call spread profile. In particular it can be shown that

$$\mathbf{E}\left[L_{\text{Tranche}}(t)|V\right] = (\mu_V - K_{\text{L}})N\left(\frac{\mu_V - K_{\text{L}}}{\sigma_V}\right) + \sigma_V n\left(\frac{\mu_V - K_{\text{L}}}{\sigma_V}\right)$$
$$- (\mu_V - K_{\text{U}})N\left(\frac{\mu_V - K_{\text{U}}}{\sigma_V}\right) + \sigma_V n\left(\frac{\mu_V - K_{\text{U}}}{\sigma_V}\right)$$

is the conditional expected loss of the tranche. The unconditional expected loss of the tranche is given by the usual expression

$$\mathbf{E}\left[L_{\text{Tranche}}(t)\right] = \int_{-\infty}^{+\infty} \mathbf{E}\left[L_{\text{Tranche}}(t)|V\right] P(V)\mathrm{d}V.$$

Once the expected tranche loss at each coupon date has been determined we can compute the fee and contingent legs using the usual expressions:

$$V_{\text{Fee}}(0) = \sum_{j=1}^{N} Z(0, T_j)\Delta_j \left[(K_{\text{U}} - K_{\text{L}}) - \bar{L}_{\text{Tranche}}(K_{\text{L}}, K_{\text{U}}, T_j)\right],$$
$$V_{\text{Cont}}(0) = \sum_{j=1}^{N} Z(0, T_j)\left[\bar{L}_{\text{Tranche}}(K_{\text{L}}, K_{\text{U}}, T_j) - \bar{L}_{\text{Tranche}}(K_{\text{L}}, K_{\text{U}}, T_{j-1})\right].$$

This model has the clear advantage of speed. The portfolio expected loss is constructed via an expression that is very simple and quick to evaluate numerically. If in addition to this a numerical quadrature scheme such as Gauss–Hermite is used to compute the unconditional expected loss then it is blisteringly fast. It is also worth pointing out that the model can easily deal with portfolios of inhomogeneous obligors (which some of the other semi-analytic methods have difficulties with). As we will see the model is also very accurate when compared to Monte Carlo and recursive methods. It is highly recommended that the reader implement a version of this model as a workhorse for more complex analysis.

6.5.2.7 Approximating the portfolio loss distribution in the large portfolio limit

An alternative to the methods outlined previously for calculating the portfolio loss distribution is the large homogeneous portfolio (LHP) approximation. If under the one-factor Gaussian copula model we make the further assumption that all the obligors are homogeneous (including their marginal default probabilities) then the conditional default probability is simply

$$q(t|V) = N\left(\frac{B(t) - \rho V}{\sqrt{1-\rho^2}}\right)$$

(dropping the i index which tags an individual obligor). We have also assumed that the coupling between the obligors and the common factor is homogeneous. Let the notional amount of each obligor in the portfolio be N and their recovery amount be δ. The conditional probability of the loss being $L \equiv L_k = kN(1-\delta)/n$ is the probability that exactly k of the n obligors default and is given by

$$p(L = L_k|V) = \binom{n}{k} N\left(\frac{B(t) - \rho V}{\sqrt{1-\rho^2}}\right)^{n-k} \left[1 - N\left(\frac{B(t) - \rho V}{\sqrt{1-\rho^2}}\right)\right]^{n-k}$$

(i.e. a binomial distribution) and the unconditional probability is given by integrating out the dependence on the variable V

$$p(L = L_k) = \int_{-\infty}^{+\infty} \binom{n}{k} N\left(\frac{B(t) - \rho V}{\sqrt{1-\rho^2}}\right)^{n-k} \left[1 - N\left(\frac{B(t) - \rho V}{\sqrt{1-\rho^2}}\right)\right]^{n-k} P(V)\mathrm{d}V.$$

This model gives the probability of loss over the period $[0, t]$, but it provides no information about when defaults occur. Also being based on the binomial distribution, it underestimates tail events. Nevertheless, the evaluation of the loss distribution using this model only requires numerical evaluation of the above integral and a few calls to the cumulative normal distribution. This model is therefore more rapid in execution than the recursive and Fourier methods for constructing the loss distribution (but comparable in speed to the normal proxy model).

The model can be simplified even further by taking the large portfolio limit as $n \to \infty$ (the LHP approximation). We are still assuming the obligors are homogeneous. In this limit the expected fraction of credits of the underlying portfolio defaulting over a specific time horizon approaches the individual default probability of the underlying credits. It can be shown that the cumulative distribution of losses becomes [Schonbucher 2003, Eberlein *et al.* 2007]

$$F(l) = N\left(\frac{\sqrt{1-\rho^2}N^{-1}(l) - B}{\rho}\right)$$

and taking the derivative of this with respect to the loss l gives the probability density function of the loss distribution

$$p(l) = \sqrt{\frac{1-\rho}{\rho}} \exp\left[\frac{1}{2}N^{-1}(l)^2 - \frac{1}{2\rho}\left(B - \sqrt{1-\rho}N^{-1}(l)\right)^2\right].$$

This final expression for the portfolio loss distribution is about as simple as it is possible to obtain. It is very quick to calculate involving only a few calls to an inverse cumulative normal distribution function. Consequently it is highly appropriate for applications where large numbers of tranche revaluations are required to be performed. One such application would be the calculation of potential future exposures within a full Monte Carlo revaluation framework. However this approximation breaks down for small numbers of obligors (typically less than 100 obligors). It is therefore not appropriate for the valuation of default baskets (which typically have of the order of five obligors in the pool).

Of all the models presented in this chapter for the construction of the loss distribution which one is the best? (There are also some other models which have not been presented for the sake of expediency, see Greenberg *et al.* [2004], Yang *et al.* [2005], and Jackson *et al.* [2007]). As always the trade-off between accuracy, ease of implementation and speed of execution determines the answer to this question. If absolute speed is required but extreme accuracy is not, the LHP model is a good choice. If the obligor pool is large and homogeneous then the model is especially appropriate. If speed is not an issue and the obligors are very inhomogeneous Monte Carlo simulation is the best choice. In general however the semi-analytic methods based on the recursive, Fourier or normal proxy methods tend to offer the best trade-off between speed and accuracy.

It is recommended that, as an exercise, the reader implement the different models for the portfolio loss distribution introduced in this chapter (remembering that once the portfolio loss is determined the tranche expected loss and fee and contingent legs can be computed). There is no better way to understand a model than actually to implement it. In terms of implementation, Figure 6.6, outlines the different steps in the valuation process for a tranche γ. The general flow is to compute the conditional obligor default probabilities, use these to compute the portfolio expected losses which in turn lead to the fee and contingent payments and finally the tranche PV. From an implementation point of view this set of calculations can be neatly incorporated into an object oriented framework (see, for example, Gamma *et al.* [1995], Joshi [2004a], Koenig and Moo [2005] and Cherubini and Della Lunga [2007] for details of modern OO programming techniques applied to financial problems). Within this sort of framework a model is constructed by plugging in different methods for calculating $q_i(t|V)$ and $\bar{L}^\gamma(T_j)$.

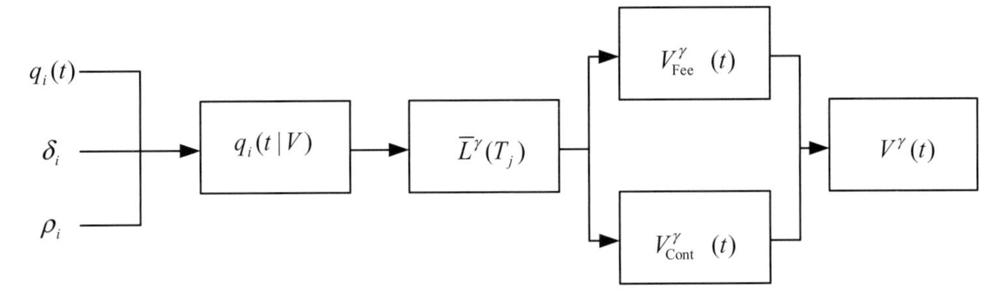

Figure 6.6 Schematic diagram showing the calculation flow and the different elements in the tranche valuation problem. See the text for a description of this model.

6.5.3 *Default baskets revisited*

Within the context of the semi-analytic recursive method we can also value default baskets. For a kth-to-default basket we can write for the fee and contingent payments

$$V^k_{\text{Fee}} = \sum_{j=1}^{N} Z(0, T_j)\Delta_j \left[1 - p_k(0, T_j)\right],$$

$$V^k_{\text{Cont}} = \sum_{j=1}^{N} Z(0, T_j)(1 - \delta)\left[p_k(0, T_j) - p_k(0, T_{j-1})\right].$$

Where $p_k(0, T_j)$ is the probability of k defaults in the period $[0, T_j]$. This quantity can be obtained using the recursive algorithm outlined.

6.6 Structural models

For the sake of completeness we briefly discuss structural models (the Fitch Vector model [Fitch], for example, is a structural model). Good descriptions of structural models can be found in Giesecke [2004] and De Servigny and Jobst [2007]). In general, structural models are Monte Carlo models for characterising the credit risk of portfolios of defaultable assets. The basic premise behind structural models as exemplified by the original Merton model [Merton 1974] is simple in principle: an obligor defaults when the value of their assets becomes less than the value of their liabilities (or debts). Structural models therefore rely upon detailed balance sheet information about the individual obligors in order to determine their survival propensity. There are various extensions to this basic premise such as Black–Cox models [Black and Cox 1976]. In this model the obligor default is treated like a barrier option and default occurs at the first crossing of the default boundary, which can occur at any time (as opposed to default being determined solely by the value of the asset at some specified final time horizon). Extensions to this model also allow the barrier to be time dependent.

To characterise defaults in the framework of structural models therefore requires two components:

- a prescription for the dynamics of the value of obligor i's assets $V_i(t)$,
- a value for the obligor specific barrier level (possibly time dependent), B_i, below which default is assumed to have occurred

For the asset dynamics a common model is to assume correlated geometric Brownian motions (GBMs) for all of the obligors where the asset value is evolved from timenode T_{j-1} to T_j according to [Glasserman 2003]

$$V_i^\alpha(T_j) = V_i^\alpha(T_{j-1}) \exp\left[\left(\mu_i - \frac{1}{2}\sigma_i^2\right)(T_j - T_{j-1}) + \sigma_i\sqrt{T_j - T_{j-1}}\phi_i^\alpha(T_j)\right].$$

(Poisson distributed jumps can be added to this model if desired, however calibrating such models to market observed spreads can be difficult.) In this expression μ_i and σ_i are the drift and volatility of the process for the value of obligor i's assets respectively. The random shock $\phi_i^\alpha(T_j)$ is computed according to $\phi^\alpha(T_j) = \mathbf{A}\varepsilon^\alpha(T_j)$ where $\varepsilon^\alpha(T_j) \sim N(\mathbf{0}, \mathbf{I})$ is a vector of uncorrelated standard normal deviates and $\mathbf{A}$ is the Cholesky decomposition of the (user-input) asset correlation matrix (note that this is the correlation between the obligor asset values). Other forms of SDE are simple to incorporate within this framework (including mean-reverting processes and jump diffusion processes). It is also possible to include other stochastic factors within the framework, such as stochastic recovery rates and obligor asset correlations (although the benefits of such additional complications are not always clear).

We have included the simulation index α in the above expression to indicate that the synthetic CDO valuation will proceed via a Monte Carlo simulation. The obligor default barrier level can be computed from the marginal obligor default probability, i.e. $B_i(T_j) = 1 - q_i(T_j)$.

The basic principle of the model is shown schematically in Figure 6.7. With these two components in place it is possible to evolve the obligor asset value processes throughout the lifetime of the trade and track the time periods when default occurs (and which obligors have defaulted). The fee and contingent legs can be computed and Monte Carlo estimators for their values determined. It is also straightforward within this framework to evaluate instruments with more complex cashflow structures, such as CDO squared trades or tranches with a combination of long and short exposures.

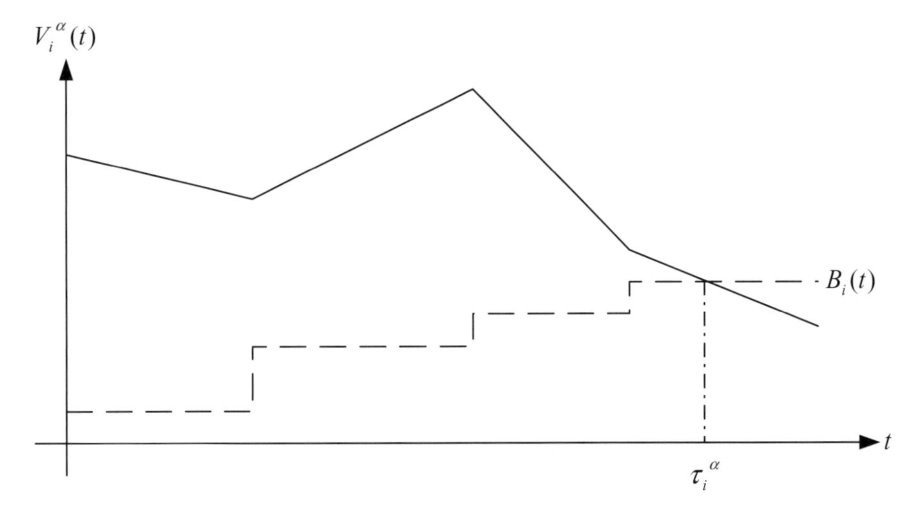

Figure 6.7 Schematic diagram showing the principle of generating defaults in the structural model. At the first crossing time of the asset value over the default threshold a default event is recorded.

When described in these terms structural models are very appealing and intuitive. However there are two downsides to the approach.

- Parameterisation and calibration: the model requires a large number of parameters to define it (more so than other approaches). Calibrating these parameters to the market observed credit spread information can be difficult.
- Computational complexity: the model is a full Monte Carlo revaluation approach. In addition to this the fee and contingent payments are path dependent (since the timing of the defaults impacts the cashflows), so the GBMs must be simulated at least at each coupon date. For a senior tranche many defaults are necessary to trigger a contingent payment – approximately 25 for the 12–22% iTraxx tranche described above – and so a large number of simulations must be undertaken (with the usual caveat that the estimators become even more difficult to obtain for sensitivities).

Despite these reservations structural models have recently enjoyed a resurgence in popularity (see, for example, Hull *et al.* [2005], Brigo and Errais [2005]). Given their simplicity of construction (if not parameterisation) a good structural model implementation is always a useful tool to have available for cross-checking other models.

6.7 Chapter review

This chapter has introduced a number of valuation models for synthetic CDOs. The model that has been adopted by the market as standard is the so-called normal copula model. This model is simple to understand and also to implement. We have presented both simulation (Monte Carlo) and semi-analytic methods for valuation. The key quantity to model is the portfolio loss distribution (at a particular time horizon corresponding to each of the coupon payment dates) and a number of different models for calculating this quantity were introduced.

As the market for synthetic CDOs has evolved the sophistication of the modelling methodology has also evolved. Extensions to the basic normal copula framework presented here (such as different copulas to capture the obligor dependence structure, random factor loadings, stochastic correlation etc.) have been developed to correct some of the perceived weaknesses of the basic model. These extensions to the basic model will be discussed in detail in Chapter 10. Nevertheless, the material presented in this chapter can be considered to be the absolute building blocks upon which modern synthetic CDO pricing is based. In the next chapter we will begin to analyse the properties of these models to develop a better understanding of the phenomenological behaviour of synthetic CDOs. Their behaviour will be found to be complex and varied.

7

Phenomenology of the standard market model

Nothing in life is to be feared, it is only to be understood.
Marie Curie

7.1 Introduction

In the previous chapter we outlined the standard market model for synthetic CDO tranche valuation and provided a number of different methods for the valuation of synthetic CDO tranches within the context of this model. In this chapter we start to analyse the properties of these models. For the sake of clarity we focus mostly on the Monte Carlo and semi-analytic recursive models for constructing the portfolio loss distribution.

In Section 7.2 the baseline scenario to be analysed will be introduced and the approximations made in the analysis described (and justified). Section 7.3 looks at a number of basic tranche simulation statistics to illustrate some of the problems that are encountered when applying Monte Carlo methods to tranche valuation. Section 7.4 analyses the portfolio loss distribution for a range of different input parameters. As we will see in subsequent chapters virtually all synthetic CDO behaviour can be explained by understanding the behaviour of the loss distribution. Following this in Section 7.5 the behaviour of the tranche present value (or equivalently the par spread) for a range of input parameters is quantified. What will quickly become apparent is that tranches have a rich and complex behaviour. Section 7.6 briefly revisits the results for default baskets obtained in Chapter 5 and provides an explanation for the phenomenology obtained. Finally in Section 7.7 we review the important points raised in the chapter and motivate our next steps.

7.2 Baseline case analysed

The baseline case to be analysed is specified below. Variations to these baseline parameters will be outlined on a case-by-case basis.

Number of obligors in pool	125
Homogeneous obligor notional	\$10 m
Capital structure	iTraxx
Upfront payments	ignore
Tranche maturity	varies (typically 5 years)
Coupon frequency	quarterly
Homogeneous obligor par CDS spread	60 bps
Homogeneous obligor recovery rate	40%
Homogeneous obligor hazard rate	1% (computed from $s = h(1 - \delta)$)
Homogeneous obligor default correlation	varies
Risk-free rate	5%
Number of simulations	100 000

The iTraxx capital structure analysed here consists of 0–3%, 3–6%, 6–9%, 9–12% and 12–22% tranches. In addition to this we will include the 0–100% tranche which is equivalent to an index position. There are a number of important simplifications that we are making in our analysis that warrant further discussion. Firstly we assume that the term structure of interest rates is approximated by a single spot rate. From the point of view of understanding the model's phenomenology this approximation is not material since the risk-free rate is only used for discounting cash-flows. If a set of discount factors constructed by blending money market, futures and swap rates together is available then this does not represent a significant change in methodology.

More important is the approximation that the term structure of par CDS spreads is represented by a single spot par CDS spread. This does have a material impact upon the tranche valuation since the par CDS spreads are used effectively to determine the obligor default times. For the current purpose of understanding the model's behaviour, this approximation is acceptable since it simplifies greatly a lot of the modelling. However, it should be remembered that in a real valuation environment the full term structure **must** be used. We are also ignoring any upfront payments.

Finally we assume that the obligors are all homogeneous. For standard index tranches the portfolio is based upon obligors with equal notional amounts (\$10 m) and assumed recovery rates (40%). However, the obligors in the actual index will have differing credit spreads. In our analysis we are assuming that the credit spreads are homogeneous. This does have a material impact upon the actual valuation, but does not impact upon the qualitative conclusions we will draw from the analysis.

7.3 Tranche loss statistics

A first useful quantity to analyse is the fraction of Monte Carlo simulations that give rise to a non-zero tranche loss. For an equity tranche the first loss to the portfolio will impact the tranche. But as we move up the capital structure each default simulation will have to generate larger and larger numbers of defaults (before the maturity of the tranche) in order for the tranche to be impacted.

To quantify the number of simulations required to impact a tranche we simulate the correlated default times of the obligors (using the Gaussian copula). Consider a pool of $i = 1, \ldots, n$ identical obligors, each with a survival curve described by $S_i(t, T) = \mathrm{e}^{-h_{i_i}(T-t)}$. The default times for simulation α are computed according to $\tau_i^\alpha = -\ln(N(\phi_i^\alpha))/h_i$ where ϕ_i^α are correlated normal deviates. The default times of course depend on the riskiness of the obligors (via the hazard rate h_i) and the obligor default correlation ρ. For a specified time horizon T (for example, 5 years) we can determine which obligors have defaulted in this simulation and calculate the total losses incurred in the collateral pool. The pool loss for this time horizon is computed from

$$L^\alpha(T) = \sum_{i=1}^{n} N_i(1 - \delta_i)\mathbf{1}_{\tau_i^\alpha < T}.$$

This loss is compared to the attachment points of the specified tranches. If the losses exceed the attachment point this is recorded as happening for this default simulation path. This process is repeated for many simulations and the fractions of paths which generate path losses for tranche γ computed from

$$q_\gamma(T) = \frac{1}{N_{\text{Sims}}} \sum_{\alpha=1}^{N_{\text{Sims}}} \mathbf{1}_{\mathrm{L}^\alpha(T) > K_{\mathrm{L}}^\gamma}.$$

Figures 7.1, 7.2 and 7.3 show the dependence of $q_\gamma(T)$ on the (homogeneous) correlation assumption, time horizon and (homogeneous) obligor spread respectively.

We would expect intuitively that as $s_i \to \infty$, $\tau_i^\alpha \to 0$, i.e. as the obligor becomes riskier, its expected default occurs sooner (from the expression for the default times). We would also expect that the equity tranche will be impacted more severely than tranches further up the capital structure, so $q_{0-3\%} \to 1$ more rapidly than for more senior tranches. It is also to be expected that as $T \to \infty$, $q_\gamma \to 1$ for all the tranches (assuming that the obligor spreads remain constant). This is indeed what Figures 7.2 and 7.3 show. Figure 7.1 is a bit more subtle. We will come back to this when we have analysed the portfolio loss distribution in more detail.

Finally we note that the algorithm outlined above can also be used to calculate the tranche default probability. The tranche default probability is computed from

$$p_\gamma^{\text{Default}}(T) = \frac{1}{N_{\text{Sims}}} \sum_{\alpha=1}^{N_{\text{Sims}}} \mathbf{1}_{L^\alpha(T) > K_{\mathrm{U}}^\gamma},$$

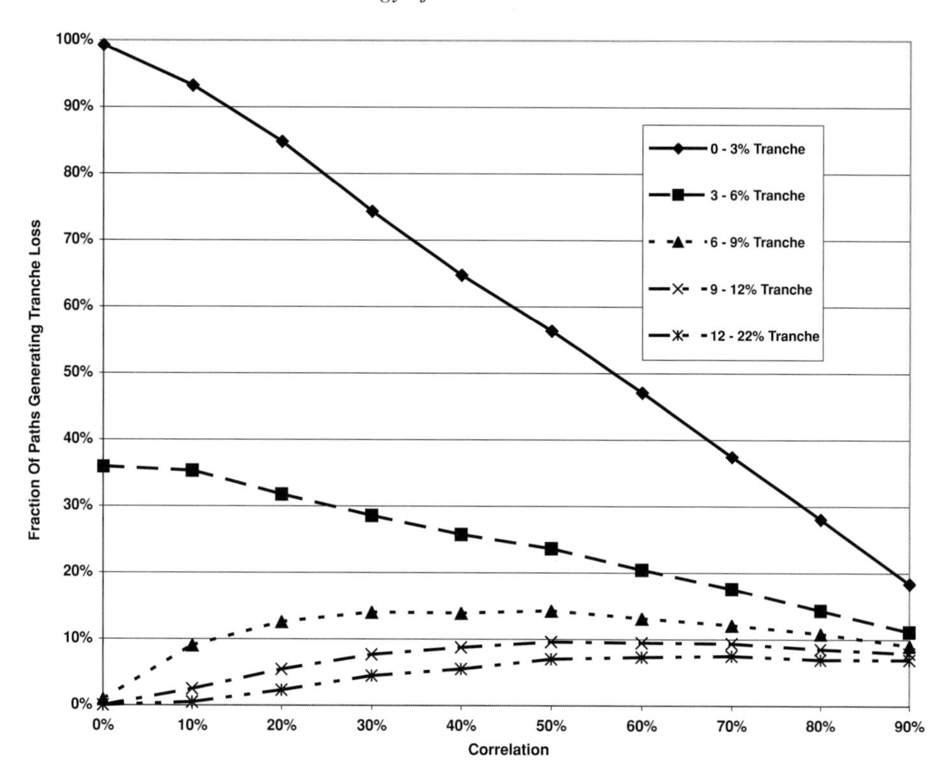

Figure 7.1 Plot of the fraction of simulation paths that generate a loss to a tranche (i.e. the total pool losses exceed the tranche attachment point) for different tranches as a function of the homogeneous correlation assumption.

i.e. the probability of tranche default is the expectation value that the losses for a particular simulation exceed the detachment point. The tranche expected loss is similarly calculated according to

$$\bar{L}_\gamma(T) = \frac{1}{N_{\text{Sims}}} \sum_{\alpha=1}^{N_{\text{Sims}}} (L^\alpha(T) - K_{\text{L}}^\gamma) \mathbf{1}_{K_{\text{L}}^\gamma \leq L^\alpha(T) \leq K_{\text{U}}^\gamma}$$
$$+ \frac{1}{N_{\text{Sims}}} (K_{\text{U}}^\gamma - K_{\text{L}}^\gamma) \sum_{\alpha=1}^{N_{\text{Sims}}} \mathbf{1}_{\text{L}^\alpha(T) \geq K_{\text{U}}^\gamma}.$$

That is, the tranche expected loss is computed from the losses in simulations where the tranche is completely wiped out (the second term) and losses in simulations where the loss is between the attachment and detachment points (the first term). Finally the tranche loss-given-default is calculated from $\text{LGD}_\gamma(T) = \bar{L}_\gamma(T)/p_\gamma^{\text{Default}}(T)$.

These quantities form the basis of the methodology for the assignment of ratings to tranches used by the ratings agency Standard & Poor's [De Servigny and Jobst

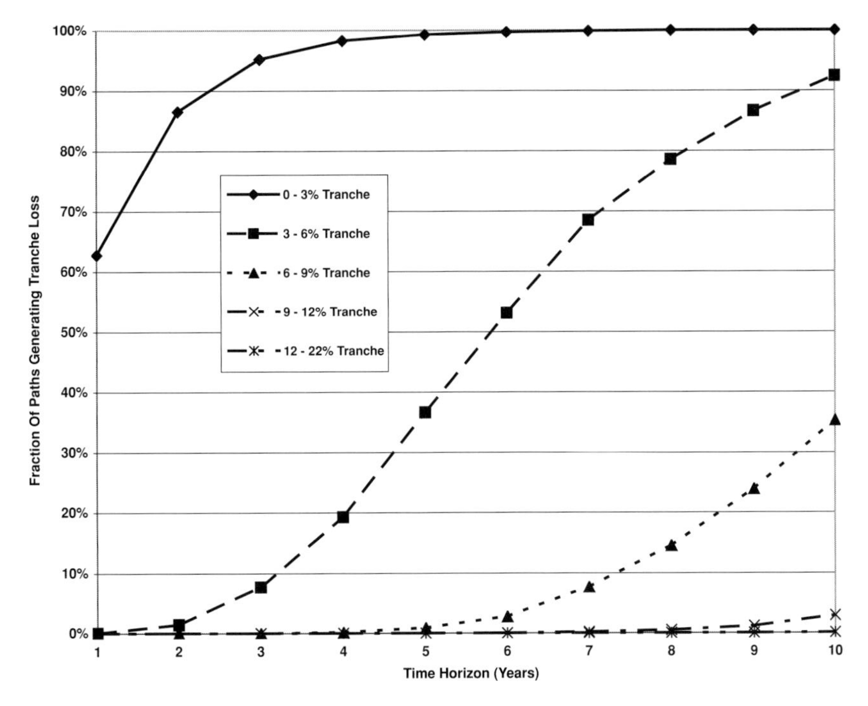

Figure 7.2 Plot of the fraction of simulation paths that generate a loss to a tranche (i.e. the total pool losses exceed the tranche attachment point) for different tranches as a function of the time horizon over which defaults are assumed to be able to impact the tranches.

2007]. Their CDO Evaluator is a Monte Carlo portfolio credit risk model for analysis of CDO transactions. It is a model for simulation of the portfolio loss distribution from which a number of different risk measures used in the CDO rating process are derived. It is not per se a model for the calculation of tranche PV (although it is of course possible to compute this once the loss distribution is known).

In common with the standard market model for synthetic CDOs, Evaluator requires the marginal default characteristics of the individual obligors to be specified and for this marginal information to be stitched together to determine the portfolio joint loss distribution. The model inputs are therefore the term structure of cumulative, marginal default probabilities for each obligor, the inter- and intra-industry correlations between the obligors and the recovery rates for the obligors Recovery rates can be constant or sampled for each simulation from a beta distribution $\delta_i^\alpha \sim \beta(a, b)$. Averaging over the ensemble of simulations we can calculate the expected portfolio loss (as described above). Using this methodology it is a simple matter to determine various parameters that characterise the tranche such as the tranche default probability, expected tranche loss, loss-given-default or the

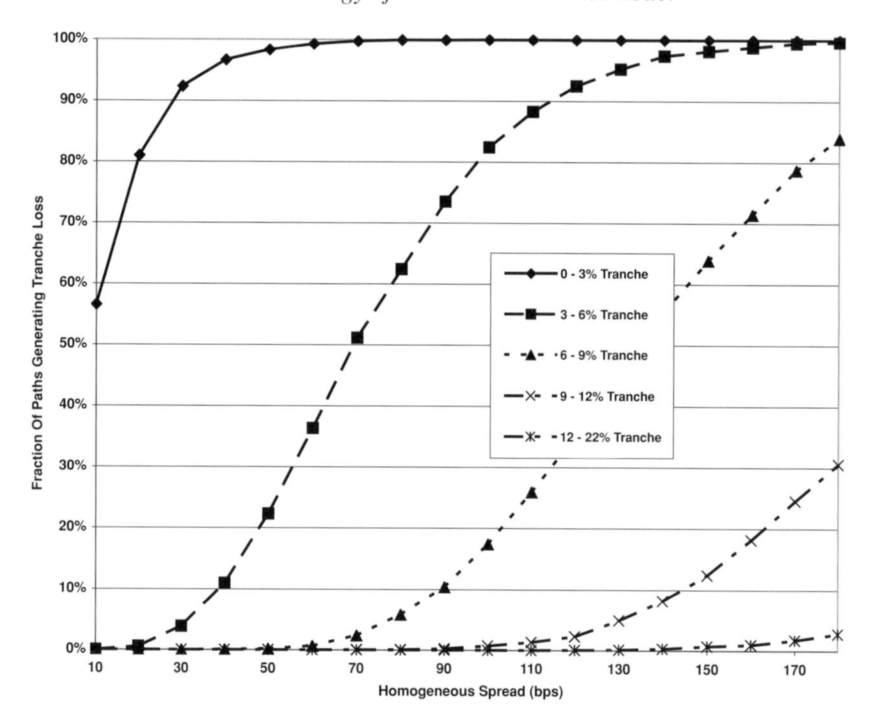

Figure 7.3 Plot of the fraction of simulation paths that generate a loss to a tranche (i.e. the total pool losses exceed the tranche attachment point) for different tranches as a function of the homogeneous CDS par spread of obligors in the pool.

scenario loss rate. The tranche expected loss can then be compared to a look-up table of losses to determine a rating for the tranche. More complex structures, such as CDO squareds, can also be accommodated in the modelling framework. In the case of a CDO squared the losses in the underlying universe of obligors are simply propagated up through the slave CDO tranches up to the master CDO tranche (see Chapter 11 or 13).

7.4 Analysis of the portfolio loss distribution

The key quantity for understanding the behaviour of tranche PVs is the portfolio loss distribution. In this section we analyse this quantity (obtained using the recursive method of valuation) for a range of different obligor properties.

7.4.1 Correlation dependence of the portfolio loss distribution

The unconditional portfolio loss distribution $p(l, \rho)$ computed from the recursive model directly impacts the expected loss of a tranche from $\bar{L}(\rho) = \sum_{l=0}^{n} p(l, \rho)L(l)$

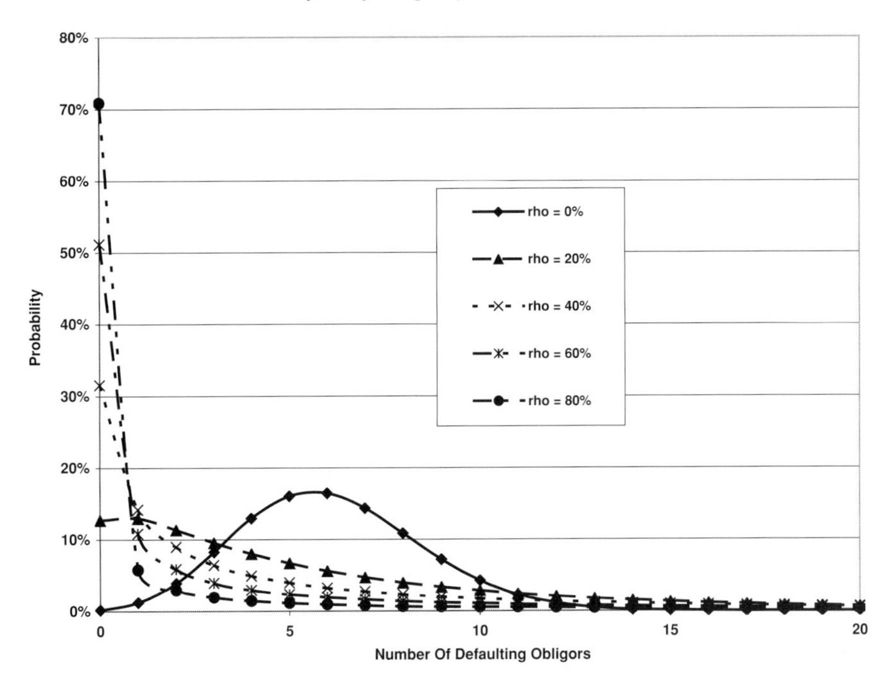

Figure 7.4 Portfolio loss distribution for a time horizon of 5 years for a variety of different correlations.

where $L(l)$ is the tranche loss for pool losses of l. This in turn determines the fee and contingent leg values. It is reasonable therefore that in order to understand the phenomenology of the model it is necessary to understand the behaviour of the portfolio loss distribution.

Figure 7.4 shows the portfolio loss distribution for a maturity of $T = 5$ years for a range of different correlation values as the portfolio loss increases. (In all of the graphs presented the portfolio loss is represented in terms of the number of defaulted obligors. We could of course also represent the loss directly as a \$ amount.) At zero correlation the loss distribution is roughly symmetrical, centred around the portfolio expected loss (in this example this is approximately $6 \times$ \$10 m $\times$ (1–40%) = \$36 m). The distribution is symmetrical because the defaults are independent, meaning it is equally likely for there to be less than the expected number of defaults as it is for there to be more. As the correlation increases the distribution ceases to be symmetrical. Increasingly the distribution becomes more concentrated, separating into either small portfolio losses or large losses. Finally at very high correlation, $1 - q_i(t)$ of the distribution is centred around zero loss and $q_i(t)$ of the distribution is centred around a very high loss (this is not apparent in the graph since we have only plotted the distribution for up to 20 defaults but the small 'bump' at high loss is quite striking at high correlations).

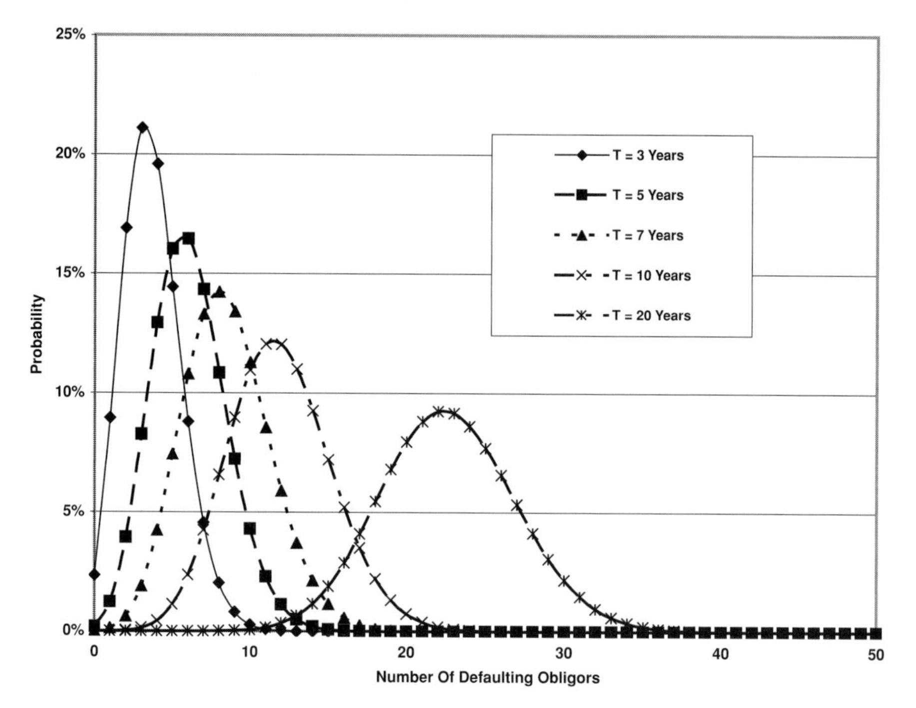

Figure 7.5 Portfolio loss distribution as a function of the number of defaulting obligors for $\rho = 0\%$ for a variety of different tranche maturities.

7.4.2 Maturity dependence

Figures 7.5–7.9 plot the maturity dependence of the loss distribution for a variety of different correlations. For $\rho = 0\%$ the loss distributions are all symmetrical about the tranche expected loss. However, as T increases the loss distribution shifts towards higher losses (as we would expect intuitively). This is because the marginal default probabilities increase as the time horizon increases.

For non-zero correlation the distributions shift towards either no losses or large losses. An interesting point to note is that as the correlation increases, the distributions for all of the different maturities converge. An expected consequence of this is that tranche PVs etc. will not be significantly impacted by the tranche maturity for high correlations.

7.4.3 Portfolio expected loss as a function of time

Finally Figures 7.10–7.12 plot the tranche expected loss as a function of time (in this case the timenodes correspond to the coupon dates) for different correlations. Also shown on these graphs is a comparison between the expected losses generated by the Monte Carlo and semi-analytic models. It will be observed firstly that there

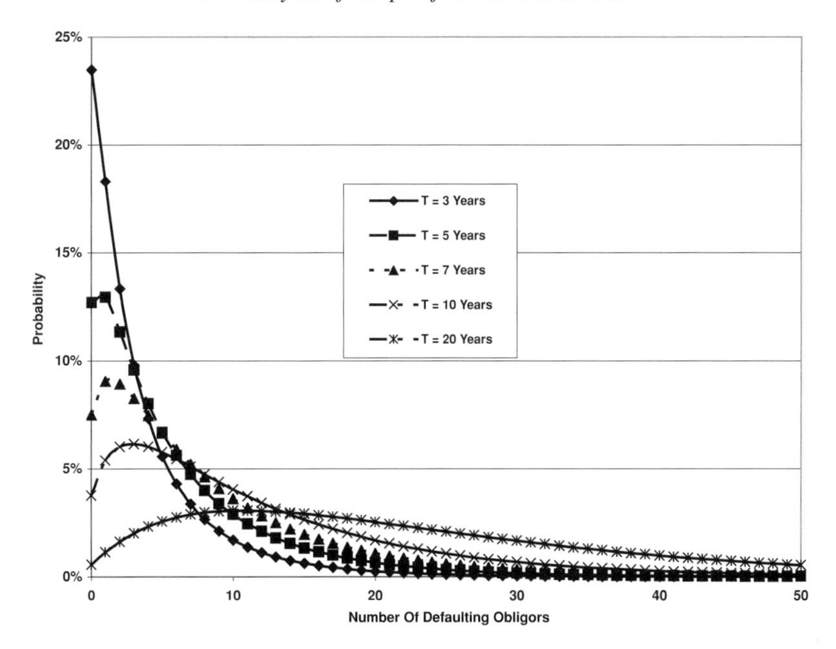

Figure 7.6 Portfolio loss distribution as a function of the number of defaulting obligors for $\rho = 20\%$ for a variety of different tranche maturities.

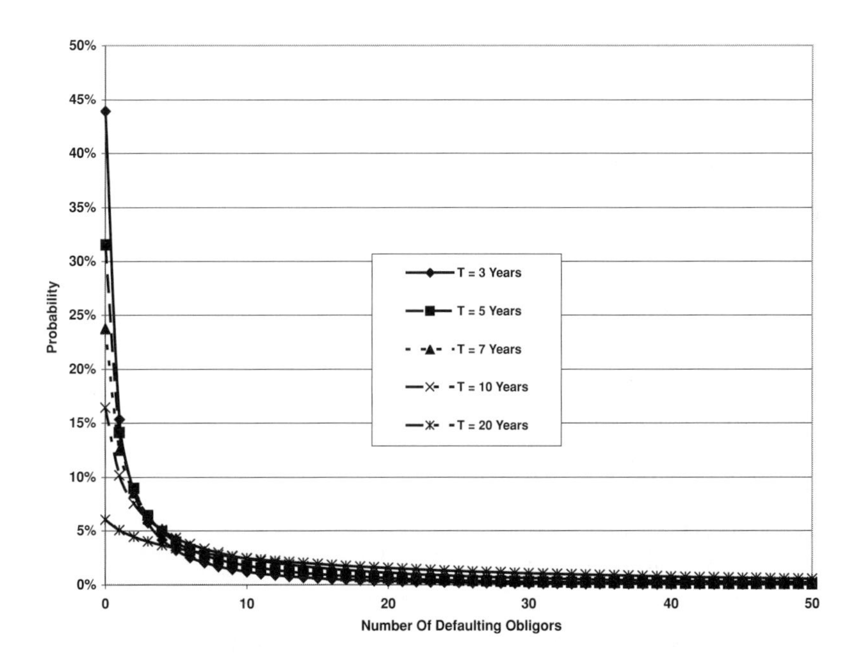

Figure 7.7 Portfolio loss distribution as a function of the number of defaulting obligors for $\rho = 40\%$ for a variety of different tranche maturities.

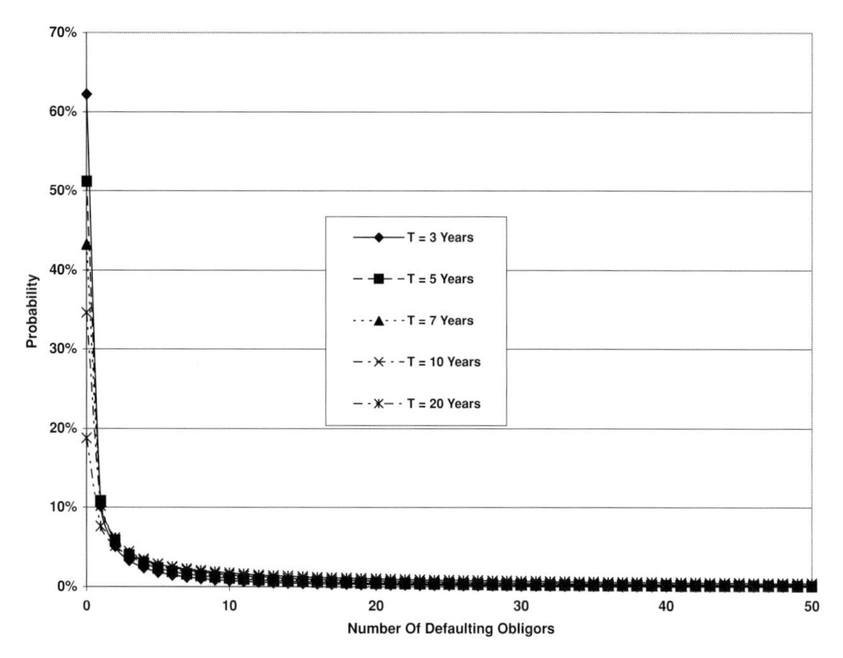

Figure 7.8 Portfolio loss distribution as a function of the number of defaulting obligors for $\rho = 60\%$ for a variety of different tranche maturities.

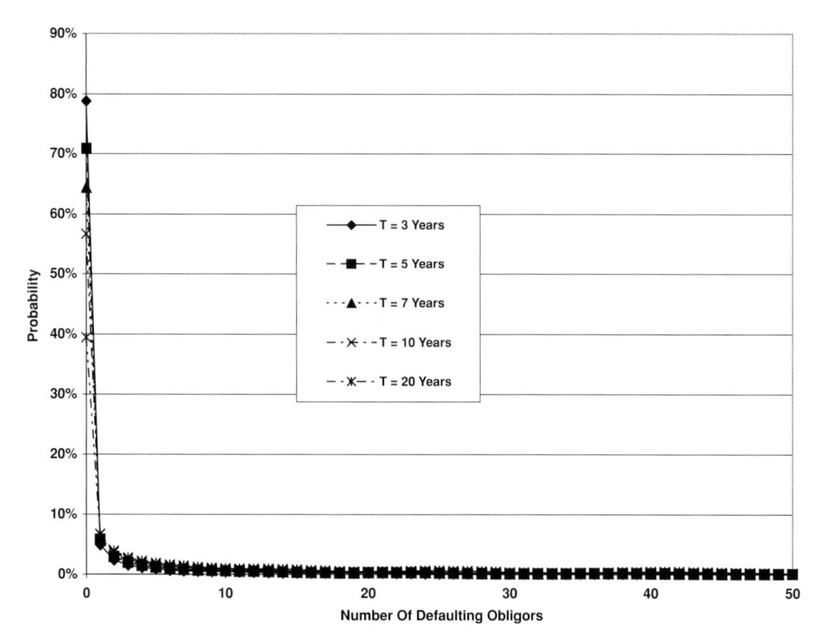

Figure 7.9 Portfolio loss distribution as a function of the number of defaulting obligors for $\rho = 80\%$ for a variety of different tranche maturities.

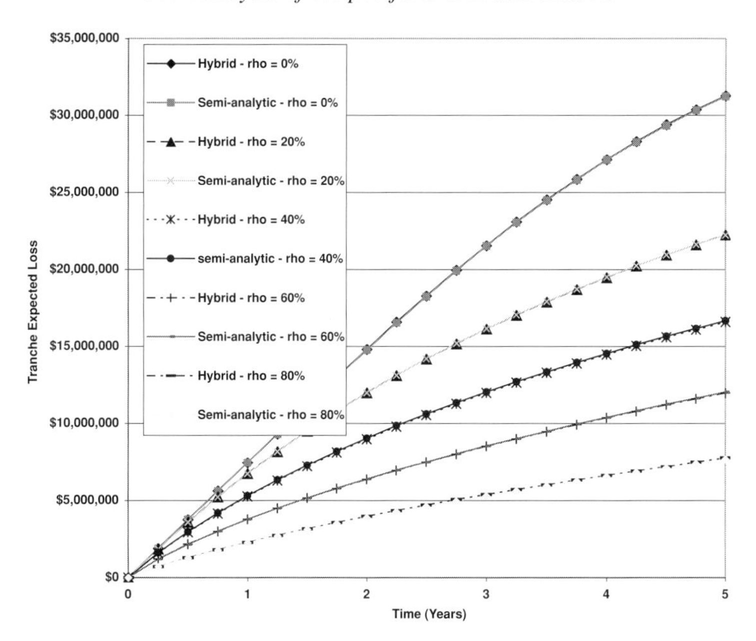

Figure 7.10 The expected loss of the 0–3% tranche as a function of time horizon for a range of different correlations. Shown on this plot are the expected losses calculated using both the Monte Carlo model as well as the semi-analytic model. There is good agreement between the two models.

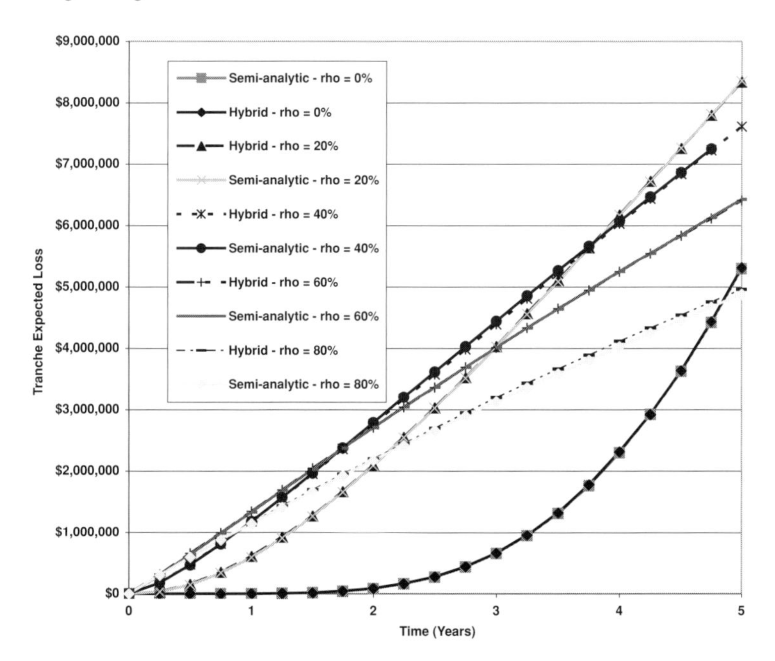

Figure 7.11 The expected loss of the 3–6% tranche as a function of time horizon for a range of different correlations. Shown on this plot are the expected losses calculated using both the Monte Carlo model as well as the semi-analytic model. There is good agreement between the two models.

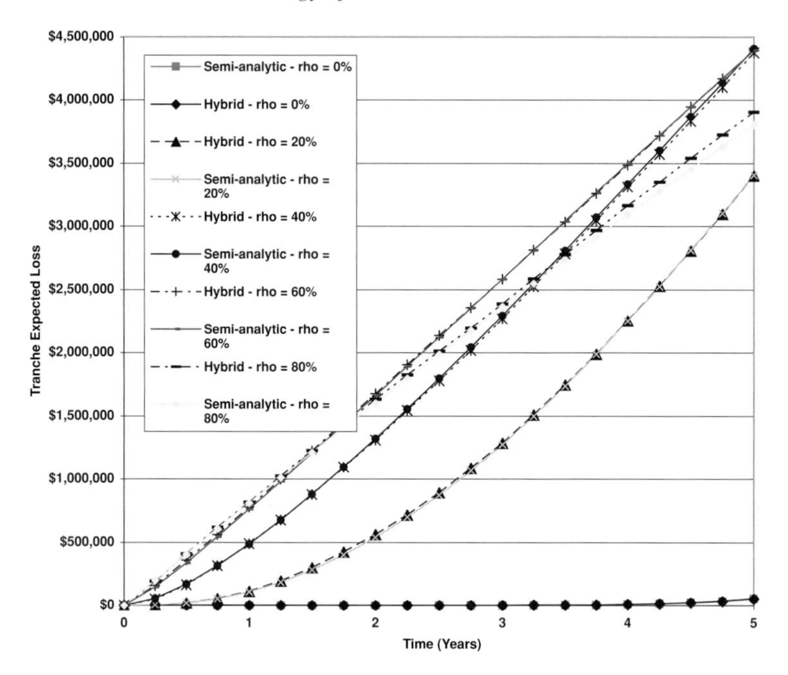

Figure 7.12 The expected loss of the 6–9% tranche as a function of time horizon for a range of different correlations. Shown on this plot are the expected losses calculated using both the Monte Carlo model as well as the semi-analytic model. There is good agreement between the two models.

is very good agreement between the two different models. As we would expect at time zero the equity tranche losses are zero (this is observed for all the tranches). As time progresses the tranche losses begin to accumulate, increasing approximately linearly with time. The tranche expected loss is greatest for zero correlation. As the correlation increases the expected losses decrease (at a particular coupon date) because the bulk of the portfolio loss distribution shifts towards zero losses. Since the expected losses decrease for increasing correlation we would expect the PV of the tranche also to decrease as correlation increases. Finally as $T \to \infty$ the tranche expected loss saturates at the total tranche notional. In contrast to the 0–3% tranche, at low default correlation it takes a finite amount of time before the pool losses begin to impact the 3–6% tranche. Also in contrast to the 0–3% tranche, as correlation increases the expected losses also increase.

At low correlation the defaults occur approximately independently. The likelihood of the multiple defaults occurring which are necessary to impact the tranche is therefore low. This results in the expected losses being low for low correlations. Conversely, as the correlation increases the likelihood of multiple defaults increases. This results in increased tranche expected losses. Similar observations also apply to tranches higher in the capital structure. We will refer back to these

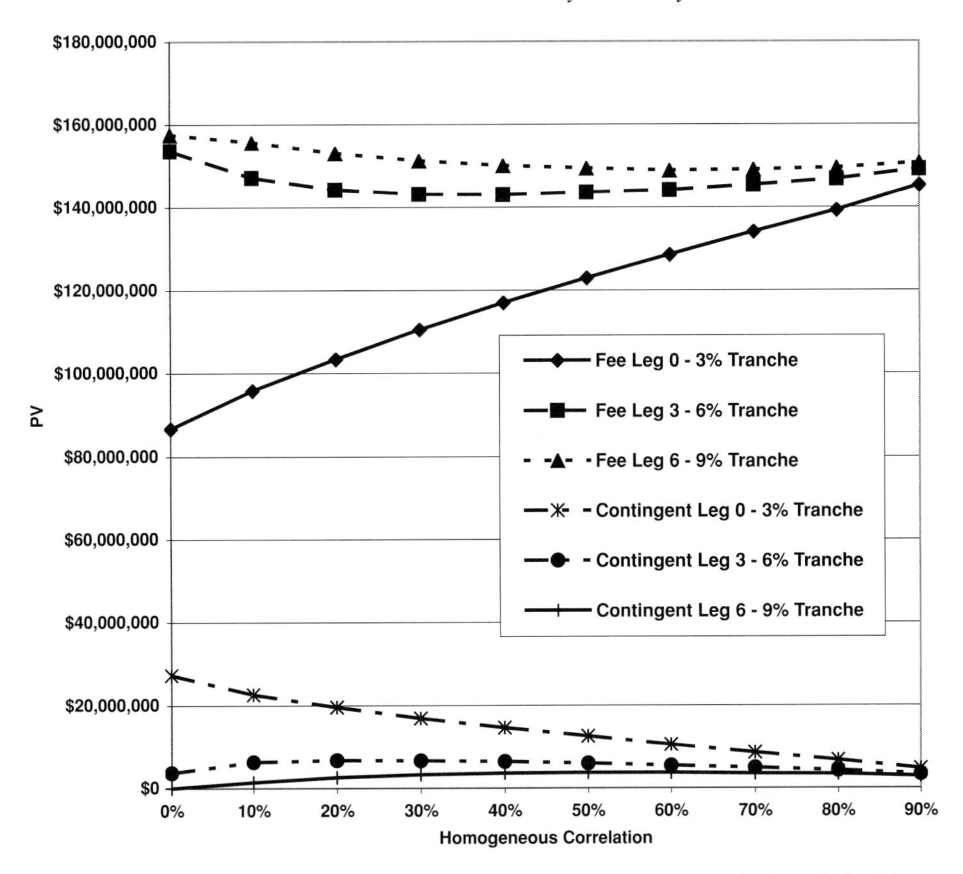

Figure 7.13 Correlation sensitivity of the fee and contingent legs for 0–3%, 3–6% and 6–9% tranches (5 year maturity).

graphs later in the chapter when we present results on the sensitivity of different tranches to various parameters.

7.5 Correlation and maturity sensitivity of the tranche par spread

Having analysed the form of the portfolio loss distribution we will now analyse one of the key model sensitivities: the homogeneous correlation (as well as assessing the maturity sensitivity of the tranches).

7.5.1 Correlation sensitivity of the fee and contingent legs

It is informative to consider the behaviour of the fee and contingent legs separately. Figure 7.13 shows the fee and contingent leg behaviours for the 0–3%, 3–6% and 6–9% tranches (for a maturity of 5 years) as the homogeneous correlation is varied. Consider first the equity tranche. As correlation increases the value of the contingent

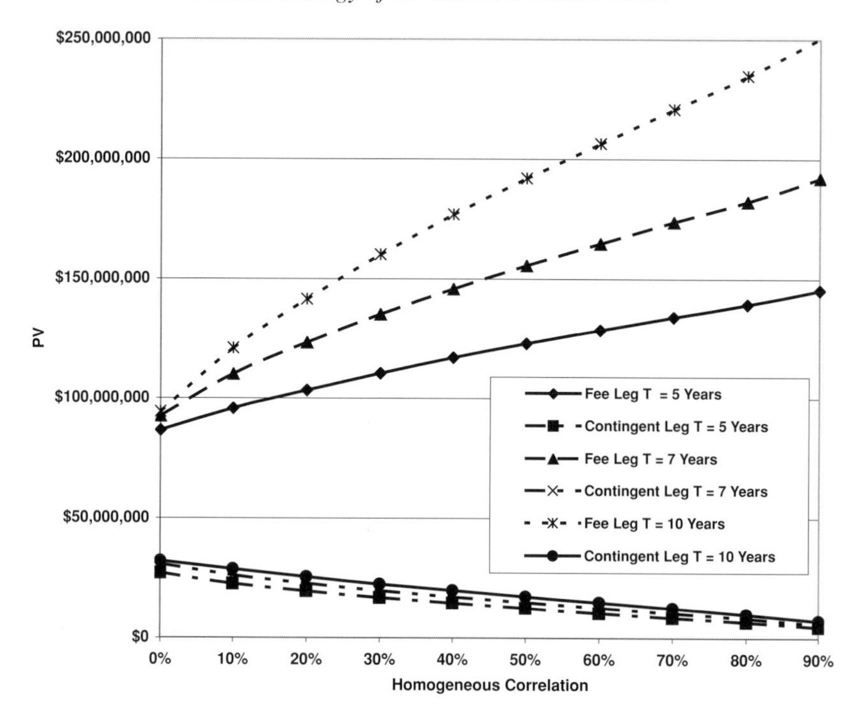

Figure 7.14 Fee and contingent legs for the 0–3% tranche as a function of correlation for a range of tranche maturities

leg decreases and the fee leg increases. Referring back to Figure 7.10 for the equity tranche expected losses, we observe that (at a fixed time horizon) the expected losses decrease as correlation increases. Since each fee leg coupon at time T_j is based on a remaining notional amount of $N^{\text{Equity}} - \bar{L}(\rho, T_j)$ we would expect the fee leg to increase as the correlation increases (since the reduction in notional from the initial value decreases). Conversely since the contingent leg is based on a notional amount of $\bar{L}(\rho, T_j) - \bar{L}(\rho, T_{j-1})$ (which is approximately constant from period to period because the expected losses increase approximately linearly), as the correlation increases the expected losses decrease and the contingent leg decreases.

For the 3–6% tranche at a 5 year horizon, the expected losses peak at around 20% correlation (in this example) and then decrease. The crucial point to note is that the behaviour is non-monotonic. This results in behaviour for the fee and contingent legs which is also non-monotonic. Similar observations apply to the 6–9% tranche.

Figure 7.14 shows the maturity dependence of the fee and contingent legs for the equity tranche (for maturities of 5, 7 and 10 years) as a function of the homogeneous correlation assumption. It is interesting to note that the contingent leg does not display too much sensitivity to the maturity of the tranche. This is to be expected since the risk of the obligors does not change with the maturity of the tranche (we are assuming that the obligor CDS term structure is characterised by a single point).

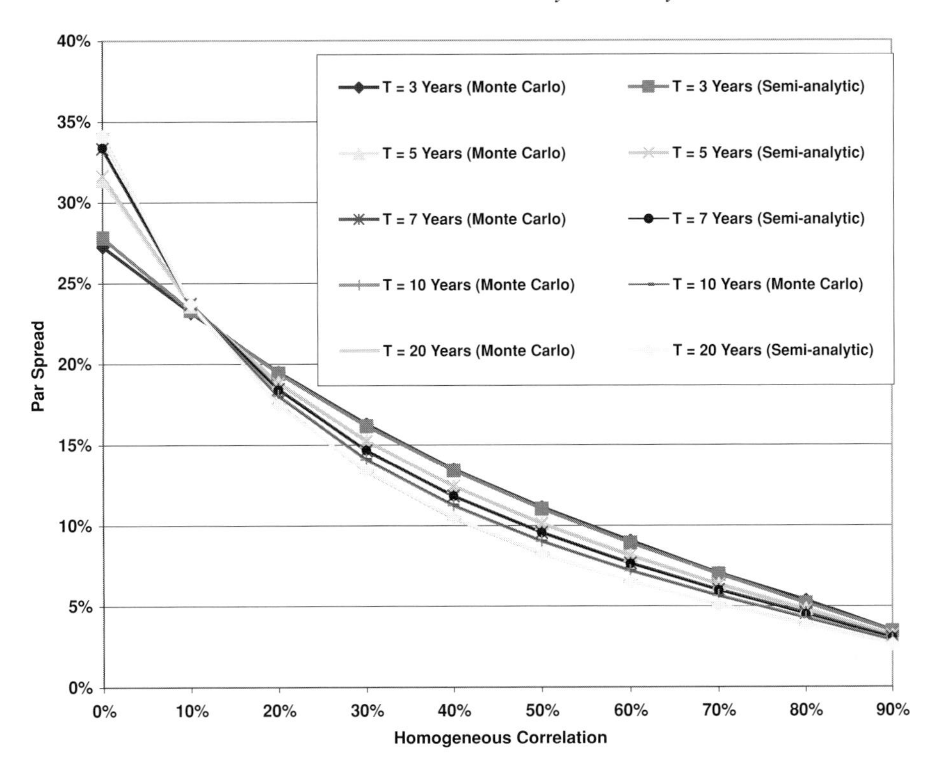

Figure 7.15 Par spread of the 0–3% tranche as a function of the homogeneous correlation for tranche maturities of 3, 5, 7, 10 and 20 years. For comparison the results obtained using the Monte Carlo and semi-analytic models are both plotted. (The fact that in this figure it is difficult to differentiate the different series reflects the relative insensitivity of this tranche to the tranche maturity.)

7.5.2 Correlation sensitivity of the tranche par spread

Figures 7.15–7.18 plot the tranche par spread for different tranches in the capital structure as a function of the homogeneous correlation for a range of different maturities. For comparison each figure plots the par spreads computed using the Monte Carlo and semi-analytic (using the recursive technique) models. In this section we merely report the results. We explain them and discuss the overall model phenomenology in the following section.

Finally out of interest, Figure 7.19 compares the result of the recursive model par spreads with that of the normal proxy model for the 3–6% tranche. It will be observed that there is very good agreement between the two models (similarly good agreement is found for the other tranches in the capital structure). The good agreement between the two models demonstrates that although the normal proxy model is simple in concept, its accuracy is certainly comparable to that of the more sophisticated recursive model (and also the full accuracy Monte Carlo model). This, coupled with the advantage it has in speed of execution, make the model very useful indeed.

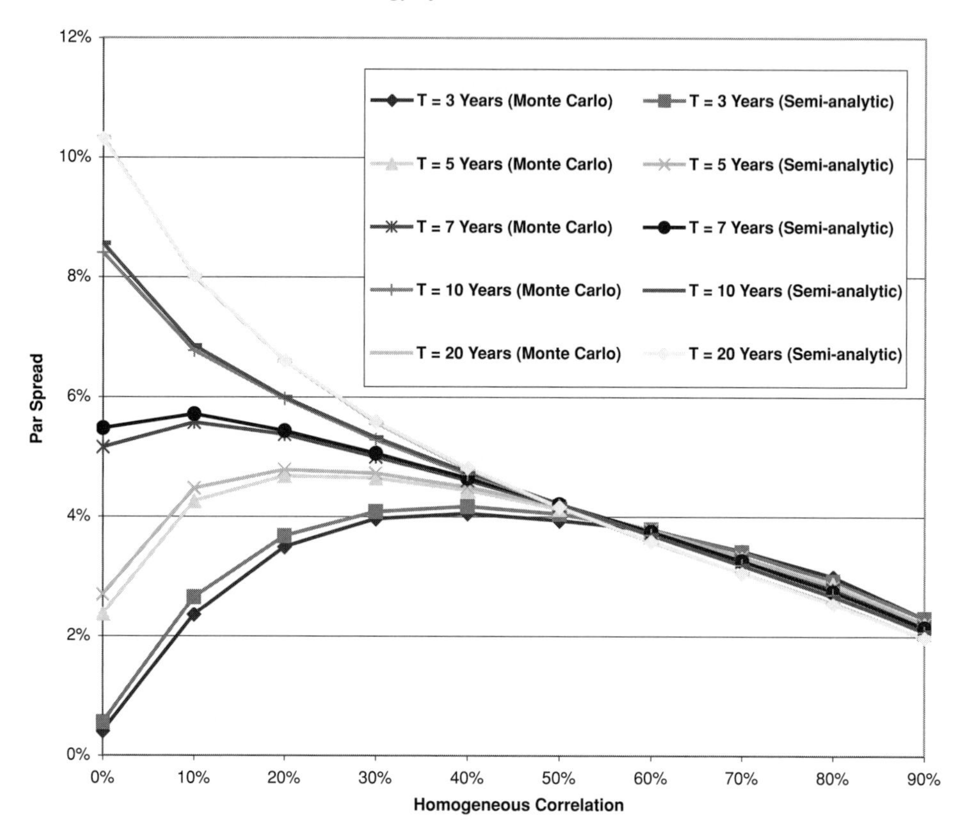

Figure 7.16 Par spread of the 3–6% tranche as a function of the homogeneous correlation for tranche maturities of 3, 5, 7, 10 and 20 years. For comparison the results obtained using the Monte Carlo and semi-analytic models are both plotted.

Note that common index tranche maturities that trade liquidly are 3, 5, 7 and 10 years. The 20 year results plotted here are for illustrative purposes only.

7.5.3 Understanding tranche phenomenology

Having quantified the tranche behaviours we will now provide an explanation. In what follows it is useful to remember the relationship between the value of the tranche fee and contingent legs and the tranche expected losses:

$$V_{\text{Fee}}(0) = \sum_{j=1}^{N} Z(0, T_j)\Delta_j[(K_{\text{U}} - K_{\text{L}}) - \bar{L}(K_{\text{L}}, K_{\text{U}}, T_j)],$$

$$V_{\text{Cont}}(0) = \sum_{j=1}^{N} Z(0, T_j)[\bar{L}(K_{\text{L}}, K_{\text{U}}, T_j) - \bar{L}(K_{\text{L}}, K_{\text{U}}, T_{j-1})].$$

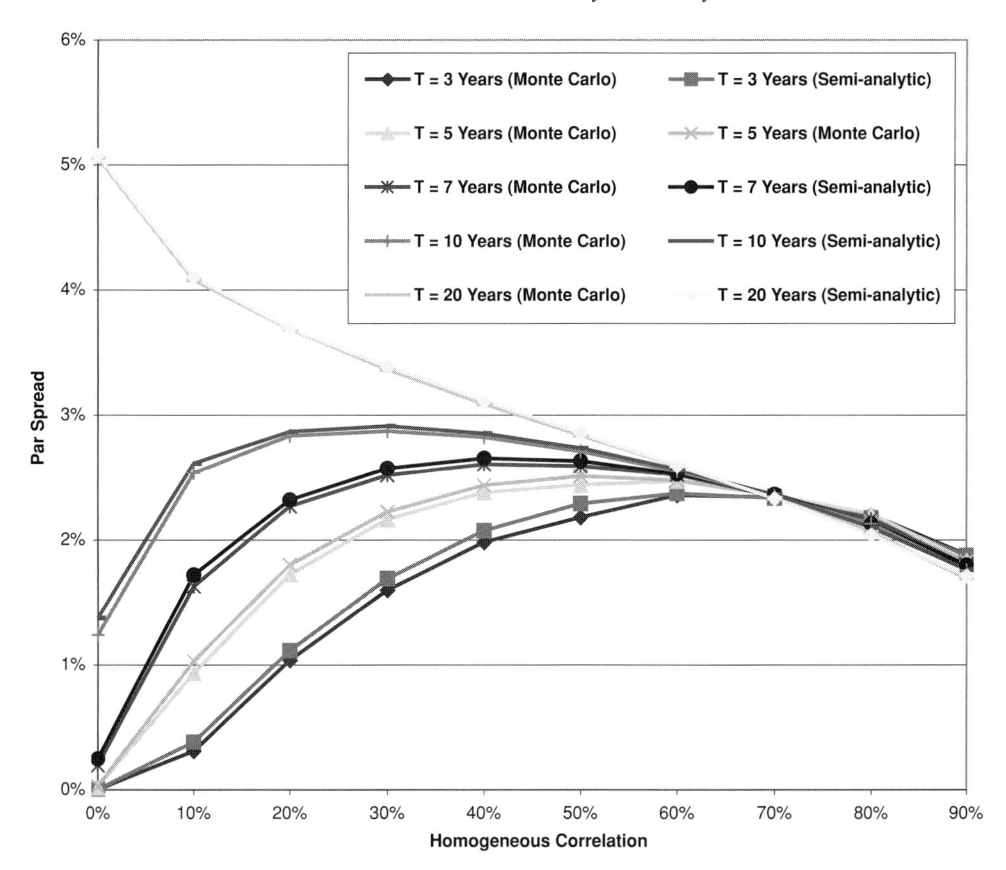

Figure 7.17 Par spread of the 6–9% tranche as a function of the homogeneous correlation for tranche maturities of 3, 5, 7, 10 and 20 years. For comparison the results obtained using the Monte Carlo and semi-analytic models are both plotted.

It will firstly be observed that for all the tranches there is very good agreement between the Monte Carlo valuation model and the results obtained using the recursive model to construct the portfolio loss distribution. As we move further up the capital structure small deviations between the two models begin to appear. However, this is due to the relatively small numbers of simulations used to obtain the fee and contingent leg estimators; more simulations would generate more defaults, providing better estimators for the more senior tranches. This comparison provides good evidence that the semi-analytic methods give accurate valuations with the clear advantage of speed of execution and quality of sensitivity estimates over Monte Carlo methods.

It is observed that the maturity sensitivity of the 0–3% tranche is not particularly significant. This should not be too surprising since the equity tranche is exposed to the first losses incurred in the portfolio. It therefore does not really matter if the

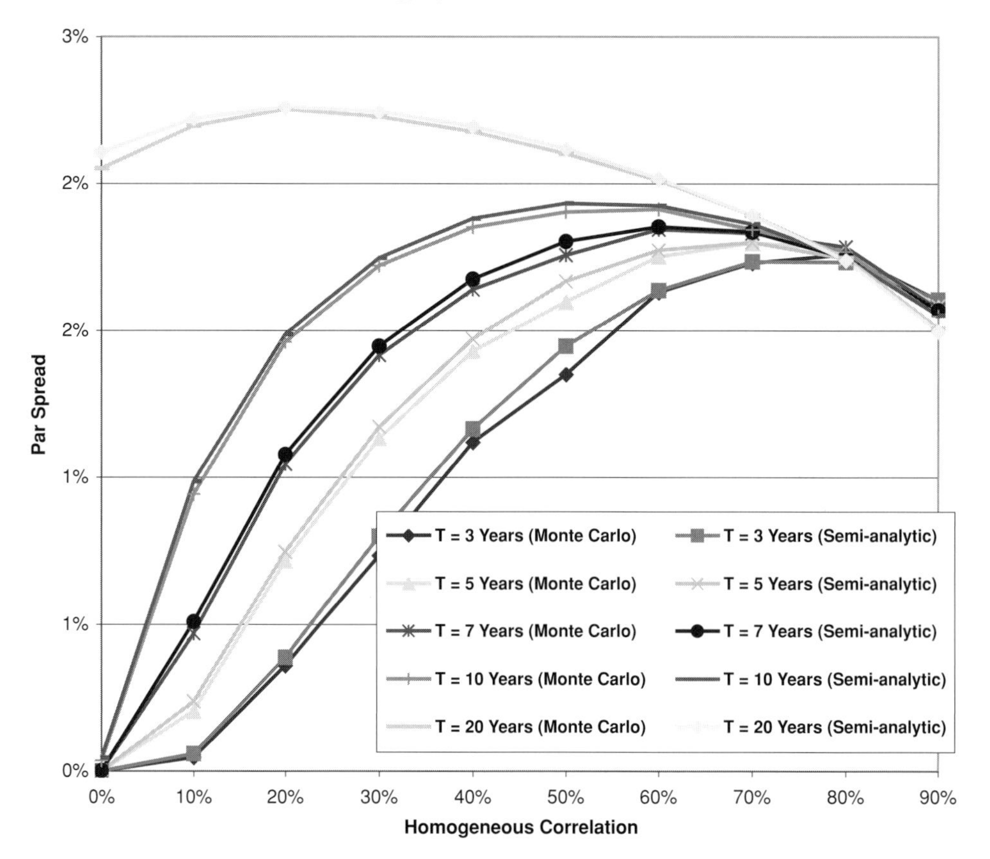

Figure 7.18 Par spread of the 9–12% tranche as a function of the homogeneous correlation for tranche maturities of 3, 5, 7, 10 and 20 years. For comparison the results obtained using the Monte Carlo and semi-analytic models are both plotted.

maturity is 3 years or 30 years; it only takes a few defaults to wipe out the equity tranche and these can occur sufficiently early to render the overall tranche maturity irrelevant. What is more interesting is the observation that the par spread decreases as the correlation increases.

At high correlations the maturity sensitivity of the 3–6% tranche is not material. From our earlier observations of the portfolio loss distributions at high correlations for varying maturities (Figures 7.4–7.9) this is to be expected. At low correlation, however, there is a clear maturity dependence of the par spread. Specifically at small maturities (less than 7 years in this case) the overall correlation sensitivity is non-monotonic; firstly the spread increases as the correlation increases, and then it decreases (with the turning point being at about 40% correlation). Above 7 year maturities the behaviour 'flips' and the overall correlation sensitivity comes to resemble that of an equity tranche (monotonically decreasing). Similar behaviour

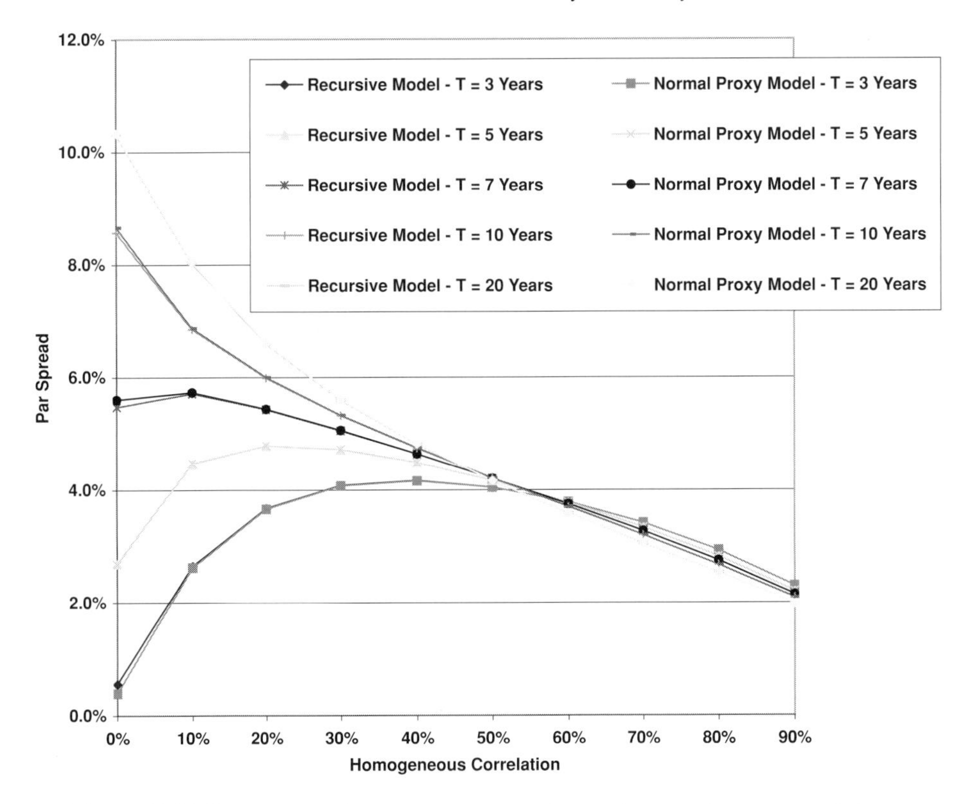

Figure 7.19 Comparison between the recursive and normal proxy model par spreads for the 3–6% tranche.

is also observed for the 6–9% and 9–12% tranches. In the 6–9% case, however, it is only tranches with maturities greater than 10 years that 'flip' from behaving like a mezz to behaving like an equity tranche, while for the 9–12% tranche the flipping is not observed at all (but the trend is beginning to appear and if the maturity were extended sufficiently it would be visible).

It is not difficult to understand qualitatively why the flipping is observed. If a tranche has a longer maturity this means there is more time for defaults to occur and for losses to accumulate. As the maturity increases therefore the relative advantages of subordination that mezz tranches have begin to be eroded, meaning that the mezz tranches become riskier and start to resemble more closely an equity tranche.

We turn now to considering just a single fixed maturity, for example 5 years. It has been observed that the equity tranche has a par spread which is a monotonically decreasing function of the correlation. Mezzanine tranches have a par spread that is at first an increasing function and then a decreasing function of the correlation. Senior tranches (with a detachment point of 100%) have a par spread that is a monotonically increasing function of the correlation. It is also worth noting that the

Table 7.1 *Portfolio loss distribution (in per cent) for different default correlations*

Number of defaulting obligors	0%	10%	20%	30%	40%	50%	60%	70%	80%	90%
0	0.19	4.81	12.71	21.87	31.53	41.35	51.18	60.99	70.82	80.95
1	1.24	8.75	12.95	14.40	14.14	12.80	10.80	8.41	5.80	3.12
2	3.93	10.68	11.33	10.39	8.96	7.41	5.86	4.36	2.93	1.56
3	8.27	11.02	9.57	7.93	6.45	5.13	3.96	2.91	1.94	0.95
4	12.93	10.42	8.01	6.28	4.95	3.87	2.96	2.16	1.45	0.72
5	16.04	9.34	6.70	5.10	3.96	3.07	2.34	1.71	1.17	0.67
6	16.45	8.10	5.62	4.22	3.26	2.52	1.92	1.41	0.97	0.65
7	14.34	6.86	4.72	3.54	2.74	2.12	1.62	1.19	0.80	0.58
8	10.84	5.72	3.99	3.01	2.33	1.82	1.39	1.03	0.67	0.46
9	7.23	4.72	3.38	2.58	2.02	1.58	1.22	0.91	0.59	0.32
10	4.30	3.85	2.87	2.23	1.76	1.39	1.08	0.80	0.55	0.22
115	0.00	0.00	0.00	0.00	0.00	0.00	0.01	0.02	0.04	0.09
116	0.00	0.00	0.00	0.00	0.00	0.00	0.01	0.02	0.04	0.08
117	0.00	0.00	0.00	0.00	0.00	0.00	0.01	0.02	0.04	0.07
118	0.00	0.00	0.00	0.00	0.00	0.00	0.01	0.02	0.04	0.06
119	0.00	0.00	0.00	0.00	0.00	0.00	0.01	0.02	0.04	0.05
120	0.00	0.00	0.00	0.00	0.00	0.00	0.01	0.02	0.04	0.06
121	0.00	0.00	0.00	0.00	0.00	0.00	0.01	0.02	0.05	0.09
122	0.00	0.00	0.00	0.00	0.00	0.00	0.00	0.02	0.05	0.11
123	0.00	0.00	0.00	0.00	0.00	0.00	0.00	0.02	0.05	0.13
124	0.00	0.00	0.00	0.00	0.00	0.00	0.00	0.02	0.06	0.17
125	0.00	0.00	0.00	0.00	0.00	0.00	0.00	0.00	0.00	0.00

magnitude of the overall sensitivity of the mezz tranches to the correlation is small compared to the equity tranche.

To understand this behaviour we need to understand the tranche portfolio loss distributions. Table 7.1 shows the portfolio loss distribution for increasing numbers of obligors as a function of the default correlation. What Table 7.1 demonstrates is that at low correlations there is zero probability of large losses. The majority of the distribution is centred around losses corresponding to less than 10 defaults. Conversely, for high correlation there appears a small, but non-zero, probability of extremely large losses. The vast majority of the distribution is concentrated at zero losses. Intuitively this is obvious: at high correlation either lots of obligors default (resulting in very large losses) or no obligors default (resulting in no losses). At low correlations the dispersion of losses is greater since the obligors can behave more independently.

For very senior tranches to experience losses (and hence have non-zero contingent payments) large numbers of obligors must default leading to large losses. From

Table 7.1 at $\rho = 0\%$ the probability of large losses is essentially zero. This means the contingent payments will be zero and hence the par spread is zero. At large values of ρ the probability of zero losses is very high, but there is also a significant probability of large losses meaning the contingent payments will be non-zero and hence the par spread will also be non-zero. **Therefore as correlation increases, the par spread of a senior tranche will increase**.

For the equity tranche the opposite behaviour is true since the equity tranche does not care about the probability of large losses in the portfolio (it only takes a small number of losses to wipe it out completely). At low correlations there is a significant probability of sufficient losses to impact the equity tranche. This means the contingent leg value will be non-zero (because the tranche is risky and the seller of protection will require compensation for the risk they are taking on) and the fee leg value will be materially reduced in value (since its notional is reduced by the expected losses). This implies the par spread will be relatively large meaning that the seller of protection will receive a large premium for providing protection. At high correlations the probability of zero losses is extremely high meaning the contingent leg is small and therefore the par spread is small. **Therefore as correlation increases the par spread of an equity tranche will decrease**.

Mezz tranches behave in a hybrid state in-between equity and senior tranches. As risk moves from an equity tranche the only place it can seep into is the mezz tranche. As risk moves from a senior tranche it seeps into the mezz tranche. Mezz tranches really are caught between the devil and the deep blue sea. This means that as correlation increases the probability of small numbers of losses increases and so the mezz tranche par spread increases (since its subordination becomes less effective). However, for very large correlation, the bulk of the portfolio loss distribution is at zero loss meaning that the par spread is small. **Therefore as correlation increases from zero the par spread of a mezz tranche initially increases (behaving like a senior tranche), but at large correlation the par spread decreases (behaving like an equity tranche)**.

It is recommended that the reader verify for themselves that the par spread is monotonically decreasing for equity tranches of any width, i.e. a 0–x% tranche, and that the par spread is monotonically increasing for any super-senior tranche, i.e. a x–100% tranche. These observations lie at the heart of the base correlation concept to be introduced in Chapter 9.

It is worth mentioning at this point that a clue to one of the most serious shortcomings of the standard models is contained in the observations above. For low levels of correlation (which are more plausible than values of e.g. 90%) the model predicted fair spread for a senior tranche will be zero. However, this is at odds with what is observed in the market where senior tranches typically trade at a few basis points. As we will see this has a serious consequence for the model when attempting to calibrate it to market observed prices.

The results presented in this section are absolutely fundamental in understanding synthetic CDO behaviour. To fully understand synthetic CDOs, the most crucial quantities are the portfolio and tranche loss distributions and the expected losses. From these quantities everything else follows.

7.6 Default baskets revisited

In light of the results obtained for synthetic CDOs we now briefly revisit the default basket results from Chapter 5. In essence a default basket is very similar to a synthetic CDO. Both products have a portfolio which will have a loss distribution and both products are essentially tranched (albeit with very thin tranches a single default wide).

Figures 5.10–5.12 show the correlation and maturity sensitivity of first-, second- and third-to-default baskets. Qualitatively the first-to-default sensitivity resembles that of an equity tranche. Intuitively this is to be expected since the first-to-default is only sensitive to the first default in the basket. As correlation increases the probability of no losses increases meaning that the cost of protection will decrease.

The second-to-default sensitivity resembles, at least superficially, a mezz tranche. The fact that it is the second default in the basket that triggers the payoff means that the basket essentially has a layer of subordination (albeit a layer only one default thick). As maturity increases the relative benefit of the subordination becomes less, meaning that the basket becomes more like a first-to-default variety. Finally the third-to-default basket resembles something between a mezz and a senior tranche. The first two defaults provide subordination for the basket. It is only for larger correlations that the probability of large losses becomes sufficiently significant so as to impact the contingent payment.

7.7 Chapter review

This chapter has analysed the properties of the standard market tranche valuation model (the normal copula model). The key quantity necessary to understand tranche behaviour is the portfolio loss distribution. In summary, as the correlation increases the probability of large losses also increases. At zero correlation idiosyncratic default risk is the most dominant factor. For perfect correlation, systemic risk dominates. In terms of tranche par spreads, as the correlation increases the par spread of the equity tranche decreases, the par spread of the senior tranches increases, and the par spread of mezz tranches first increases and then decreases. The correlation sensitivity of the different tranches lies at the heart of much of the market for synthetic CDOs. In the next chapter we will analyse further the parameter sensitivities of the standard market model and introduce a number of important measures for quantifying the risk of synthetic CDOs.

The quotation at the beginning of this chapter was meant to reassure the reader that with the application of rigorous, scientific reasoning and logic, no problem, however complex it may appear, can be understood. When viewed in its totality, the behaviour of a set of synthetic CDO tranches can initially appear daunting and complex. However, by breaking the problem down into smaller more easily understood components, insight and understanding can be gained. Once simple problems have been understood, more complex problems can be tackled incrementally. This is a guiding principle in all scientific analysis.

8

Risk quantification of synthetic CDOs

8.1 Introduction

In this chapter we continue our analysis of synthetic CDOs and their valuation within the standard model by quantifying the sensitivity of tranche PVs to fluctuations in input parameters. Several different risk measures commonly used in the marketplace will be introduced to facilitate this quantification.

Section 8.2 provides an overview of the different risk factors that synthetic CDOs are sensitive to and Section 8.3 describes the baseline case to be analysed (and justifies any simplifying assumptions made in the analysis). Broadly speaking risks can be classified as either market or credit variety. In Sections 8.4, 8.5 and 8.6 we introduce different measures to quantify the extent to which a position is sensitive to market and credit risk factors (specifically sensitivities to spreads, correlation and default respectively). These risk measures are similar to those previously introduced for single-name CDSs. However, the fact that we are now dealing with multiple obligors introduces additional complexities into the characterisation of risk. Section 8.7 analyses the time decay of a tranche as it's time-to-maturity decreases (as we will see this is an interesting quantity to analyse).

Value-at-risk (VaR) based measures are the subjects of Sections 8.8 and 8.9. Section 8.8 looks at credit spread VaR. This measure characterises the P/L uncertainty of a tranche as obligor credit spreads evolve over time. Section 8.9 introduces default VaR. This measure quantifies the impact of multiple obligor defaults over a specified time horizon on tranche P/L. Finally Section 8.10 reviews the key points introduced in the chapter.

8.2 Synthetic CDO risk factors

Taking a position in a synthetic CDO obviously entails an element of risk. Consider an investor selling protection (receiving periodic protection payments from the protection purchaser in exchange for covering losses) on an equity tranche of a standardised index. If the purpose of the investment is to hold it until maturity then the daily MtM fluctuations of the tranche in the traded market are not the most important concern; what is of concern are the occurrence of outright defaults in the underlying pool. Conversely if the position is entered into with the expectation of actively trading it before maturity, the daily MtM fluctuations induced by fluctuations in the values of the market observed input variables become very important since these are what will crystallise P/L when the position is exited. In addition to this the position is still sensitive to outright default of individual obligors and so this must also be taken into consideration.

Synthetic CDOs therefore exist in a superposition of states (borrowing terminology from quantum mechanics) between being a pure credit instrument and a pure market instrument depending on the strategy and horizon of the investor. Adding in the fact that the overall PV is sensitive to the behaviour of a pool of obligors (meaning we also have to consider the effects of idiosyncratic and systemic behaviour) renders the risk characterisation of synthetic CDOs quite complex. As an aid to the quantification of risk it is helpful to characterise risks as being of either the market or credit variety. A good discussion on the risk management of synthetic CDOs can be found in De Servigny and Jobst [2007] and Douglas [2007].

8.2.1 Market risk factors

Market risk factors refer to those risks which drive daily changes in the P/L of tranches. These risks are especially important under MtM accounting practices where the daily P/L reported by the bank must be calculated according to the current market conditions (and not just the original book value at trade origination). As with other derivative instruments the value of a synthetic CDO will have some sensitivity to common risk factors such as the levels of interest rates and FX rates. However, the sensitivity of synthetic CDOs to these factors is far less important than the sensitivity to fluctuations in credit spreads and correlations.

Perhaps the single most important factor in driving the daily changes in tranche MtMs is the market perceived credit quality of the underlying pool. This is reflected in the par CDS credit spreads of the debt that trades on each of the obligors in the pool (assuming the pool is a relatively standardised one). The other important class of market risk factor is sensitivity to fluctuations in default correlations. There are other risk factors. Recovery rate risk refers to the fact that CDO valuation models have to make an assumption as to the expected recovery rate of an obligor in the event of an actual default. Typically most models assume a value of 40% for this. Some models treat the recovery rate as a random variable itself typically drawn from a beta distribution. Little consideration has been given, from a modelling perspective, to the actual sequence of events within the marketplace post a large default (large defaults have become relatively rare in recent times). This relates to the actual settlement mechanisms in place in the vanilla CDS market. When an obligor defaults all those speculators who had sold protection on that obligor will now find themselves obligated to compensate the protection purchaser. This could lead to a run in trading in the distressed debt of the defaulted obligor, thereby pushing down the actual recovered amount substantially below the assumed value (typically 40%). This is an issue that has thus far received little attention, but is clearly of concern.

There are also settlement and legal risks as well as documentation risks. The documentation of credit derivatives has to be worded very precisely to define explicitly the circumstances under which contingent payments will be triggered. Recent history has demonstrated, however, that no amount of documentation can account for all possible eventualities. Residual risk remains which is difficult to mitigate against. Finally there is also model risk. This refers to the potential dangers introduced by using a particular model to value an instrument. If the model is a good one which captures well the behaviour of the instrument under distressed market conditions, then a hedging strategy based on this model will perform better (lead to fewer losses) than a poor model. Quantifying model risk is not always clear cut. In Chapter 14 we will introduce the technique of hedging simulation as a means to potentially quantify the impact of model risk.

8.2.2 Credit risk factors

Credit risk factors refer to those risks arising from outright default of an obligor in the underlying pool. When a default occurs and the total cumulative loss in the underlying pool is sufficient to exceed the attachment point of a particular tranche, then the investor who has sold protection on that tranche is obligated to make a protection payment to the protection purchaser; these payments continue until the tranche notional is exhausted. The protection seller is therefore concerned not just with individual defaults (as is the case for a single-name CDS which terminates upon the arrival of the first default), but also with the occurrence of multiple defaults within the pool. Issues such as the degree to which default events occur collectively or individually thus become important. We expect factors including the tranche subordination and the tranche thickness to influence the severity of this type of risk (since a thin tranche will be wiped out more quickly by multiple defaults than a thick tranche).

8.3 Baseline case analysed

To introduce and quantify these concepts we will consider again the baseline case introduced in Chapter 7, but with the following modifications.

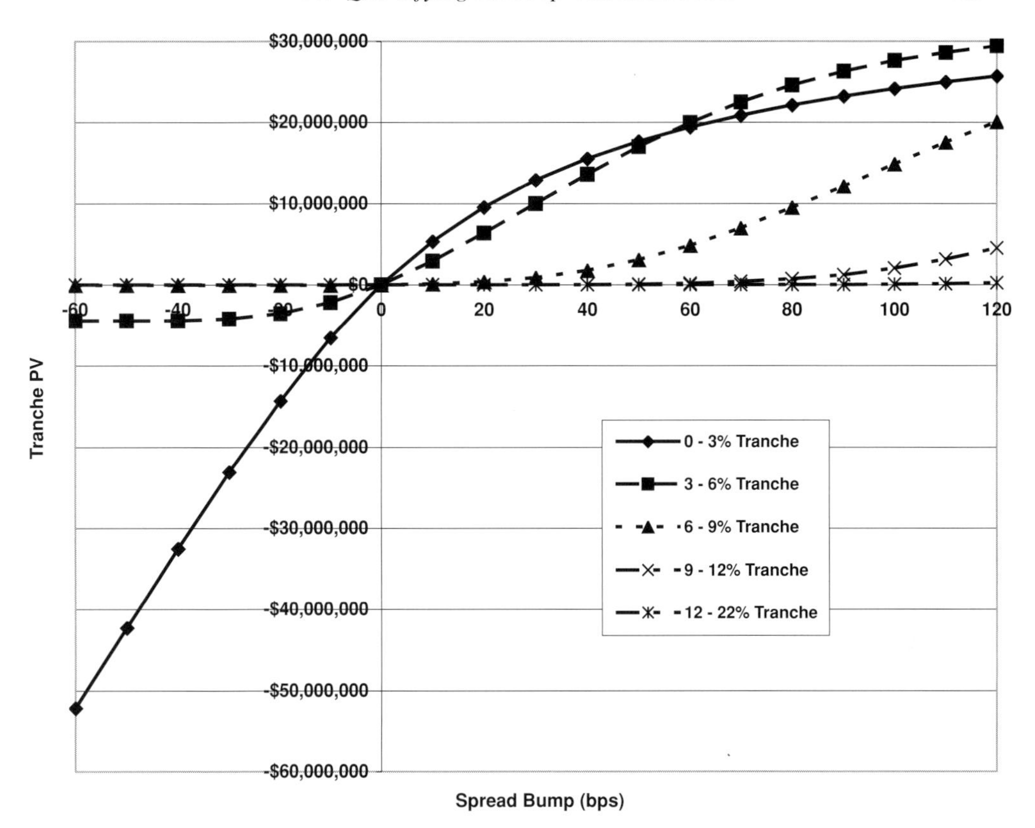

Figure 8.1 The tranche PVs as a function of (homogeneous) bumps applied to the par CDS spreads (for a correlation of $\rho = 0\%$).

Rather than being homogenous we will in some cases assume that the obligors in the underlying pool have inhomogeneous credit spreads distributed uniformly within the range $[s_{\text{Min}}, s_{\text{Max}}]$ (when this is the case it will be stated clearly). Unless otherwise stated all of the PV results reported in this chapter will be given from the perspective of the protection purchaser. That is, the PV reported for a tranche γ is calculated according to $V^\gamma = -s_{\text{Contract}} V^\gamma_{\text{Fee}} + V^\gamma_{\text{Cont}}$ (protection purchaser pays the coupon, and receives the contingent payments in return).

8.4 Quantifying credit spread sensitivities – CS01

8.4.1 Par spread sensitivity

Synthetic CDOs are clearly influenced by the levels of the underlying par CDS spreads. Figures 8.1–8.4 plot the PVs of the different tranches as a function of the par CDS spreads of the obligors in the underlying pool. Each graph plots the tranche PVs for different homogeneous correlations. Note that the scales are identical for all

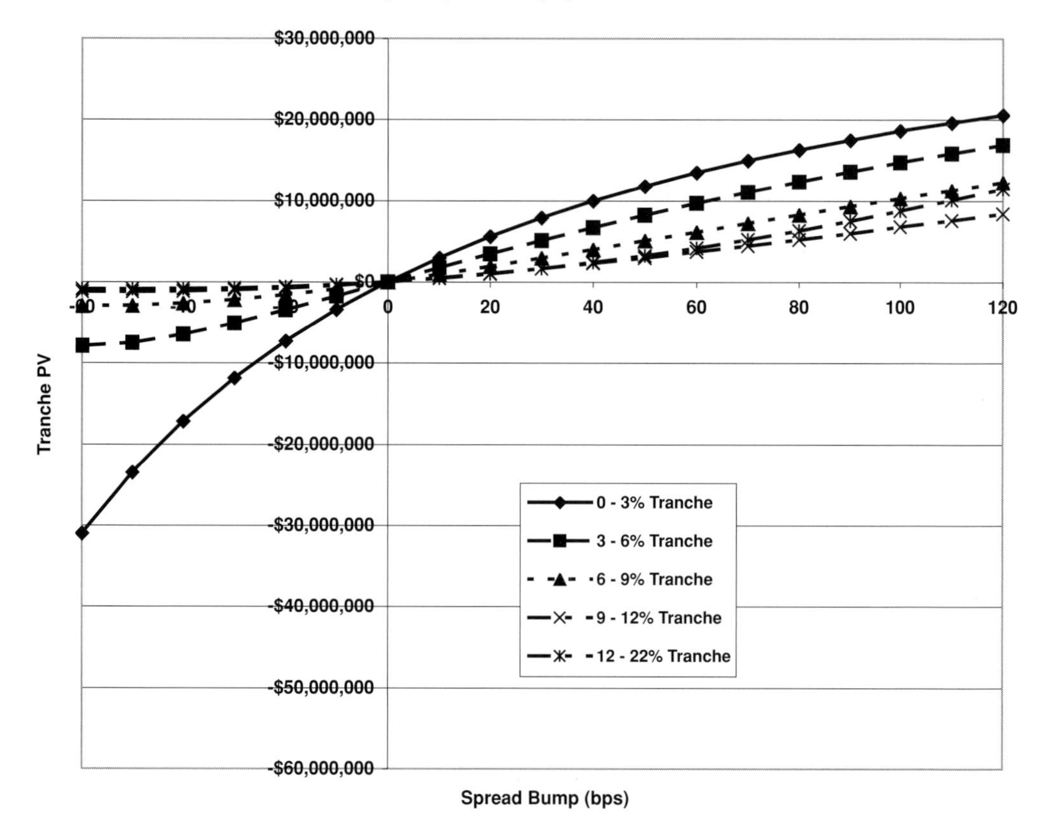

Figure 8.2 The tranche PVs as a function of (homogeneous) bumps applied to the par CDS spreads (for a correlation of $\rho = 20\%$).

the graphs to demonstrate the relative sensitivities of the tranches as the correlation increases.

The procedure followed to obtain these graphs is the same as for default baskets. For a par CDS spread of 60 bps for each obligor, the par spread is calculated for each tranche; the contractual spread is then set to this value (hence the PV at zero spread bump will be identically equal to zero for all tranches). The spreads are then simultaneously bumped for all obligors and the new tranche PVs determined. Note that these results are presented from the perspective of the tranche protection purchaser. To understand the PV sensitivity from the perspective of the protection seller we simply reflect the results in the x-axis. The tranche PVs display marked convexity as a function of the magnitude of the spread bump (evident for all of the tranches). To understand this phenomenology it is instructive to consider the behaviour of the fee and contingent legs separately. These are plotted for the different tranches for a default correlation of 0% in Figures 8.5 and 8.6. It is simplest to explain this behaviour in terms of a Monte Carlo simulation model. At a spread bump

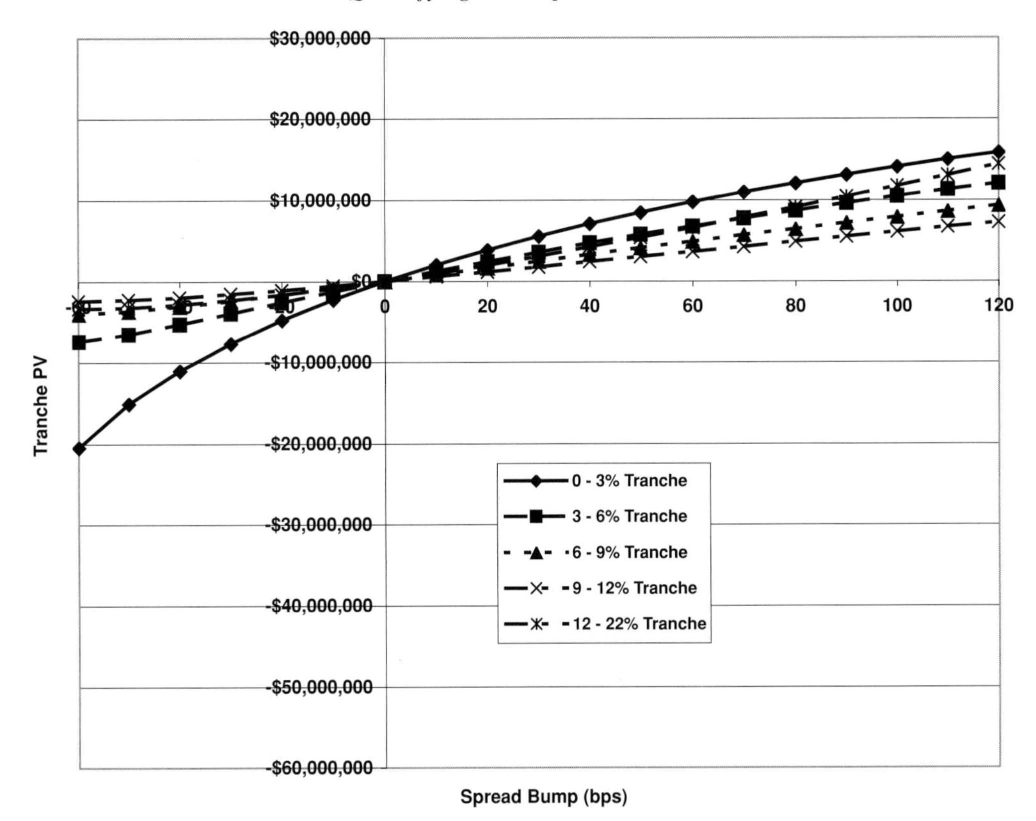

Figure 8.3 The tranche PVs as a function of (homogeneous) bumps applied to the par CDS spreads (for a correlation of $\rho = 40\%$).

of −60 bps the contingent payments are zero (as expected since the hazard rates are now identically equal to zero and there are no losses in the underlying pool). As the spread bump increases non-zero contingent payments begin to appear. These impact the low lying tranches first (hence the equity tranche almost immediately begins to be hit), followed by the more senior tranches. In this particular example, as the spreads begin to become very large, the simulated default times become very small for all of the obligors leading to large losses in the underlying pool. The equity tranche gradually becomes saturated at its maximum loss (close to the notional of the tranche, \$37.5 m). In the graph the 3–6% tranche is just beginning to become saturated as well.

At −60 bps spread bump there are no (or at most few) losses in the underlying pool and therefore the fee leg takes its maximal value (the tranche notional is not reduced due to any losses). As the spread bump increases more losses occur and the fee leg value decreases (because the notional of the tranche is being reduced due to the losses eating into it). This happens first for the low lying tranches and then

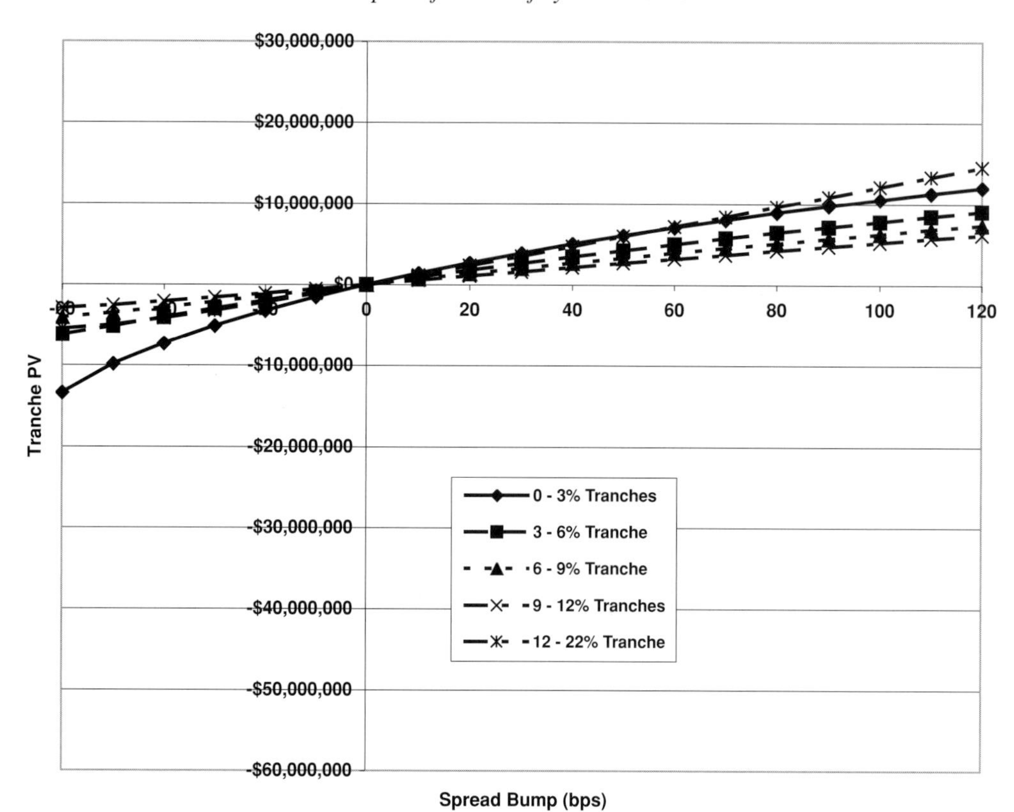

Figure 8.4 The tranche PVs as a function of (homogeneous) bumps applied to the par CDS spreads (for a correlation of $\rho = 60\%$).

gradually begins to impact senior tranches. In this particular example the 12–22% tranche is completely unaffected by the effects of the spread bumps, because of the large amount of subordination it has. Eventually we would see that as the spread was bumped to an infinite level, the fee leg would become zero.

The convexity (non-linearity) of the PV as a function of the par spread, as was observed for default baskets, is therefore a direct consequence of the saturation of the losses that the tranche suffers as the obligors in the underlying pool become more risky. As the obligors become more risky, greater contingent payments are made thereby reducing the notional of the tranche (and hence the value of the fee leg). Obviously, because of the existence of the detachment points, the maximum value of the contingent payments is bounded by the total notional of the tranche.

8.4.2 Marginal CS01

The marginal CS01 of a tranche is defined as the change in PV of the tranche for a 1 bp move in an underlying credit spread. Because of the tranching the impact of fluctuations in credit spreads for the same obligor will impact different tranches

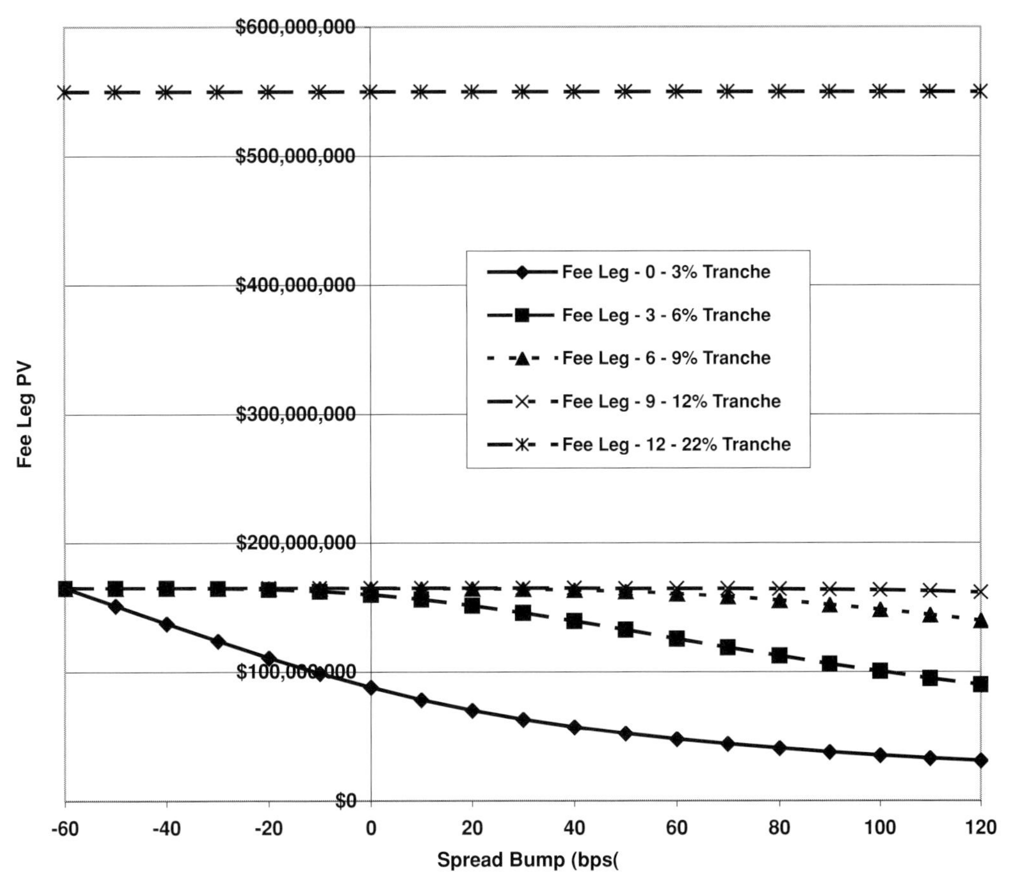

Figure 8.5 The fee leg PV of the tranches as a function of (homogeneous) bumps applied to the par CDS spreads (for a correlation of $\rho = 0\%$).

in different ways. The equity tranche is susceptible to any small fluctuations in the credit spread of an obligor, but the senior tranche may not be influenced at all due to the layers of subordination below it (this was apparent from the last section).

We define the marginal CS01 of tranche γ for obligor i as being

$$\Delta_i^\gamma = \frac{\Delta V^\gamma(\{s_i\})}{\Delta s_i}$$

where $\Delta V^\gamma(\{s_i\})$ is the change in tranche PV for a 1 bp bump in the spread of obligor i. Standard forward, backward or central differences [Hull 1999, Wilmott 2000] may be used to calculate the CS01. We can also compute marginal gammas according to

$$\Gamma_i^\gamma(t) = \frac{\partial^2 V^\gamma}{\partial s_i^2} \approx V^\gamma(s_i + \delta s_i, t) + V^\gamma(s_i - \delta s_i, t) - 2V^\gamma(s_i, t)$$

for a 1 bp shift up and down (using a central difference approximation).

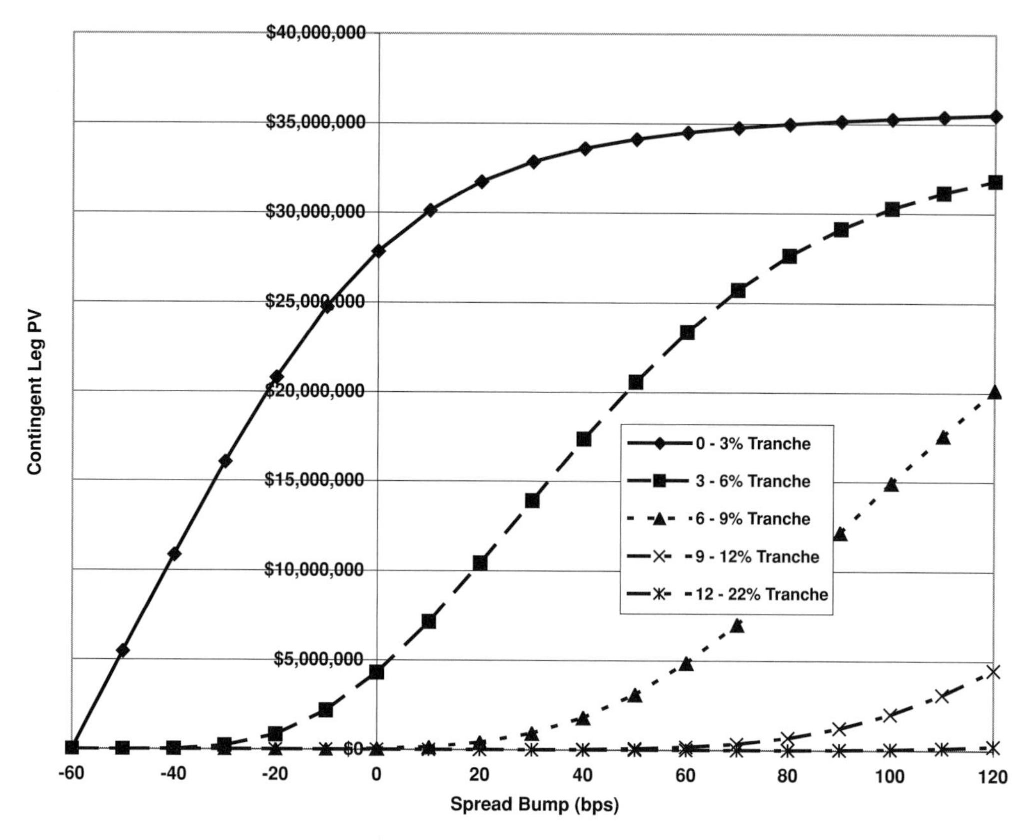

Figure 8.6 The contingent leg PV of the tranches as a function of (homogeneous) bumps applied to the par CDS spreads (for a correlation of $\rho = 0\%$).

Figure 8.7, plots the delta value for three different obligors for different tranches. In this example the obligors in the underlying pool are inhomogeneous with spreads uniformly distributed in the range $s_i \in [20\,\text{bps}, 100\,\text{bps}]$ (if the obligors were all homogeneous their marginal CS01 for a particular tranche would be identical). As expected, the same obligor has a different delta value for different tranches. In particular, the delta value for the equity tranche is larger than for other tranches higher up in the capital structure. It is simple to understand why this is the case: the equity tranche is sensitive to the first defaults in the pool. Note that these results were obtained for the case where $\rho = 0\%$.

Should Δ_i^γ depend on the correlation? Intuitively we would expect not since we are only bumping a single obligor's spread and seeing the impact on the tranche of the bump. If an obligor becomes riskier but has no correlation with any others then the impact would be limited. But if the obligor is correlated with others, then increasing its risk will also impact upon other obligors. Also, as we have seen, the correlation directly impacts the shape of the portfolio loss

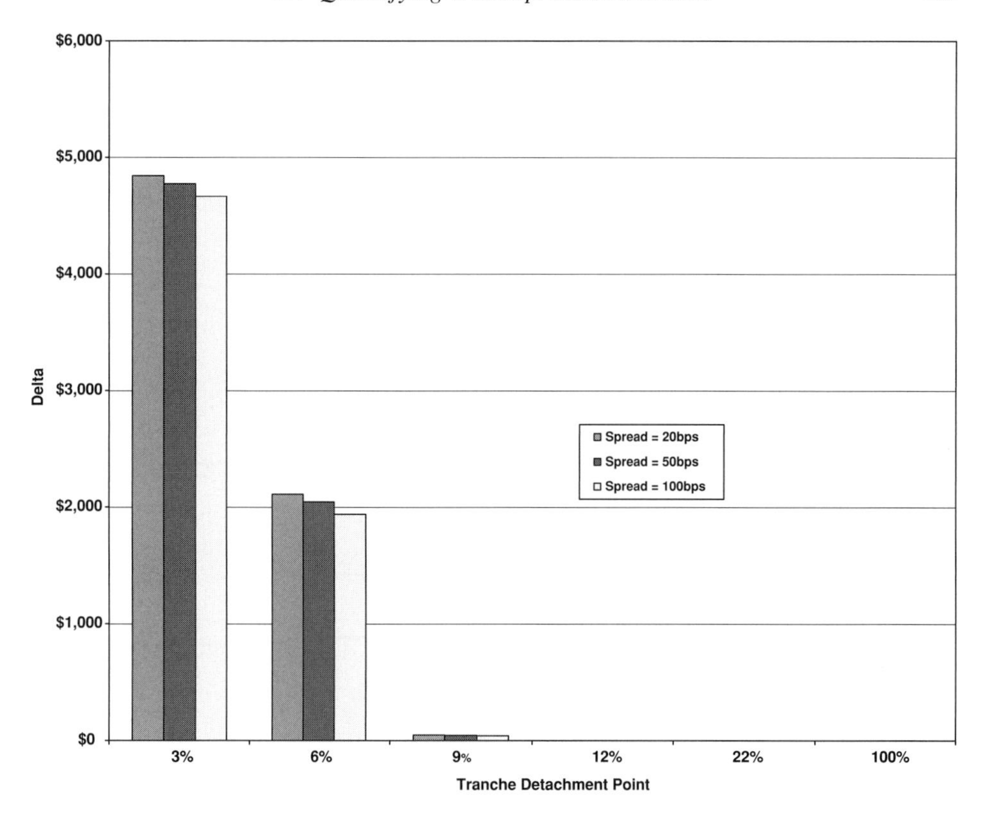

Figure 8.7 Histogram of marginal CS01s (for three representative obligors) across the capital structure.

distribution so it is reasonable to expect that the correlation plays a role in the marginal CS01.

Figure 8.8 shows the delta values for the different obligors at a correlation of 40%. This figure shows that as the spread of an obligor increases its equity tranche delta value increases. Conversely, for the same obligor as we move up the capital structure its delta value decreases.

Finally, recall the discussion of Chapter 5 regarding the magnitude of the spread bump and the impact of this upon Monte Carlo estimators of sensitivities. In that case, using Monte Carlo, it was desirable to use as large a spread bump as possible to compute the delta values, to minimise the effects of simulation noise. In the current context, however, we are using a semi-analytic model for the valuation, so it is not necessary to be concerned with simulation noise. However, due to the convexity of the PV with respect to the spread, it is important not to use a large spread bump to avoid introducing non-linear effects into the delta values. It is recommended, therefore, even though simulation noise is not an issue here, to consider carefully the magnitude of spread bump used to compute sensitivities.

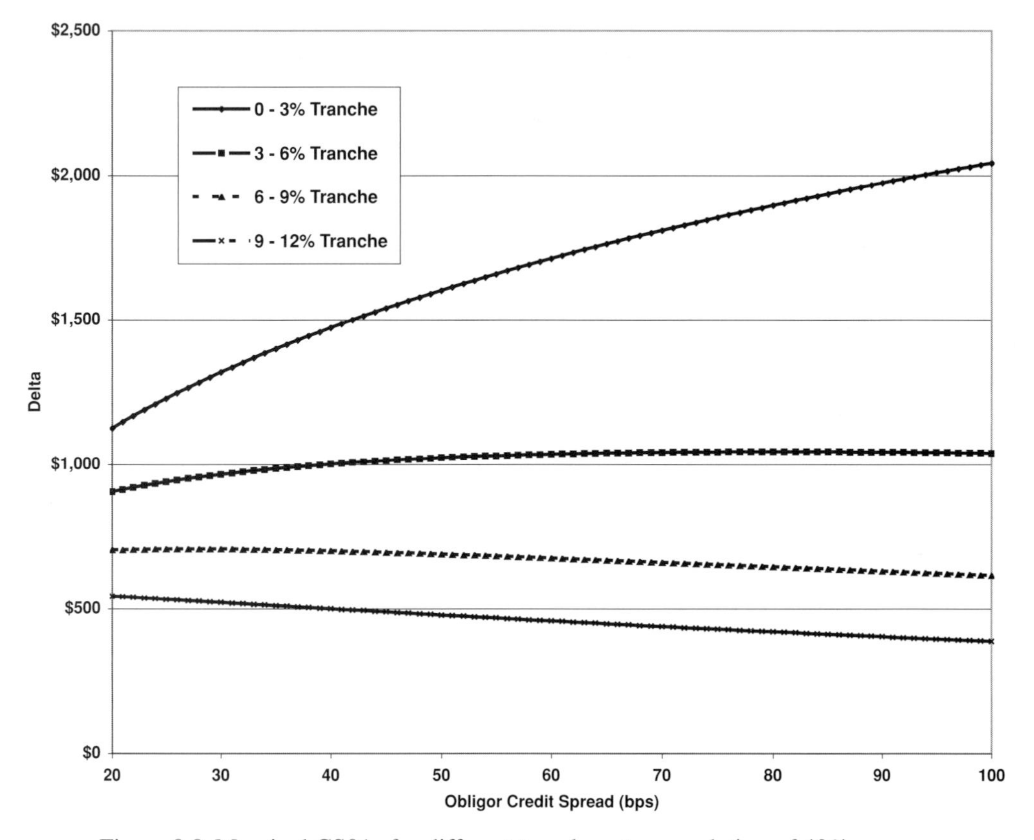

Figure 8.8 Marginal CS01s for different tranches at a correlation of 40%.

8.4.3 Multiple CS01

Marginal CS01 is an important spread sensitivity measure, but it is not enough for synthetic CDO risk capture. This is because of the presence of spread convexity (the change in PV of a tranche is not a linear function of the magnitude of the spread bump), but in addition to this we have to consider not only the spread movements of single obligors, but also the synchronous spread movements of many obligors (possibly the entire pool widening or tightening simultaneously).

Because of this the risk of the tranche will now also include some exposure to the degree to which credit spreads fluctuate together. If all the obligors' spreads widen together at the same time this will have a commensurately larger impact upon the tranches than if all of the spreads fluctuate independently. Figure 8.9, records the impact on the equity tranche PV of multiple spreads widening synchronously (for a default correlation of 0%). In this figure we chart the impact of bumping 1, 10, 50 and 125 obligors simultaneously by the amount indicated on the horizontal axis. The vertical axis is the 0–3% tranche PV. The contractual spread is set such that the PV at zero spread bump (a baseline value of 60 bps) is zero.

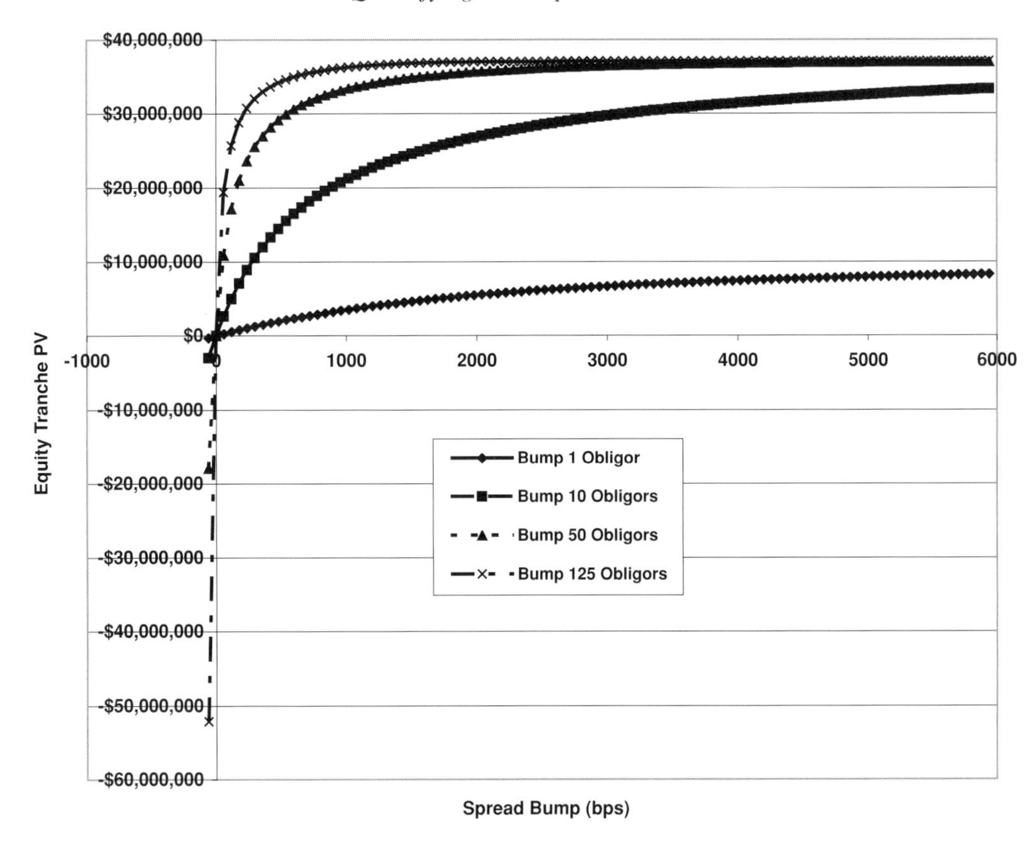

Figure 8.9 The impact of multiple spread widening on the 0–3% tranche PV.

Consider the case where all the obligors are bumped simultaneously. Fairly rapidly, as the spread bump increases, the tranche PV saturates to a value very close to the notional value of the tranche, \$37.5 m. Remember that the tranche PV is being reported as $V = -s_{\text{Contract}}V_{\text{Fee}} + V_{\text{Cont}}$, i.e. from the perspective of the protection purchaser. In the limit as $s_i \to +\infty$ for all i, the simulated default times for each path behave as $\tau_i^\alpha \to 0+$. This means that the simulated portfolio loss distribution is clustered in the first time bucket (with a value close to $125 \times \$6\,\text{m} = \$750\,\text{m}$) and the equity tranche expected loss is $\bar{L}(T_0) \approx \$37.5\,\text{m}$ and zero for all other time buckets.[1] From the general formulas for the fee and contingent legs we therefore expect $V_{\text{Fee}} \approx 0$ and $V_{\text{Cont}} \approx \$37.5\,\text{m}$. By similar arguments we would expect to see similar behaviour for the cases where 1, 10 and 50 obligors are bumped simultaneously. This time, however, the overall impact is reduced (since increasing the spread of a single obligor indefinitely does not increase indefinitely the impact upon the tranche; at some point the spread is large enough to ensure that $q_i(t) \approx 1$

[1] Being a probabilistic simulation, there is of course a small but non-zero chance that a few default times will 'leak' out of the first bucket, but this back-of-the-envelope calculation is sufficient to explain the general form of the results obtained (the risk-free rate is also non-zero).

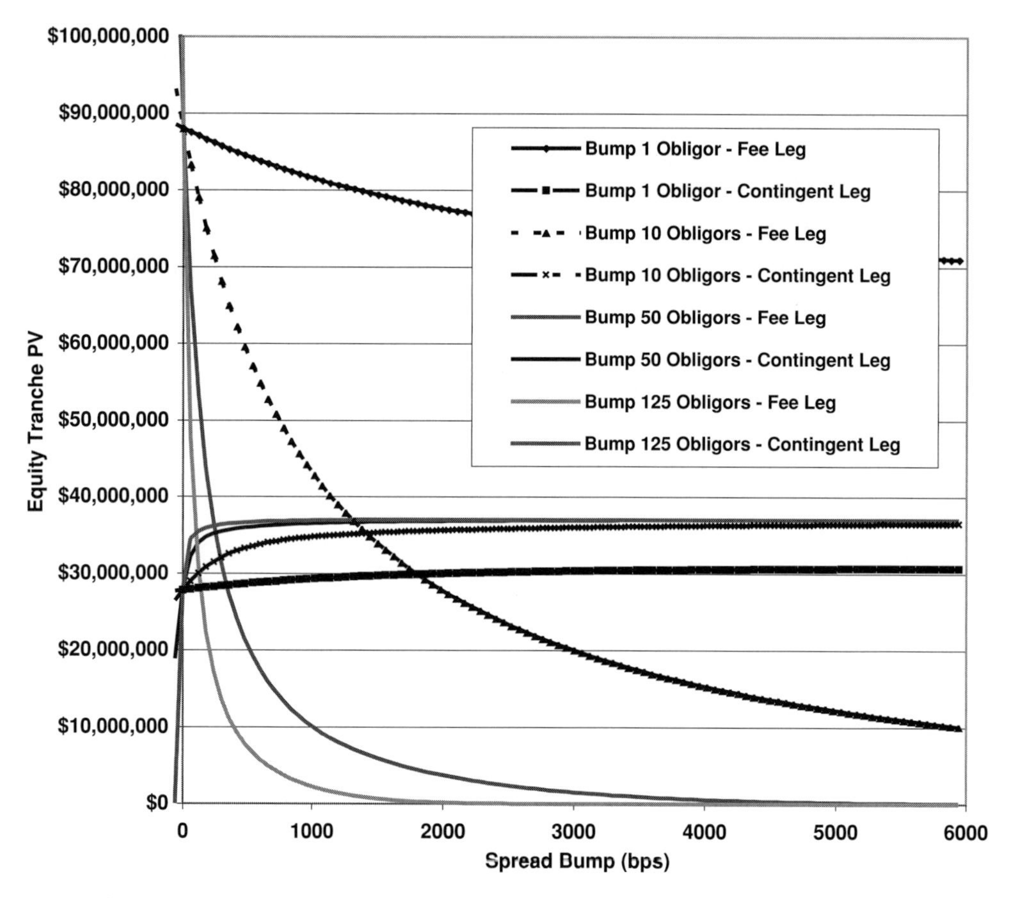

Figure 8.10 The impact of multiple spread widening on the equity tranche fee and contingent legs.

and so to all intents and purposes that obligor has defaulted; it can therefore no longer adversely impact the tranche). Figure 8.10 re-plots the previous graph, but this time separating out the fee and contingent legs. It confirms the intuition just outlined. Similar plots for the other tranches reveal identical results.

8.5 Quantifying correlation sensitivities – correlation vega

The most important input into the standard model, along with the par CDS spreads, is the default correlation. In general the correlation matrix will be composed of $n(n-1)/2$ distinct entries, and we can define a correlation vega matrix as

$$\nu_{ij}^{\gamma} = \frac{\partial}{\partial \rho_{ij}} V^{\gamma}(\{\rho_{ij} : i, j = 1, \ldots, n\}).$$

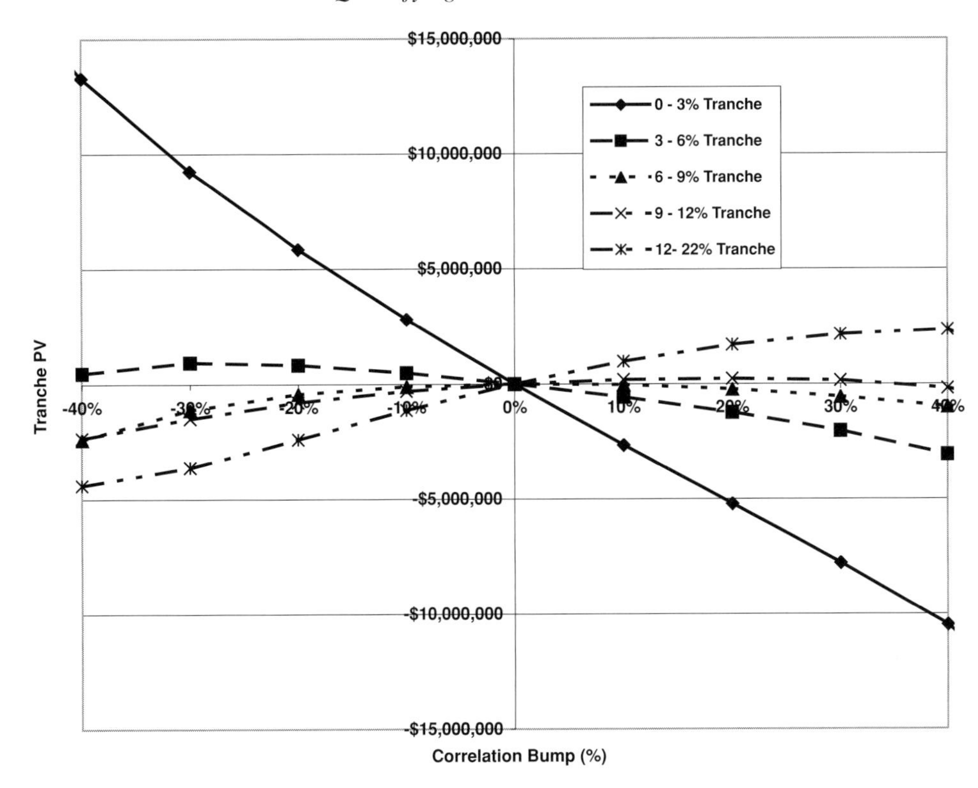

Figure 8.11 Correlation sensitivity of the tranches. The baseline correlation is assumed to be 50% and the horizontal axis represents shifts from this value.

We expect that the correlation vega is tranche γ dependent since the different tranches display different correlation sensitivities. Within the context of the standard model we assume that $\rho_{ij} \equiv \rho$ and this expression simplifies considerably to just a single correlation vega sensitivity. In practice, for a bespoke pool of obligors it may be extremely difficult to obtain accurate estimates for the correlation entries. In this case worrying about sensitivities to each and every entry in a full correlation matrix is not justified. On the other hand, making some assumptions about the gross structure of the correlation matrix (e.g. reducing it to a block diagonal form representing obligors in different industrial classification sectors) may provide some guidance as to likely segments of the underlying pool that the position is extremely sensitive to. As always it is recommended to perform a few numerical experiments to get a feel for likely behaviours.

Figure 8.11 shows the correlation vega of the tranche PVs (from the perspective of the protection purchaser) to fluctuations in the homogeneous correlation assumption. In this figure the baseline correlation is assumed to be 50% and the par spreads for the different tranches calculated at this correlation value are taken to be the contractual spreads (hence the PV at a correlation of 50% is zero for all the tranches). The usual comments regarding convexity and the implications of this on the calculation of linear sensitivities of course apply. For an equity protection seller the investor is long correlation, i.e. the investor will gain if the correlation of the underlying pool increases and will lose if it decreases. A senior tranche protection seller is short correlation (loses if the correlation increases). We will return to the tranche correlation sensitivity and analyse it in more detail in Chapters 12 and 13.

8.6 Quantifying default risk sensitivities – value-on-default (VoD)

A synthetic CDO position is exposed not just to the outright default of a single obligor in the underlying pool, but also to multiple defaults. Consequently the impact of multiple defaults upon a tranche's PV must also be considered. To quantify these risks we can define a number of different risk measures.

8.6.1 Marginal VoD

The marginal VoD [De Servigny and Jobst 2007, Douglas 2007] characterises the impact of the instantaneous default of an individual obligor upon the tranche PV. We expect the VoD for a particular obligor to be tranche dependent. This is because an individual default will impact the low lying tranches immediately, but will have a smaller impact upon tranches with greater levels of subordination (the default however does have some impact on the higher tranches by reducing the level of subordination they have). The marginal VoD for tranche γ due to obligor i defaulting is defined as

$$\mathrm{VoD}_i^\gamma = V^\gamma(\{s_j : j = 1, \ldots, n, j \neq i\}, s_i \to \infty) - V^\gamma(\{s_j : j = 1, \ldots, n\}).$$

The notation here is quite clumsy, but basically the marginal VoD is the change in PV of tranche γ brought on by the instantaneous default of obligor i. In the notation above the default of an obligor is represented as $s_i \to \infty$, i.e. the credit spread of the obligor blows up. In this limit the marginal default probability becomes $q_i(t) \to 1$. Since the marginal default probability is an input into the calculation of the conditional marginal default probability and hence the portfolio loss distribution (in the recursive semi-analytic model), introducing a default into the underlying pool will impact the shape of the loss distribution (the VoD can also be computed by revaluing the tranches with less obligors and modifying the attachment and detachment points accordingly). Figure 8.12 shows the portfolio loss distribution as progressively more defaults are introduced into the portfolio. Consistent with intuition, adding more defaults into the portfolio has the effect of shifting the loss distribution to the right by the loss given default of the defaulted obligor (in this case the difference between the distributions for two and four defaulted obligors is

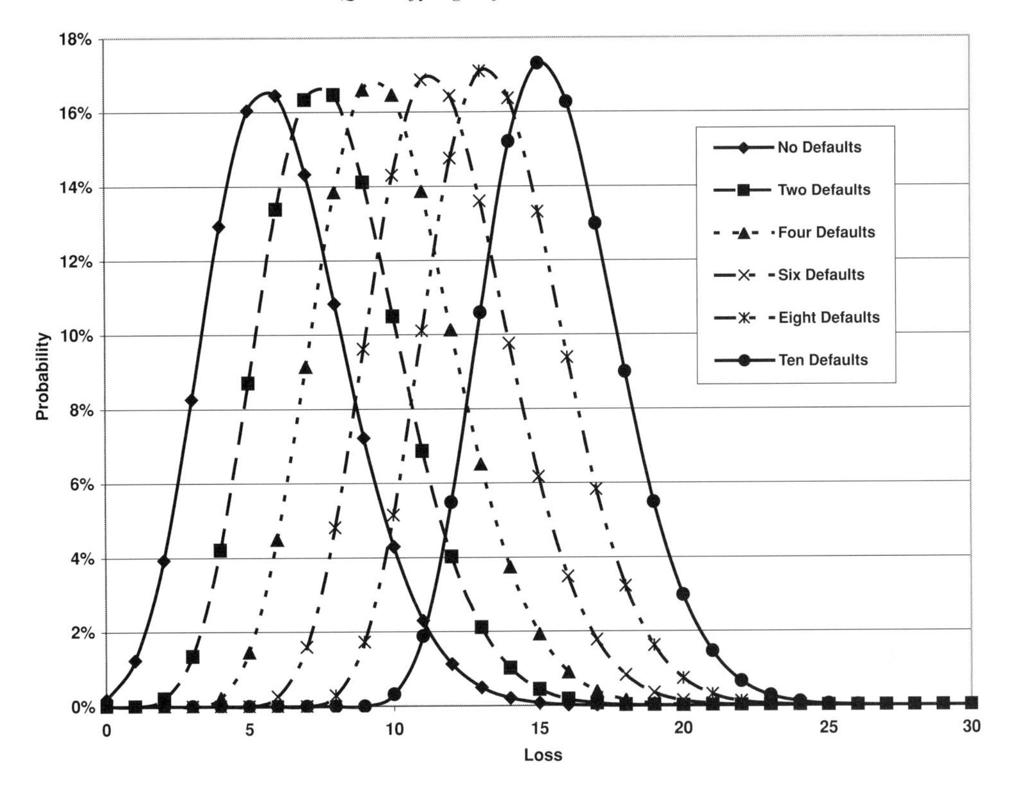

Figure 8.12 The impact on the portfolio loss distribution of adding progressively more defaulted obligors into the portfolio.

approximately 2 × \$6 m since we have assumed a homogeneous recovery rate of 40%).

The marginal VoD will impact different tranches in different ways. For example, a seller of protection on the 3–6% tranche will have a negative VoD if a single name defaults despite the fact that a single default is not enough to wipe out the 0–3% equity tranche subordination. This is because the default impacts the equity tranche immediately and reduces the overall cushion that the seller of mezzanine tranche protection enjoys. It is a simple matter to check this (and is left as an exercise for the reader).

8.6.2 Multiple or running VoD

The marginal VoD provides useful information about the impact on a tranche of a single obligor defaulting. However, it does not provide information on the impact of multiple synchronous, or near synchronous defaults (and as we saw previously

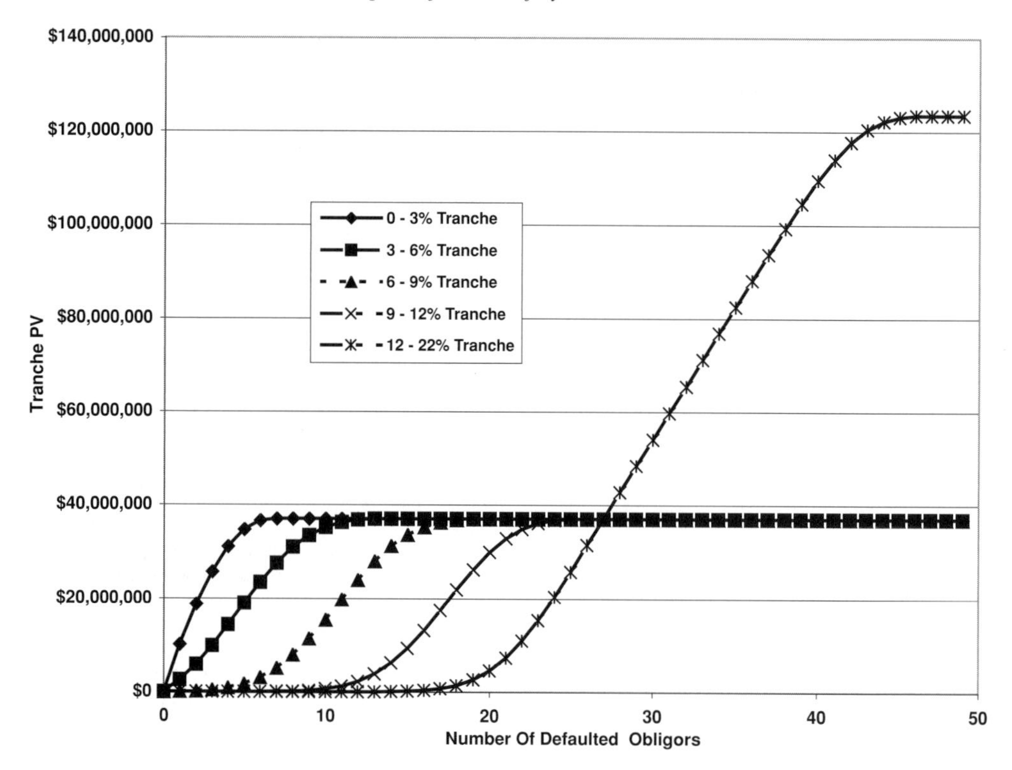

Figure 8.13 The running VoD for the tranches as a function of the number of defaulting obligors.

the behaviour of a synthetic CDO is influenced by the degree to which spreads move synchronously). To capture this a *running VoD* can be defined.

The running VoD, VoD_X is the change in value of the tranche when X obligors in the portfolio are simultaneously placed into default. The most obvious way to choose which obligors to default is to order them in terms of decreasing credit spread, i.e. the obligor with the largest credit spread is defaulted first. If the obligor notional amounts are homogeneous this will define the worst possible loss case. The total loss amount has an upper bound of the tranche notional less any upfront payments. Figure 8.13 shows the running VoD as progressively more obligors are defaulted for the different tranches. In this example all the obligors have identical spreads and all the tranches are priced at par. Note that we assume the default of an obligor has no impact on the spreads of other obligors. This is clearly not what will happen in practice.

The form of this graph should be no surprise. For the equity tranche, placing an obligor into default has an immediate impact. Increasing the number of defaulted obligors increases the losses to the equity tranche until eventually the losses saturate at the total notional of the tranche ($37.5 m). The same behaviour is observed for the

3–6%, 6–9% and 9–12% tranches, except that more defaults are necessary before the tranches begin to be impacted by the losses. The 12–22% tranche has a much larger (negative) PV because the fee payments are based on a much larger notional than for the preceding four tranches in the capital structure. It is important to remember that for a portfolio of inhomogeneous obligors (which will be the case in reality), the running VoD will be dependent on the order in which obligors are defaulted.

8.7 Tranche time decay

Sensitivity to time decay of the tranche value is defined by its theta as

$$\Theta = -\frac{\partial}{\partial t} V(t).$$

In the current context, by tranche time decay we do not explicitly mean theta. What is referred to is how the tranche value amortises over time as the remaining life of the tranche reduces (and not the rate of change of tranche value with the passage of time). The tranche time decay is one of the more interesting sensitivities due to the way in which risk shifts around in the capital structure as the time-to-maturity of the tranche approaches. Intuitively we would expect that as $t \to T$ risk will move out of tranches with subordination. This is because as the remaining life of the tranche reduces there is less time left for multiple defaults to occur and wipe out the subordination. Because of this we would expect that $V_{\text{Cont}} \to 0$ more quickly for a mezz tranche than for an equity tranche.

There are two different ways the maturity of the tranche can be rolled down. One way is to compute the value of a tranche with a reduced maturity of $T - t$ including only the cashflows in the period $[0, T - t]$ in the calculation of the fee and contingent legs. The other way is to include only the cashflows in the period $[t, T]$. If the coupon schedules are regular (in particular identical first and last coupons) then the two methods should yield similar results.

In the following we employ the latter method. In particular, consider a tranche γ with coupon dates $\{T_j : j = 1, \ldots, N\}$ and maturity $T = T_N$. If we want to value the tranche at some intermediate coupon date $T_k \in [0, T]$ when the remaining tranche maturity is $T - T_k$ the fee and contingent legs are computed according to

$$V^{\gamma}_{\text{Fee}}(T_k) = \sum_{j=k}^{N} Z(T_k, T_j)\Delta_j \left[N^{\gamma} - \bar{L}^{\gamma}(T_{j-k})\right]$$

$$V^{\gamma}_{\text{Cont}}(T_k) = \sum_{j=k}^{N} Z(T_k, T_j) \left[\bar{L}^{\gamma}(T_{j-k}) - \bar{L}^{\gamma}(T_{j-k-1})\right].$$

That is, we assume the tranche does not experience losses in the period $[0, T_k]$ and shift the expected losses such that there is zero expected loss at time T_k.

The parameters used for this analysis are as follows. The tranche expected losses

Number of obligors in pool	125
Capital structure	iTraxx
Tranche maturities	5 years
Coupon tenor	quarterly
Homogeneous obligor notional	\$10 m
Homogeneous obligor recovery rate	40%
Homogeneous (flat) obligor par CDS spread	60 bps
Homogeneous (default) correlation	varies
Risk-free rate	5%
Number of simulations (Monte Carlo model)	100 000

are computed using the Monte Carlo model. The procedure for generating the results to be discussed is as follows.

- Specify the obligor parameters (par CDS spread, recovery rate and notional).
- Specify the tranche parameters (attachment/detachment points, maturity T, correlation ρ).
- Set $t = 0$ and calculate the tranche par spread.
- Set the tranche contractual spread equal to the par spread. The PV at $t = 0$ will therefore be zero. The contractual spread is kept fixed at this value throughout the rest of the analysis.
- Set $t = t_1$ such that the remaining maturity of the tranche is given by $T - t_1$ and calculate the new tranche fee and contingent legs and hence PV.
- Repeat this procedure for a discrete set of times up until time T.

The above procedure is repeated for a range of correlations. At each correlation the tranche par spread is recalculated and the contractual spread set to this value. For all cases considered, all the contract and obligor parameters (such as the obligor spreads) are kept static; the only parameter that varies is the time from contract inception $0 \leq t \leq T$.

Since the tranches are priced at par at $t = 0$, the tranche PVs are zero. If the obligor par CDS spreads remain static we might expect, all other things being equal, that $V^{\gamma}(t) = 0$ for all $t > 0$. This is what we would observe for a simple CDS. It will be observed that the tranche behaviour is rather more subtle.

Figure 8.14 shows the 0–3% tranche fee leg at different times to maturity and for varying values of the homogeneous correlation. As we would expect, as $t \to T$, $V_{\text{Fee}} \to 0$. At low correlation the fee payment is smaller than for higher correlation. This is because as correlation increases the equity tranche becomes less risky (at high correlation the distribution of losses becomes concentrated around either no losses or large losses). Figure 7.10 showed that as correlation increases, the expected

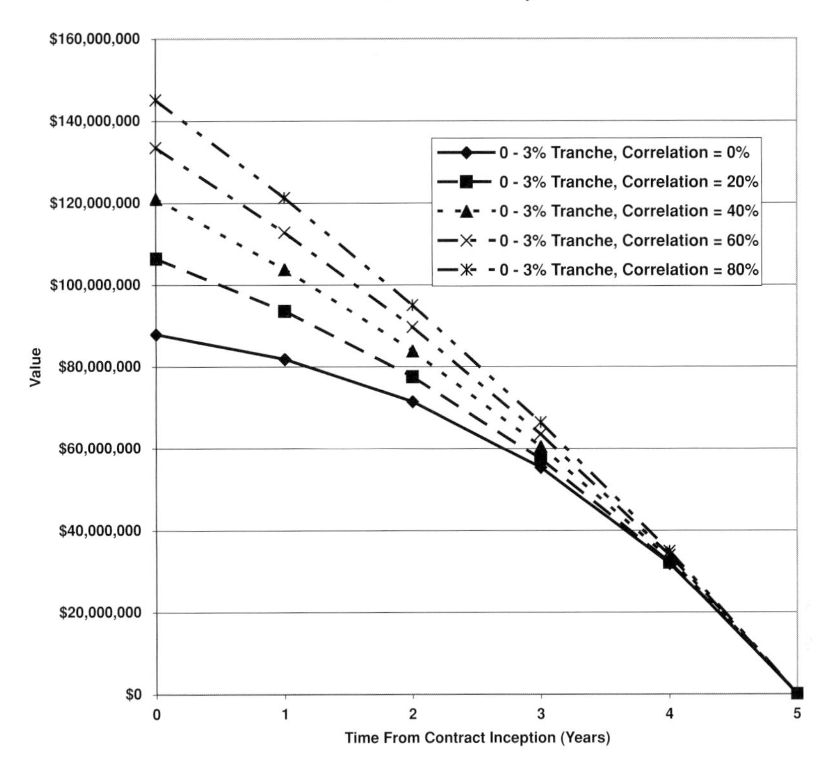

Figure 8.14 Value of the 0–3% tranche fee leg as time to maturity rolls down for various correlations.

losses for the equity tranche decrease (at a particular coupon date T_j). Because the fee payment at each coupon date T_j is proportional to $N^\gamma - \bar{L}^\gamma(T_j)$, if the expected losses decrease then the fee payment increases.

Figure 8.15 shows the 0–3% tranche contingent leg at different times to maturity and for varying values of the homogeneous correlation. As for the fee leg, the contingent leg approaches zero as the time-to-maturity approaches zero. This is consistent with intuition since as the remaining tranche life decreases the time available for defaults to impact the tranche decreases and so the expected value of the contingent leg decreases. In contrast to the fee leg, however, as the correlation increases the contingent leg decreases. The reason for this is as follows. Referring again to Figure 7.10, if we define the gradient of the expected loss as $\Delta\bar{L}(\rho, T_j) = \bar{L}(\rho, T_j) - \bar{L}(\rho, T_{j-1})$, then each contingent payment is given by $V_{\mathrm{Cont}}(\rho, T_j) = Z(0, T_j)\Delta\bar{L}(\rho, T_j)$. Figure 7.10 shows that as ρ increases $\Delta\bar{L}(\rho, T_j)$ decreases, i.e. the gradient of the expected loss decreases as correlation increases. This implies that $V_{\mathrm{Cont}}(0\%, T_j) < V_{\mathrm{Cont}}(20\%, T_j)$, for example.

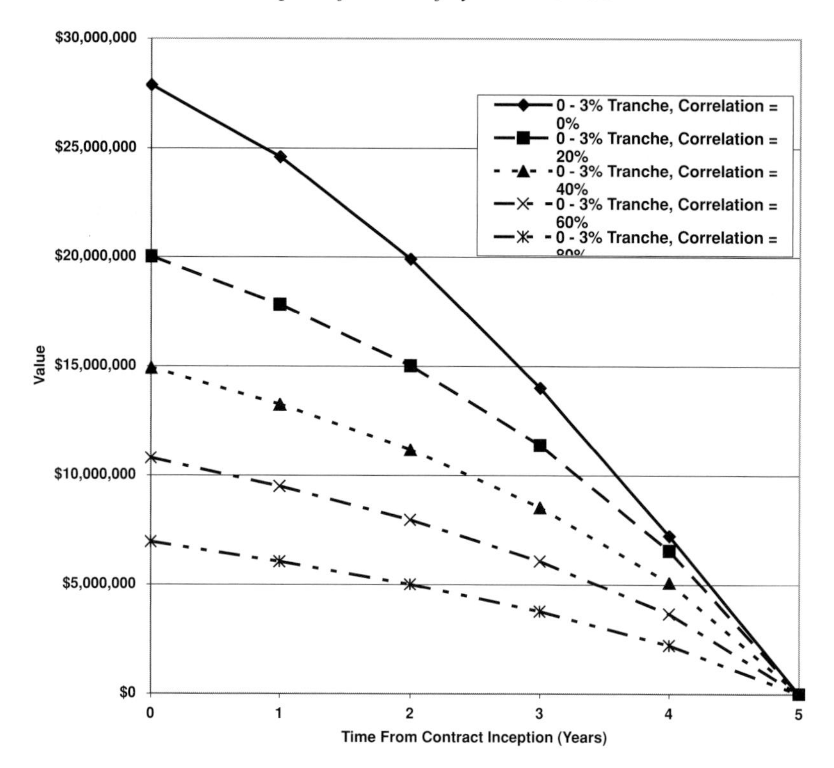

Figure 8.15 Value of the 0–3% tranche contingent leg as time to maturity rolls down for various correlations.

Now that we understand the time decay of the fee and contingent legs, Figure 8.16 shows the 0–3% tranche PV at different times to maturity and for varying values of the homogeneous correlation. As expected at $t = 0$ and $t = T$ the PV is zero. However, for intermediate times the PV is non-zero (although it should be noted that the magnitude of the PV as a fraction of the tranche notional is not huge). This is slightly counter-intuitive since none of the tranche valuation parameters, such as the obligor CDS spreads, are varying. The reason for this behaviour is in the non-linear nature of the tranche expected loss as $t \to T$ (see Figure 7.10 showing the equity tranche expected loss as a function of time).

Specifically, a coupon payment at time T_j for the fee leg is proportional to $N^\gamma - \bar{L}^\gamma(T_j)$ and for the contingent leg it is $\bar{L}^\gamma(T_j) - \bar{L}^\gamma(T_{j-1})$. Although both payments are linear functions of the expected loss, their functional form is different. The contingent payment depends on the gradient of the tranche expected loss as a function of time; the fee payment depends only on the level of the expected loss. Because the gradient of the expected losses is not constant the values of the fee and contingent payments drift away from their initial balance imposed at $t = 0$. This means the net of the fee and contingent payments will not remain zero.

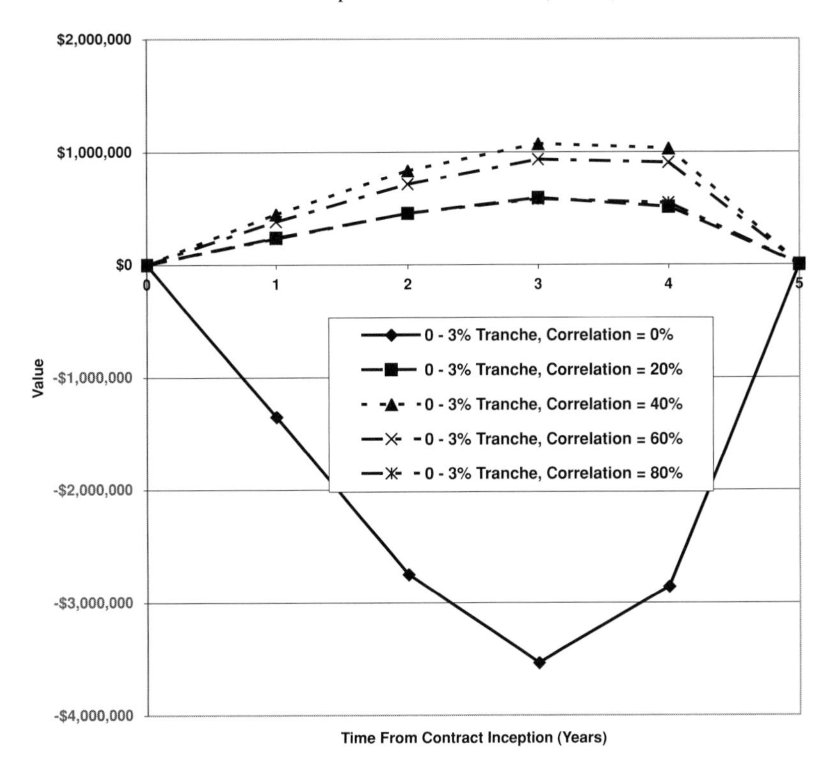

Figure 8.16 Value of the 0–3% tranche PV as time to maturity rolls down for various correlations.

It can also be observed that for $\rho = 0\%$ the PV is negative, whereas for $\rho > 0\%$ the PV is positive. Figure 8.17 plots the contingent legs and total fee legs (the fee leg multiplied by the tranche par spread to give the $ value of the fee payment) for the cases $\rho = 0\%$ and $\rho = 20\%$. For $\rho = 0\%$ the fee leg is greater than the contingent leg for all times, leading to a positive PV. However, for $\rho = 20\%$ the fee leg is less than the contingent leg, leading to a negative PV. As is usually the case, this behaviour arises due to the redistribution of losses as the correlation increases. The analysis for the equity tranche can be carried out for the other tranches in the capital structure. The results obtained are consistent with the behaviour observed for the equity tranche.

8.8 Credit spread value-at-risk (CVaR)

The MtM value of a synthetic CDO is sensitive to the day-to-day fluctuations in the credit spreads of the underlying obligors. It makes sense to ask how much the value of a tranche can possibly deteriorate over a specified time horizon at a certain

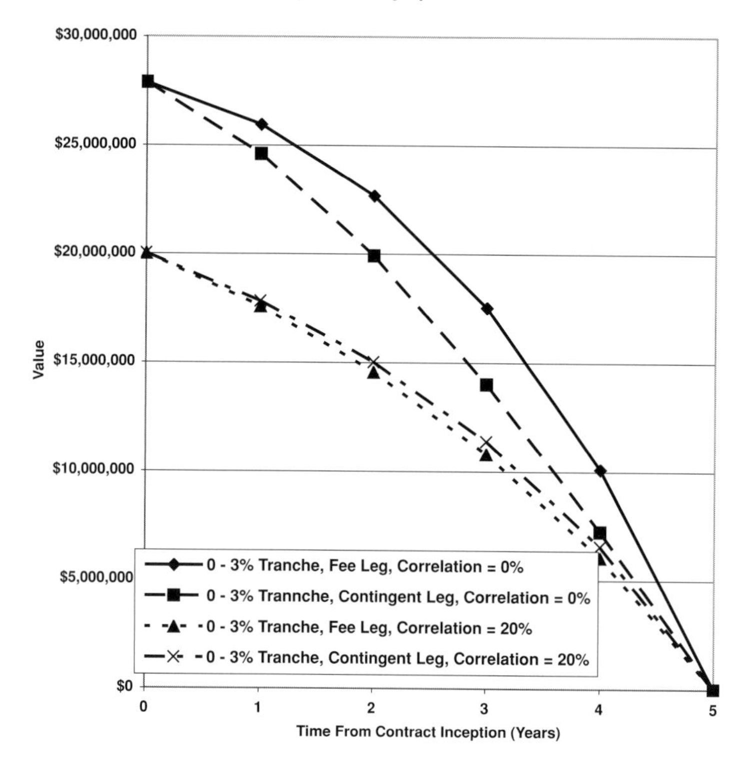

Figure 8.17 Value of the 0–3% tranche fee and contingent legs for 0% and 20% correlations.

confidence level due to the fluctuations in the credit spreads. This can be quantified with a credit spread based VaR measure.

VaR is defined as the maximum possible loss (to a position, portfolio) that can be observed over a certain time horizon at a specified confidence level [Hull 1999]. It is a statistical measure that is commonly utilised for the calculation of short time horizon (typically of the order of days) risk statistics.

To calculate the VaR for a position we need to specify the time horizon and the confidence level. We assume that the temporal horizon for the calculation of CVaR is T_{CVaR} and that the confidence level we wish to calculate it at is q. A general procedure for calculating CVaR is as follows.

1. At $t = 0$ (with remaining tranche maturity T and CDS par spreads $\{s_i(0) : i = 1, \ldots, n\}$) calculate the tranche PV, V.
2. Evolve the par CDS spreads of the obligors over the horizon $[0, T_{\text{CVaR}}]$, $\{s_i^\alpha(T_{\text{CVaR}}) : i = 1, \ldots, n\}$ (how this can be done is discussed below).
3. Calculate the new tranche PV $V^\alpha(\{s_i^\alpha(T_{\text{CVaR}})\})$ (with remaining maturity $T - T_{\text{CVaR}}$) and hence the change in tranche MtM over the horizon $[0, T_{\text{CVaR}}]$ $\Delta V^\alpha = V^\alpha - V$.
4. Repeat this for a large number of simulations $\alpha = 1, \ldots, N_{\text{CVaR}}$, each simulation corresponding to a different spread scenario.

5. Calculate the VaR from the appropriate quantile of the change in MtM distribution [Bouchaud and Potters 2000]

$$1 - q = \int_{-\infty}^{\text{VaR}} P(\Delta V)\text{d}\Delta V$$

where $P(\Delta V)$ is the simulated distribution of tranche values under the different spread scenarios.

The VaR is equal to the area under the left tail of the constructed distribution of tranche values. Typical confidence levels are $q = 99\%$ and horizon 1, 10 or 30 days.

The obligor par spreads are typically evolved in one of two different ways:

- using historically observed data,
- by simulation.

8.8.1 Evolving the spreads using historical data

Evolving the spreads using historical data has the clear advantage that we are making no assumptions about the parametric form of the underlying distribution of spreads. This is particularly useful for corporate credit spreads which can exhibit significant fluctuations over short time periods (due for example to adverse news, LBO approaches etc.). These fluctuations make it difficult to justify an assumption such as log-normality for the distribution of spread returns. On the other hand we are assuming that the historically observed pattern of spread movements is a good indicator of future spread movements. For equities this may not be too bad an assumption, but for credit spreads that can fluctuate violently over short time periods (even inter-day) the particular choice of dataset can have a significant impact upon the results obtained. Another potential drawback of historical simulation is that we need enough historical data in order to sample sufficient historical paths to build up a distribution of simulated tranche values. Techniques such as bootstrapping (sampling with replacement) [Wilmott 2000] can be useful in generating large numbers of sample paths from relatively small datasets (although introducing spurious auto-correlation into the sample paths must be avoided).

8.8.2 Evolving the spreads using simulated data

Generating simulated data has the advantage that we can construct as much synthetic data as we wish. For example we could assume that the spread dynamics follow a geometric Brownian motion (GBM)

$$\frac{\text{d}s_i^\alpha(t)}{s_i^\alpha(t)} = \mu_i \text{d}t + \sigma_i \text{d}W_i^\alpha$$

and sample from the exact distribution according to [Glasserman 2003]

$$s_i^\alpha(T_{\text{CVaR}}) = s_i^\alpha(0) \exp\left[\left(\mu_i - \frac{1}{2}\sigma_i^2\right) T_{\text{CVaR}} + \sigma_i \sqrt{T_{\text{CVaR}}} \varepsilon_i^\alpha\right]$$

where $\varepsilon_i^\alpha \sim N(0, 1)$ for all i and each α. In this example we are evolving the spreads in one shot from time 0 to time T_{CVaR}. To specify the model fully it is necessary to obtain estimates for the drift and volatilities and correlations of each of the spreads. A common approximation for short time horizon VaR applications is to neglect the drift of the stochastic process. A further advantage of simulated data is that additional components can be added to the SDE as required. For example a jump diffusion component can easily be added to the above GBM in order to quantify the impact of instantaneous spread jumps on the VaR.

The disadvantage of using simulated data of course is that we are making an assumption about the distributional properties of the underlying spreads, typically that they are log-normally distributed. As seen in Chapter 4, the distribution of returns of the iTraxx index is non-normal. It can therefore be dangerous to make assumptions about the distribution without first analysing empirically the properties of the data.

8.9 Default value-at-risk (DVaR)

Although the running VoD introduced in Section 8.6 is a useful measure of potential default scenarios, it does not provide a probabilistic assessment of what the distribution of portfolio losses is likely to be. This is because we specify exogenously what the default scenario is in terms of choosing the names to default without assigning a likelihood to that particular scenario.

To define a VaR figure associated with obligor defaults we need to study many possible default scenarios where each is generated according to a probabilistic rule (default VaR is also discussed in De Servigny and Jobst [2007] and Douglas [2007]). Each default scenario can then be translated into a tranche VoD and the empirical distribution of VoD values built up. From this distribution the required tail statistic representing VaR at a particular confidence level (and a particular time horizon) can be computed. Consider calculating a default VaR estimate for a time horizon of T_{DVaR} years (DVaR will typically be calculated over a much longer time horizon than CVaR) at a confidence level of q. We denote by $\mathbf{D}^\alpha$ the $n \times 1$ vector indicating which obligors have defaulted before T_{DVaR} for simulation α. $D_i^\alpha = 1$ if obligor i has defaulted and $D_i^\alpha = 0$ if they have not. We assume that the correlation matrix $\mathbf{C}$ constructed from a single homogeneous value ρ is available, and that its Cholesky decomposition has been computed from $\mathbf{C} = \mathbf{A}\mathbf{A}^{\text{T}}$. The DVaR algorithm is as follows:

set $\mathbf{D}_0 = \mathbf{0}$	initially no defaults
calculate $V_\gamma^0 = V_\gamma(\{s_i\}, \{\delta_i\}, \rho_{\text{Def}}, \mathbf{D}_0)$	for each tranche γ the baseline, no default valuations
for $\alpha = 1$ to N_{Sims}	
set $\mathbf{D}_0 = \mathbf{0}$	no defaults initially
generate $\varepsilon^\alpha \sim N(\mathbf{0}, \mathbf{I})$	uncorrelated $N(0, 1)$ deviates
calculate $\phi^\alpha = \mathbf{A}\varepsilon^\alpha$	correlated $N(0, 1)$ deviates
generate $\tau_i^\alpha = -\dfrac{1}{h_i} \ln N(\phi_i^\alpha)$	correlated default times of obligors
set $D_i^\alpha = \mathbf{1}_{\tau_i^\alpha < T_{\text{DVaR}}}$	default vector
calculate $V_\gamma^\alpha = V^\alpha(\{s_i\}, \{\delta_i\}, \rho_{\text{imp}}, \mathbf{D}^\alpha)$	new tranche PVs
calculate $\text{VoD}_\gamma^\alpha = V_\gamma^\alpha - V_\gamma^0$	VoD for tranche γ and simulation α
end α.	

This procedure generates a distribution of VoD values from each default scenario. From this distribution we can now calculate the DVaR at the required quantile level.

To illustrate DVaR we now consider a simple example. The parameters are as follows.

Number of obligors in tranches	125
Tranche capital structure	iTraxx
Tranche maturities	5 years
Homogeneous obligor notional	\$10 m
Homogeneous obligor recovery rate	40%
Homogeneous obligor par spreads	varies (typically 60, 120 or 180 bps)
Default correlation for tranche revaluation	varies
Default correlation for default scenarios	same as revaluation correlation
DVaR calculation horizon	1 year
Number of default time simulations	1000

The tranche revaluations are performed using the recursive model (due to its speed advantage over Monte Carlo valuation methods). Figure 8.18 shows the distribution of default scenarios as a function of the correlation for generating the obligor default times. At high correlations the majority of the loss distribution occurs for zero defaults, with a small but non-zero probability of extremely large losses. As correlation decreases more of the probability mass becomes dispersed among small numbers of defaults. These results are consistent with those found earlier for the portflio loss distribution in Chapter 7.

Turning now to the DVaR, Figure 8.19 plots the VoD of the 0–3% tranche for $\rho = 0\%$. This figure plots the cumulative VoD generated from the 1000 default

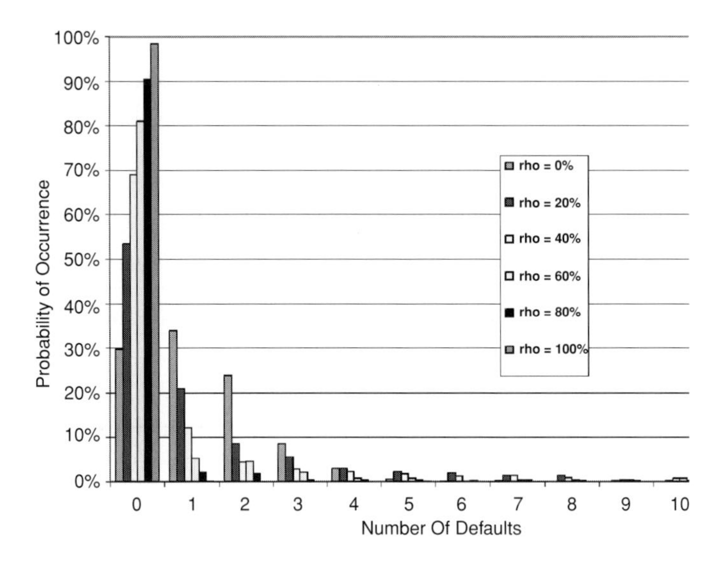

Figure 8.18 Distribution of default scenarios as a function of the default correlation.

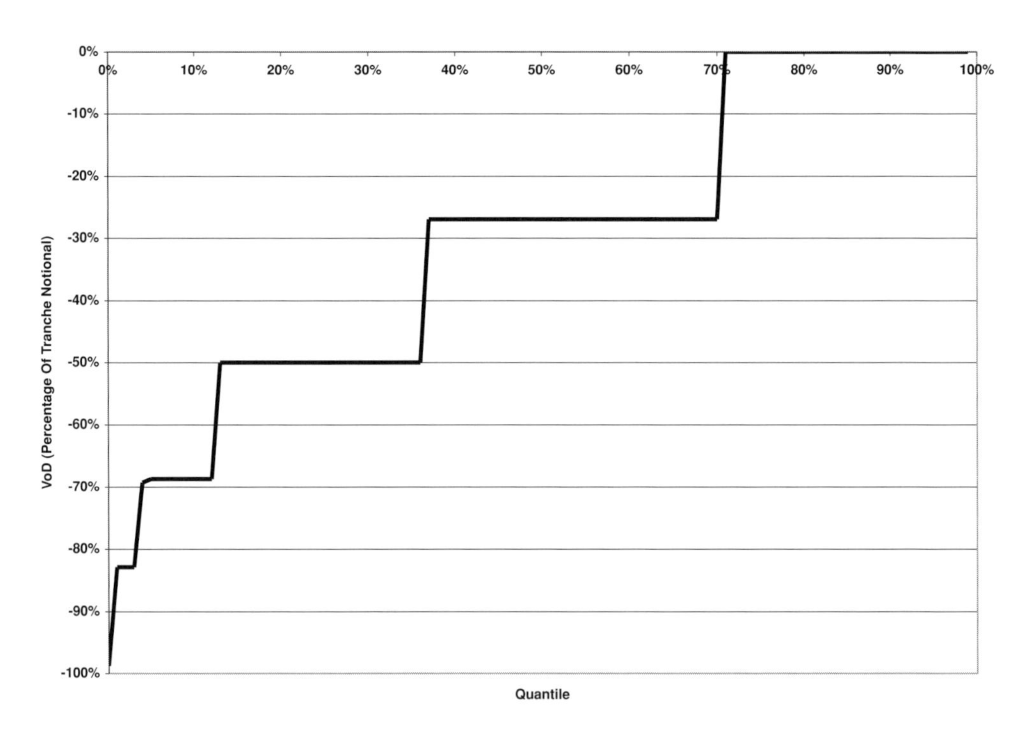

Figure 8.19 Cumulative VoD distribution for the 0–3% tranche ($\rho = 0\%$).

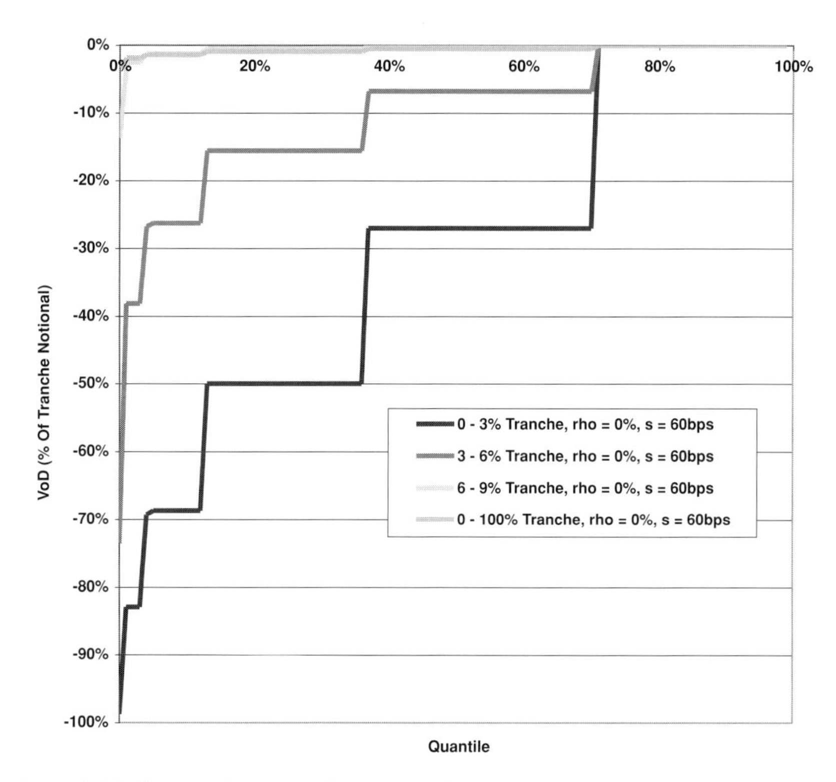

Figure 8.20 Cumulative VoD distribution for the different tranches ($\rho = 0\%$).

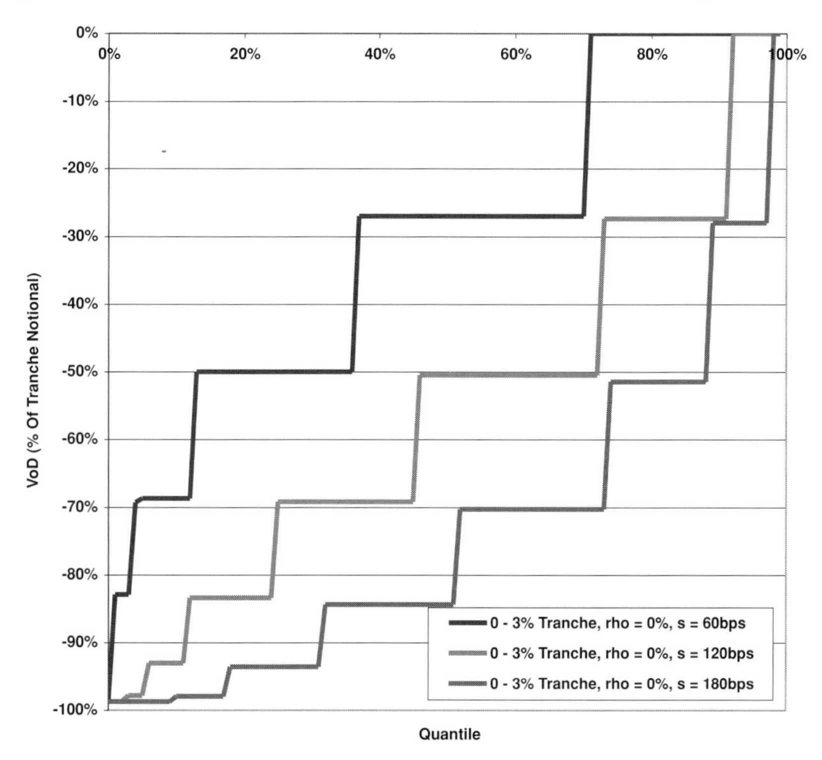

Figure 8.21 Cumulative VoD distribution for the 0–3% tranche for differing levels of risky obligor ($\rho = 0\%$).

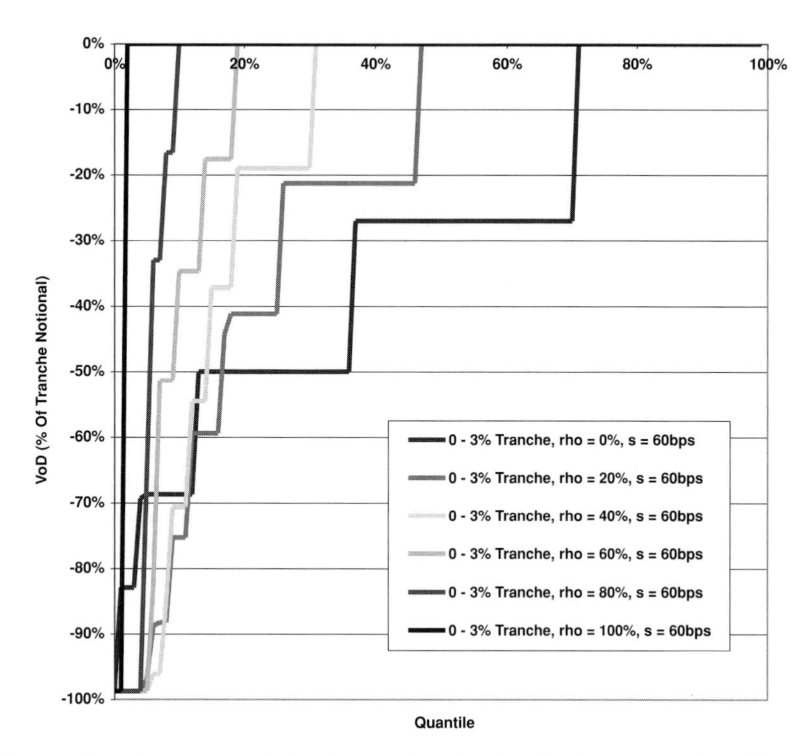

Figure 8.22 Cumulative VoD distribution for the 0–3% tranche for differing correlations.

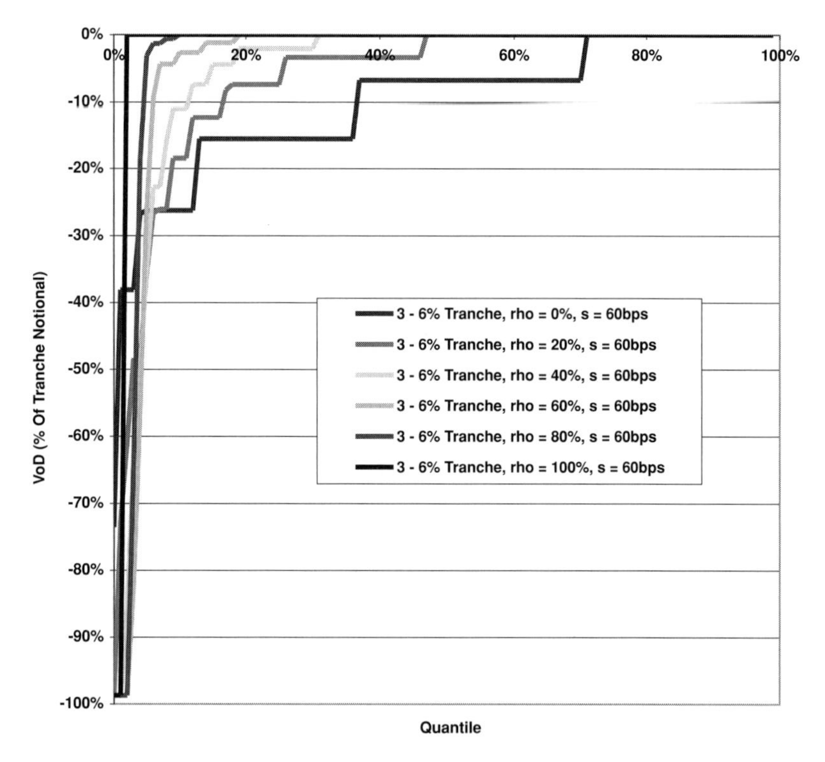

Figure 8.23 Cumulative VoD distribution for the 3–6% tranche for differing correlations.

simulations. The first point to note is the piecewise continuous nature of the plot. This is because the number of defaults is a discrete integer number (hence the VoD takes only a discrete number of values). In this particular example aproximately 30% of the simulations result in no defaults (hence a VoD of zero) and 0.2% result in seven defaults (enough to wipe out the equity tranche). From figures such as this we can read off a DVaR figure at the required quantile.

Figure 8.20 plots the cumulative VoD for the different tranches. As we would expect, the VoD decreases as we move up the capital structure due to the enhanced levels of subordination. Figure 8.21 plots the cumulative VoD for the 0–3% tranche for increasingly risky obligors. As expected, as the obligors become riskier, the number of defaults per simulation increases and so does the average VoD. Finally Figures 8.22 and 8.23 plot the cumulative VoD for the 0–3% and 3–6% tranches as a function of the correlation. For the 0–3% tranche we see that as the correlation increases, the cumulative VoD decreases. This is consistent with all that has been observed for equity tranches (as correlation increases the tranche becomes less risky).

Finally we note that in contrast to CVaR, the DVaR risk measure is only really meaningful over relatively long time horizons. For example, it is meaningful to ask how many obligors may default over a one-year time horizon, but it is not meaningful to consider a one-day horizon.

8.10 Chapter review

In this chapter we have analysed further the sensitivities of synthetic CDOs to fluctuations in their input parameters. Synthetic CDOs have a complex risk profile, being susceptible (simultaneously) to both market and credit risk factors. Several different measures were introduced to quantify the risk sensitivities including marginal CS01s, value-on-default, running value-on-default and correlation vega. The characteristics of these different risk measures were analysed as was the tranche time decay behaviour. Finally in this chapter we also introduced VaR measures based on credit spread and default scenarios to characterise potential tranche losses.

An important point to note is that we have used the semi-analytic model (based on the recursive method of portfolio construction) for most of the analysis in this chapter. Because of this choice of model we have not had to consider the effects of simulation noise upon the risk sensitivity calculations (which would have been a major concern had we been using the Monte Carlo valuation method). For vanilla products (with simple cashflow structures) the use of this model is perfectly acceptable. However, it should be noted that it may only be possible to value more exotic products with simulation methods. In these circumstances great care must be taken when estimating and interpreting sensitivities.

9

Implied and base correlations

9.1 Introduction

In previous chapters we have introduced and analysed the standard market model for valuing synthetic CDOs. In this and the next chapter we examine the natural consequences of what happens when a bespoke, illiquid market becomes standardised and highly liquid. Specifically with the introduction of standardised CDS indices and documentation specifying the terms and conditions of credit derivative contracts, the trading of credit exposure ramped up enormously. The subsequent introduction of exposure to tranched portions of these standardised portfolios also proved to be immensely popular with investors. At some point the nature of the market changed and it became possible to observe quoted prices for the standardised indices and tranches provided by the brokers and market makers.[1]

Once a market becomes sufficiently liquid to allow efficient price discovery for 'vanilla' instruments, the rationale for a model changes. In an illiquid market where there is no price guidance a model is the only means for determining an instrument's value. In a liquid market, however, the market observed prices must act as a target that a model is calibrated to. If a model does not accurately reproduce the market observed prices of vanilla instruments then there is a real danger someone using the model to price instruments will be arbitraged. The model, once calibrated, can then be used to provide price guidance for bespoke tranches. This is done by attempting to map bespoke tranches to standardised index tranches and interpolating prices accordingly (this is, in fact, the major challenge in the credit markets).

The structure of this chapter is as follows. In Section 9.2 we briefly outline the market standard conventions for quoting the price of tranche protection on the standard CDS indices. Section 9.3 then introduces the key observation that is at the heart of the portfolio credit derivative market: the correlation smile. This section also defines the implied correlation and provides some commentary as to why

[1] Indeed, standard index tranche prices are available on Bloomberg.

Table 9.1 *Illustrative quotations for standardised iTraxx tranches*

Tranche	Upfront payment	Running spread (bps)	Par spread (bps)
0–3%	30%	500	—
3–6%	—	—	200
6–9%	—	—	30
9–12%	—	—	15
12–22%	—	—	5
0–100%	4 bps	—	25

the correlation smile exists at all and why it represents a fundamental modelling problem. Having described the problem, the next step is to describe in Section 9.4 the solution that has currently been adopted by the market – base correlation. As we will see this is a solution that removes a number of the problems introduced by implied correlation, but this is at the expense of introducing other complications. Section 9.5 summarises the important points introduced in this chapter.

9.2 Market quoting conventions

As described in Chapter 6 the cashflows in a synthetic CDO contract include the fee payments and the contingent payments. In addition to these recurring, semi-periodic payments some tranches also trade with a one-off upfront payment and fixed running spreads. The convention in the market place is for the equity tranche of the standardised indices to trade with an upfront payment (calculated as a percentage of the initial – with no defaults – tranche notional) and a 'running spread' of 500 bps. The running spread is applied to the remaining tranche notional at each coupon payment date.

Table 9.1 shows a series of illustrative quotes for tranches defined on the iTraxx main index. The upfront payment for the equity tranche acts as compensation for the seller of tranche protection. This is because the equity tranche is immediately impacted by any losses in the underlying portfolio. Protection sellers therefore require additional compensation over and above a periodic coupon payment for providing default protection.

Mezz tranches tend to have tight bid-ask spreads due to strong demand for them. Senior and super-senior tranches tend to have wide bid-ask spreads (despite the fact that the expected losses for these tranches are almost certainly going to be very small). This is because there is low demand for these tranches and hence not much guidance is available from the market.

9.3 The correlation smile and implied correlation

Synthetic CDOs are valued by specifying as data inputs (as opposed to contractual inputs such as the maturity of the tranche) the following variables:

- marginal obligor survival curves (constructed from the term structure of par CDS spreads of each of the obligors),
- the obligor recovery rates,
- the obligor default correlation.

All of these inputs are of crucial importance to the valuation. The marginal survival curves determine when an individual obligor is likely to default and the recovery rates determine the loss-given-default. But it is the default correlation that glues all of the marginal behaviour together to determine the overall portfolio behaviour. The marginal survival curves can be constructed from market observed data. For a wide range of corporate exposures (particularly those corporate entities referenced in the standard iTraxx and CDX indices) these data are liquid and readily available from standard sources such as Markit Partners. The marginal recovery rates can be estimated from historical data although it is typically assumed that this parameter has the value of 40% for all the obligors. The default correlation, however, is altogether more difficult to estimate from historical data and represents a source of considerable uncertainty. Different market participants will have differing views on the likelihood of default contagion amongst the obligors in a portfolio.

Proceeding on the assumption that the spreads and recovery rates are standardised inputs that all market participants agree on, the only unknown model parameter left in the valuation is the default correlation. If we have market observed quotes (as outlined in the previous section) for the different index tranches, we can run the standard model in reverse. That is, we can compute the value of the default correlation which when input into the standard model reproduces the market observed prices. The value of the default correlation that gives a match between the market and model prices is what is known as the *implied correlation* (which is sometimes also referred to as the *compound correlation*). The correlation can be thought of as a 'dial' that is used to tune the model to the correct market value. The process is shown schematically in Figure 9.1. In the market for correlation products the implied correlation is used to communicate correlation views amongst market participants and can be thought of as the market's view of the shape of the portfolio loss distribution (since the correlation directly influences the portfolio loss distribution as seen in Chapter 7). It represents a mechanism whereby investors and traders can communicate these views (in much the same way implied volatility communicates the market's view of volatility in the Black–Scholes world [Hull 1999, Wilmott 2000]).

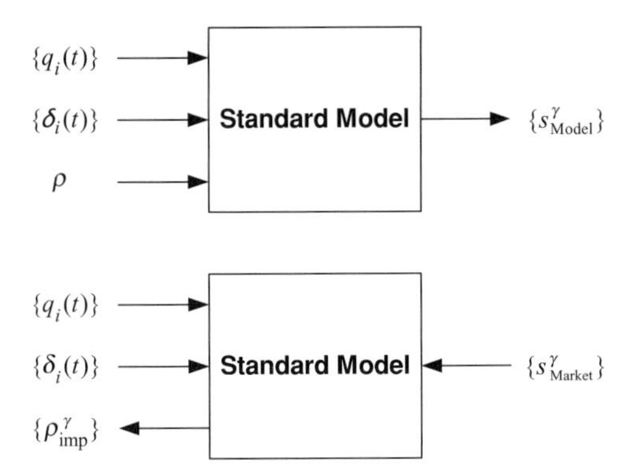

Figure 9.1 Illustration of the concept of implied correlation. In the top figure the single correlation ρ is an input into the model. The model output is a set of par spreads (and upfront payments) for each of the tranches γ. In the lower figure the process is run in reverse. The market observed par spreads (and upfront payments) are an input to the model. The output of the model is a set of implied correlations $\rho^{\gamma}_{\text{imp}}$, a separate correlation for each tranche in the capital structure.

The implied correlation is computed for each of the tranches independently. Figure 9.2 plots typical results of this calculation for the iTraxx capital structure. Also shown in this figure is the base correlation curve (which we will come back to later). Clearly the implied correlation is not constant across the capital structure. This means that the standard model cannot simultaneously price all tranches with a consistent/single value for the default correlation. Similar plots for different maturities yield the same basic phenomenology. Generally speaking the correlation smile has a 'hockey stick' structure. This observation provides a first insight into how to explain the origin (and perhaps future evolution) of the correlation smile as we will discuss shortly.

The key observation from Figure 9.2 is that the implied correlation decreases when moving from equity to mezz tranches, and then increases again when going from mezz to senior. The senior tranches require a higher correlation input than the equity or mezz tranches to reproduce the market observed spreads. As we saw previously in Chapter 7, as correlation increases this has the effect of increasing the mass in the tails of the portfolio loss distribution (increasing the probability of no losses and more importantly increasing the probability of large losses, thus impacting senior tranches). A higher implied correlation therefore implies that the market believes there is greater risk of losses eating into the senior tranches than the vanilla Gaussian copula model is predicting.

Is this result, that the correlation is not constant across the entire capital structure, unexpected? Yes. Does it represent a major modelling issue? Yes. Irrespective of the particular tranche and instrument maturity, the propensity for simultaneous

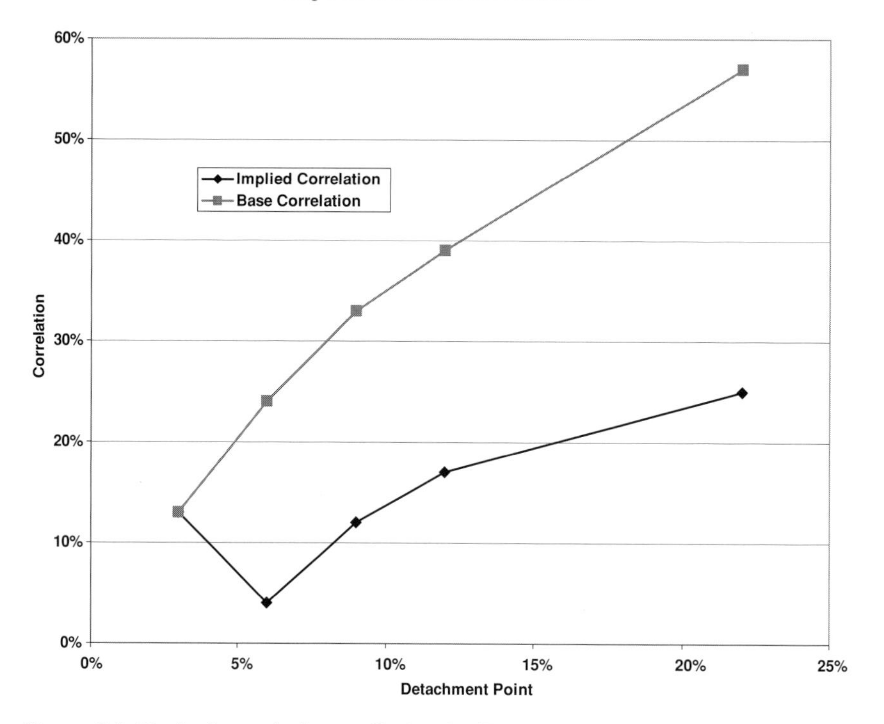

Figure 9.2 Typical correlation smile for the iTraxx capital structure.

defaults amongst the obligors in the pool should be a function of the interactions/dependencies between the obligors themselves and the coupling of these interactions to the prevailing economic environment. The correlations should not be a function of the particular tranche. This means that if the pricing model were a perfect representation of how to value synthetic CDOs, the implied correlations recovered from market observed prices using the standard market model should be the same for all tranches. Note however that implied correlations of tranches based on the different underlying portfolios of the different indices will not be the same. Also we would expect tranches defined on the same index but of different maturities to have different implied correlations since the market's view of default correlation may vary depending upon the term considered.

From the expression for the conditional default probability of an obligor in the single factor model[2]

$$q_i^{\text{Implied}}(t|V) = N\left(\frac{B_i - \rho_i^{\text{Implied}} V}{\sqrt{1 - \left(\rho_i^{\text{Implied}}\right)^2}}\right),$$

[2] ρ is the homogeneous obligor default correlation, V is the systemic factor and $B_i(t) = N^{-1}(q_i(t))$ is the default barrier level (q_i is the marginal default probability of the obligor which is time dependent).

(remembering that this quantity is the fundamental building block of the portfolio loss distribution in the semi-analytic models), we can see that the correlation smile implies the loss distribution of the pool will vary from tranche to tranche. This clearly cannot be the case since it is the same pool that is used to define the various tranches; the pool loss distribution cannot be a function of the different attachment/detachment points. Calculating the implied correlation is tantamount to uncovering the market's view of the shape of the portfolio loss distribution. What the correlation smile is saying is either that the market view of the portfolio loss distribution varies from tranche to tranche, or that the standard model does not accurately reflect how the market is pricing tranches. It is this latter possibility that seems more plausible.

9.3.1 Why does the correlation smile exist?

An initial explanation for the existence of the correlation smile focused on market supply/demand dynamics, specifically market inefficiencies and segmentation of investor preferences across the capital structure. With regard to investor preferences, investors in different tranches will hold different views on parameters such as default correlation. Supply/demand imbalances might be sufficient to explain some of the variation of the implied correlation across the capital structure, but it is hard to believe it could explain the magnitude of variation observed (or the persistence of the smile through time).

The most likely explanation for the existence of the correlation smile is that the market is not pricing the tranches using the same assumptions of the standard market model. For example, the standard model makes assumptions about deterministic, homogeneous default correlations, deterministic recovery rates, decoupling of default/recovery/interest rates etc. If these assumptions are not factored into the market's pricing of the tranches then there will inevitably be a mismatch between the model prices and the market prices. The question then arises as to how the standard model can be modified so as to move it more into line with the market's 'model' of pricing tranches (this will be the subject of the next chapter).

9.3.2 What are the problems with implied correlation?

If we accept that the correlation smile exists (and the empirical evidence shows clearly that it does exist) how can we make use of implied correlation? Although a simple concept to understand, there are a couple of fundamental problems with it. For an equity tranche, as ρ varies in the range $[0\%, 100\%]$, the tranche par spread varies in the range $[0, \infty)$. Additionally, for each value of ρ there is a single value for the par spread (since the par spread is a monotonically decreasing function

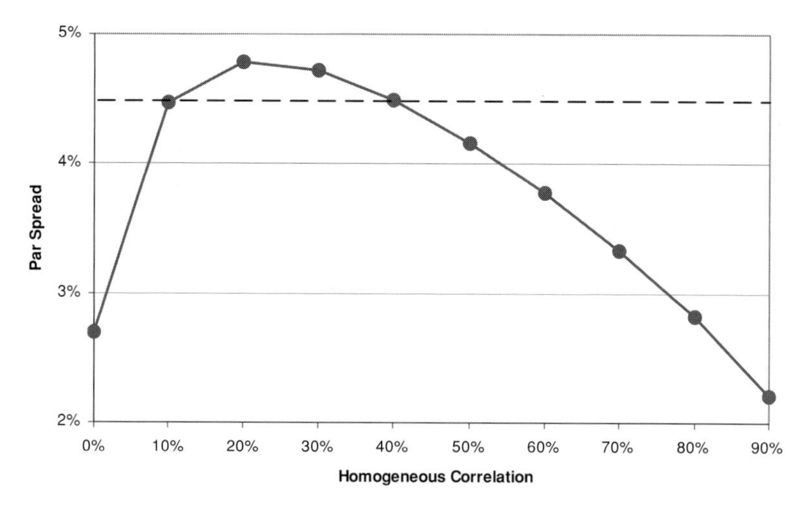

Figure 9.3 Graph showing the fundamental problems with implied correlation. For market observed spreads greater than approximately 4.75%, there is no solution for implied correlation. At a spread of approximately 4.5%, there are two correlation solutions, one at 10% and one at 40%.

of the correlation). In the language of implied correlation, whatever the market observed spread level of the equity tranche, we will always be able to transform this into a unique value for the implied correlation. Similar observations also apply for senior tranches (except in this case the par spread is an increasing function of the correlation). The difficulties arise with mezzanine tranches and are illustrated in Figure 9.3 (focusing on the 5 year maturity tranche).

The problems shown in Figure 9.3 for a 3–6% tranche are twofold. Firstly for market observed spreads greater than approximately 4.75% (in this particular example), the model will fail to return an implied correlation. For spreads less than this the model returns not one but two possible implied correlations which can recover the market observed spread. It is not clear which correlation would be appropriate. Due to the relatively shallow nature of the mezz tranche's spread correlation sensitivity, the implied correlation can also fluctuate by significant (and non-linear) amounts as the par spread fluctuates. In addition to these problems, which are fundamental artefacts of the nature of the tranching, there is the practical problem of how to value bespoke tranches using implied correlations. For example, if we have a 4–7% tranche referencing the iTraxx portfolio what would be an appropriate implied correlation to use as an input to value this tranche? The non-linear smile of the correlation dependence on detachment point makes it difficult to interpolate a correlation accurately, to say nothing of how to determine the correlation which is appropriate for the attachment point.

9.3.3 Calculating the implied/compound correlation

It is a simple matter to determine the implied correlation from market observed data. For tranche γ the PV is given by $V^{\gamma}(\rho) = -s^{\gamma} V^{\gamma}_{\text{Fee}}(\rho) + V^{\gamma}_{\text{Cont}}(\rho)$ where we have explicitly included the dependence of the fee and contingent leg values on the correlation parameter input. These are given by the usual expressions (ignoring for simplicity the upfront payment on the equity tranche):

$$V^{\gamma}_{\text{Fee}}\left(\rho^{\gamma}_{\text{imp}}\right) = \sum_{j=1}^{N} Z(0, T_j)\Delta_j\left[\left(K^{\gamma}_{\text{U}} - K^{\gamma}_{\text{L}}\right) - \bar{L}^{\gamma}\left(\rho^{\gamma}_{\text{imp}}, T_j\right)\right],$$

$$V^{\gamma}_{\text{Cont}}\left(\rho^{\gamma}_{\text{imp}}\right) = \sum_{j=1}^{N} Z(0, T_j)\left[\bar{L}^{\gamma}\left(\rho^{\gamma}_{\text{imp}}, T_j\right) - \bar{L}^{\gamma}\left(\rho^{\gamma}_{\text{imp}}, T_{j-1}\right)\right].$$

The correlation dependence arises through the conditional default probabilities for each obligor (which in turn arises from the formulation of the factor model).

Given an input for the market observed tranche par spread s^{γ} the implied correlation $\rho^{\gamma}_{\text{imp}}$ is defined such that $V^{\gamma}(s^{\gamma}, \rho^{\gamma}_{\text{imp}}) \equiv 0$, i.e. the correlation that prices the tranche at par. An iterative root-finding procedure may be applied to determine the value of $\rho^{\gamma}_{\text{imp}}$ which satisfies this condition. The bisection method is the simplest to use in this case (although not necessarily the quickest) [Burden and Faires 1989, Press *et al.* 2002]. A simple (and practical) test of the implementation of an implied correlation algorithm is to re-price the tranche using the computed value of the implied correlation; the tranche par spread should be equal to the original market input (to within machine precision).

It is important to note that the implied correlation is calculated in isolation for each tranche in the capital structure. That is, the values for the senior tranches are not derived from the mezz tranches which in turn are not derived from the equity tranche. This is in contrast to base correlation which we now discuss.

9.4 The market solution – base correlations

Implied correlation has a number of drawbacks, most notably that there are occasions when an implied correlation cannot be determined for certain mezz tranches and that it is difficult to interpolate values for tranches with non-standard attachment/detachment points due to the correlation smile. The first solution proposed to rectify these problems was the *base correlation* concept [McGinty and Ahluwalia 2004, McGinty *et al.* 2004, St Pierre *et al.* 2004, Willemann 2004, O'Kane 2005]. It has rapidly been adopted by The Street as the market standard methodology for quoting correlations.

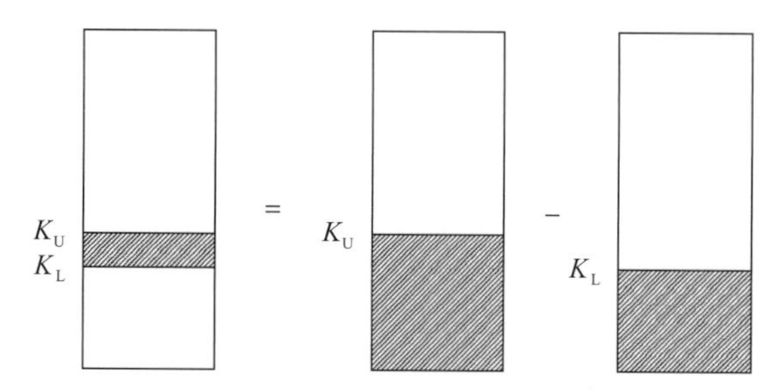

Figure 9.4 Schematic representation of a tranche's payoff as a call-spread option.

The base correlation concept implicitly accepts that there is no single correlation that simultaneously can fit all tranches and all maturities for a particular set of index tranches. The method treats non-equity tranches like a call spread option on the underlying portfolio loss. That is the payoff of a tranche with attachment/detachment points (K_L, K_U) where the cumulative loss is l is given by $L(l) = (l - K_U)^+ - (l - K_L)^+$.

This is represented schematically in Figure 9.4. The tranche payoff can be viewed as the difference between an equity tranche with a detachment point of K_U and an equity tranche with a detachment point of K_L. Base correlations are defined as the correlation inputs required for a series of equity tranches (also referred to as the base tranches) that give tranche values consistent with the observed spreads for traded tranches (which, of course, are not a series of equity tranches). Because base correlations are defined with respect to equity tranches with ever increasing levels of subordination, the limitations of implied correlations are immediately overcome. Specifically, for any level of market observed tranche spread it will always be possible to back out a unique base correlation (because the correlation sensitivity of the par spread of an equity tranche, irrespective of the width of the tranche, is always monotonically decreasing).

9.4.1 Computing base correlations

The base correlations can be computed via an iterative bootstrapping procedure (determining the base correlation for each detachment point in ascending sequence) as follows. Consider that we want to compute the base correlation curve $\{\rho_{\text{Base}}(K_U)\}$ for a capital structure with a set of detachment points $\{K_U\}$. For the iTraxx capital structure the detachment points of the liquidly traded tranches are 3%, 6%, 9%, 12% and 22%. For a tranche with attachment/detachment points (K_L, K_U) the PV of the tranche expressed as the difference between two tranches with attachment/detachment points $(0, K_L)$ and $(0, K_U)$ is given by [Parcell and

Wood 2007] (see also Torresetti *et al.* [2006] for further discussion and numerical examples)

$$V(K_L, K_U) = -s\,[V_{\text{Fee}}(0, K_U, \rho_U) - V_{\text{Fee}}(0, K_L, \rho_L)] + V_{\text{Cont}}(0, K_U, \rho_U) - V_{\text{Cont}}(0, K_L, \rho_L) - u$$

where we have included the upfront payment u in the calculation. The fee and contingent legs are computed from the usual expressions in Chapter 6 (and of course depend upon the portfolio and tranche loss distribution).

For the 0–3% tranche at inception the tranche is priced at par. The above formula therefore reduces to

$$0 = -s\left[V_{\text{Fee}}\left(0, 3\%, \rho_{3\%}^{\text{Base}}\right) - 0\right] + V_{\text{Cont}}\left(0, 3\%, \rho_{3\%}^{\text{Base}}\right) - 0 - u$$

(we are writing the correlation with the superscript 'Base' to distinguish it from implied correlation). The premium s is known (500 bps for the iTraxx equity tranche) and the upfront u is observed from the market. We can therefore solve this equation numerically to obtain $\rho_{3\%}^{\text{Base}}$. By construction we expect that $\rho_{3\%}^{\text{Base}} = \rho_{3\%}^{\text{Implied}}$.

For the 3–6% tranche we need to solve

$$0 = -s\left[V_{\text{Fee}}\left(0, 6\%, \rho_{6\%}^{\text{Base}}\right) - V_{\text{Fee}}\left(0, 3\%, \rho_{3\%}^{\text{Base}}\right)\right] + V_{\text{Cont}}\left(0, 6\%, \rho_{6\%}^{\text{Base}}\right) - V_{\text{Cont}}\left(0, 3\%, \rho_{3\%}^{\text{Base}}\right)$$

(the 3–6% tranche has no upfront payment). As before, the par spread for this tranche is known from the market so we can compute numerically the value for $\rho_{6\%}^{\text{Base}}$. The bootstrapping nature of the process is evident here: to compute $\rho_{6\%}^{\text{Base}}$ it is first necessary to have computed $\rho_{3\%}^{\text{Base}}$.

This process is continued until we have computed the base correlations for all the tranches for which there are market observed quotes. Re-pricing an index tranche with the base correlations ensures the model reproduces the market observed prices. Because the market observed par tranche spreads evolve on a daily basis, the base correlation curve also fluctuates daily.

The bootstrapping procedure requires that we only calibrate the model to the price of a single tranche in the capital structure at a time. This is to be contrasted with more general skew models (to be analysed in the next chapter) which attempt a simultaneous calibration of a model to all the tranche prices observed in the market. Base correlation implicitly accepts the existence of an inhomogeneous implied correlation (and therefore implicitly accepts the limitations of the standard model without trying to fix it). The base correlation curve is stripped from

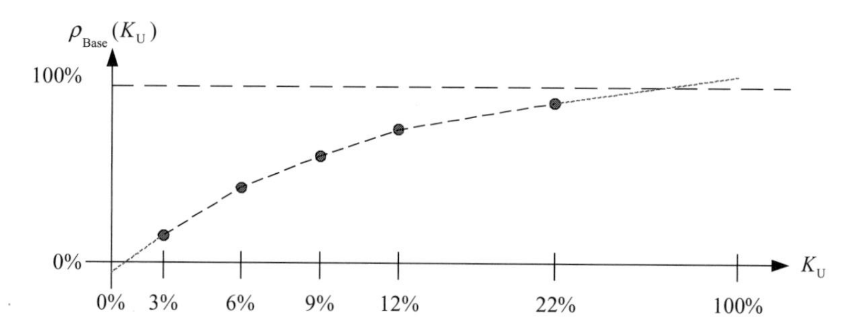

Figure 9.5 Interpolation of the base correlation curve (within the range $K_U \in [3\%, 22\%]$) and extrapolation outside of this range.

the market observed prices of index tranches. A 'steep' base correlation curve corresponds to more losses in the equity tranche, compared to higher up in the capital structure. A 'shallower' curve means that losses have shifted up the capital structure.

9.4.2 Valuation of bespoke tranches using the base correlation curve

Now the problems begin! The base correlation concept is very simple to apply to the standard tranches to obtain the base correlation curve. In principle it should be equally simple to value an index tranche with non-standard attachment/detachment points. Since the base correlation curve is monotonically increasing, the simplest approach will be to interpolate between the calibrated points.

For bespoke tranches with detachment points in the range 3–22% then linear interpolation between the calibrated points can be used. For detachment points outside this range some assumptions need to be made. For example we could assume that correlations below 3% and above 22% are simply extrapolated from the known values. This essentially assumes that the gradient of the base correlation curve continues without any discontinuity. This is shown schematically in Figure 9.5 This figure demonstrates that by simple extrapolation we may obtain $\rho^{\text{Base}}(0\%) < 0\%$ and $\rho^{\text{Base}}(100\%) > 100\%$ which is clearly incorrect. Other alternatives include artificially setting $\rho^{\text{Base}}(0\%) = 0\%$ and $\rho^{\text{Base}}(100\%) = 100\%$ and dividing the intervening range into an arbitrary number of intervals (much like how a temperature scale is defined), or artificially capping the base correlation in the range [0%, 100%]. The former choice is quite arbitrary and the latter choice will introduce a discontinuous change in the gradient of the base correlation curve (which as we will see shortly can have serious consequences).

Following Parcell and Wood [2007] we assume that we use linear interpolation and extrapolation to obtain the base correlation curve. At some intermediate detachment point k the base correlation for the $(0, k)$ tranche can be interpolated according to

$$\rho^{\text{Base}}(k) = \rho^{\text{Base}}(K_i) + \frac{k - K_i}{K_{i+1} - K_i}\left[\rho^{\text{Base}}(K_{i+1}) - \rho^{\text{Base}}(K_i)\right]$$

for $K_i \leq k \leq K_{i+1}$ where $K_i = 3\%, 6\%, 9\%, 12\%, 22\%$ for the iTraxx capital structure. We can write down a similar expression for the extrapolated values.

Valuing thin tranchelets at various points in the capital structure can highlight the problems that base correlation introduces. That is, we consider valuing a tranche with attachment/detachment points $(K_\text{L}, K_\text{L} + \delta K)$ giving a tranche of thickness δK. Remember that a sufficiently thin tranchlet (e.g. 1% thick) will straddle the calibrated base correlation control points (e.g. we can have a tranche with attachment point 2.5% and detachment point 3.5%).

The base correlations for the tranches $(0, K_\text{L})$ and $(0, K_\text{L} + \delta K)$ can be determined from the base correlation curve. The tranche can be valued using these base correlations and the expressions above for $V(K_\text{L}, K_\text{U})$ and the par spread obtained. Following this procedure it is observed [Parcell and Wood 2007] that for most of the tranchelets $s_{\text{Par}}(K_\text{L}, K_\text{L} + \delta K) > s_{\text{Par}}(K_\text{L} + \delta K, K_\text{L} + 2\delta K)$ etc. That is, as the amount of tranchlet subordination increases the par spread decreases as we would expect since the tranchlets become less risky as we move up the capital structure. However, there are occasions where this is observed not to be the case. These cases are typically observed around the calibrated control points when the tranchlet straddles the control point. This implies that a tranchlet with greater subordination is riskier than a tranchlet with less subordination! This is clearly incorrect and indicates a problem with the methodology (and not an arbitrage opportunity).

The reason for this inconsistency is nothing to do with the tranche valuation model or the base correlation concept. It is simply an artefact of the interpolation procedure used to obtain the base correlations. The interpolation matters because the gradient of the base correlation curve changes discontinuously between the control points. This is demonstrated schematically in Figure 9.6. Because the value of a $(0, k)$ tranche depends on the interpolated base correlation there may be occasions when the tranchlet straddles a control point leading to inconsistencies in pricing. A more sophisticated interpolation procedure, such as spline based procedures [Press *et al.* 2002] which ensure that the base correlation curve has continuous first and second derivatives, can potentially remove these problems. These methods tend to be good for $K_\text{U} \in [3\%, 22\%]$ (for the iTraxx index), but there are still problems outside this range. In general there is no a priori reason why any particular interpolation routine will guarantee a no-arbitrage base correlation curve (even smooth curves can lead to arbitrage opportunities).

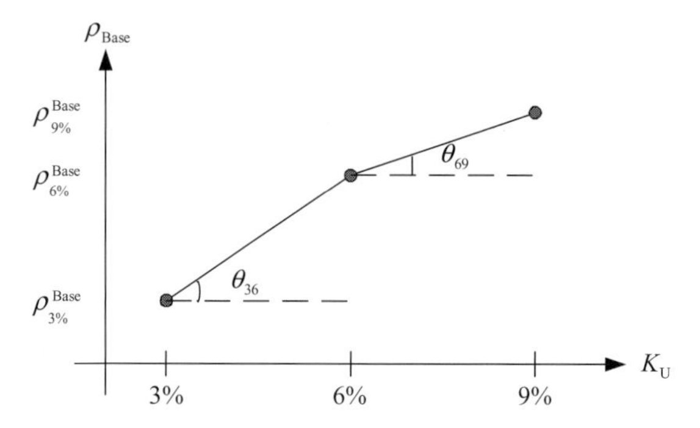

Figure 9.6 Schematic diagram showing how the gradient of the base correlation curve changes discontinuously ($\theta_{36} \neq \theta_{69}$) from control (detachment) point to control point.

In addition to the problems observed above (which are typically observed for mezz tranches), linear interpolation of base correlations can also lead to par spreads for tranchlets which are negative! Clearly this is also incorrect. See Parcell and Wood [2007] for further discussion of these problems with base correlation.

The reason why all of this matters is not for the valuation of index tranches. For index tranches with standardised pools of obligors (which remain static for 6 month periods at a time), maturities (3, 5, 7, and 10 years) and attachment/detachment points, good price discovery of market observed prices render a model's predictions irrelevant. The objective of a model is to ensure it replicates these observed prices. The standard model based on the Gaussian copula does not achieve this. Base correlation is a 'rescaling' of the problem to ensure that the standard model (with no modifications) can replicate market prices. The next chapter will assess more fundamental approaches to capturing the correlation skew. The problems arise when attempting to price bespokes. The general approach to bespoke valuation is to use information from the index market to determine a fair price for the bespoke. This mapping from index to bespoke is typically the biggest headache for managing synthetic CDO correlation books.

The base correlation approach outlined so far has fit a curve for a single tranche maturity. The approach must be extended to calibrate base correlation curves simultaneously for the different tranche maturities of 3, 5, 7 and 10 years. This results in a calibrated base correlation surface (see also Andersen [2006] and Sidenius [2006] for an alternative approach 'chaining' together tranches across maturities to resolve the term structure issue and also Pain *et al.* [2005]). This surface is used to interpolate prices for bespoke tranches (and also provide hedges). The issue now is to decide how to map a bespoke tranche onto an index tranche. An additional complication is in the fact that at contract inception we may be able to identify a suitable proxy base correlation surface (e.g. the iTraxx surface) which fits the bespoke well. As time progresses the base correlation curve evolves, and is used to provide an MtM for the bespoke tranche. This does not represent too much of a problem until the index rolls. At this point the on-the-run index becomes off-the-run. If the bespoke is continued to be hedged with the off-the-run index then problems with liquidity and observability (and hence the reliability of the base correlation curve) may become important.

In summary, the pricing of bespoke tranches in a manner that is consistent with standardised index tranches is an issue that currently has no market accepted solution. For more discussion on this topic the reader is recommended to consult what little research material is publicly available, for example, see Prince [2006] and Torresetti *et al.* [2007].

9.5 Chapter review

In this chapter we have introduced the concepts of implied and base correlations. Backing out implied correlations that fit market observed prices from the standard market model reveals the existence of a correlation smile. The correlation smile is a clear indication that the standard model has limitations. Implied correlation itself also has a number of shortcomings. The introduction of the base correlation concept has gone a long way towards resolving these issues. Base correlation has been so successful that it has become a market standard methodology (base correlations for standard index tranches are quoted on Bloomberg). However, it too has some limitations (mostly related to the details of its numerical implementation). The base correlation method calibrates tranches sequentially to market observed prices via a bootstrapping process. In the next chapter we will consider alternative models for capturing the correlation skew observed in the market where all the tranches are calibrated simultaneously to the market observed prices.

Calibration to market observed prices is important since the spreads of index tranches is the only information available for pricing bespoke instruments. Given that the vast majority of an institution's exposure is to bespokes tranches, determining how to best consistently utilise the market information available is incredibly important. This chapter has only briefly touched upon this important and difficult topic.

10

Extensions of the standard market model

10.1 Introduction

The existence of transparently observable, liquid index tranche prices means that valuation models now have a calibration target. If this target is not hit by a model then someone relying on that model for pricing will be susceptible to being arbitraged. However, the plain vanilla incarnation of the Gaussian copula model introduced and analysed in previous chapters is incapable of hitting this target. The model must therefore be extended to meet this new challenge. Some of the different methods that have been proposed for calibrating synthetic CDO tranches to the market are the topic of this chapter.

In Section 10.2 we briefly recap what the fundamental shortcomings of the standard Gaussian copula model are and then discuss how the basic model can be extended to overcome these shortcomings. There are two basic approaches for doing this. One approach is to change the dependence structure used to characterise the coupling between obligors (but still within the framework of the single factor model). Other approaches are more radical in their deviation from the standard market model, removing the assumption of deterministic correlation (stochastic correlation models) or using an entirely different framework to the copula models (gamma models). Section 10.3 then briefly discusses portfolio loss models. These models take the approach of treating the portfolio loss as the fundamental stochastic variable to be modelled. They represent the current state-of-the-art in synthetic CDO modelling. Because they are cutting edge, these models have yet to gain widespread acceptance amongst practitioners. That is not to say they will not in the future, but the Gaussian copula framework currently adopted in the marketplace has become firmly entrenched. We will therefore only outline these models and instead refer the reader to the original research papers for further details where they can form their own opinion as to the models' potential. Finally in Section 10.4 we review the important concepts introduced in the chapter.

10.2 Extending the standard market model

The fundamental shortcoming of the standard market model is that the Gaussian framework (with its flat correlation assumption) does not generate sufficient default events to value senior tranches accurately as well as undervaluing the equity tranche and overpricing mezz tranches. Because of the liquidity of the STCDO index market the rationale for a model is no longer to price tranches per se, but rather to reproduce the prices of liquidly traded tranches (using a single set of model parameters applied across the whole capital structure) and to understand the sensitivity of the price to fluctuations in the parameters so that the positions can be risk managed. The model, once it is calibrated, is then used for the valuation and risk management of bespoke tranches and these are priced consistently with the vanilla products that are their natural hedges.

One solution to the calibration problem is to increase the conditional marginal default probabilities until the model matches the market. An obvious way in which to facilitate this is to increase the default probabilities by modifying the distribution of the latent variable in the one-factor model. The marginal distributions of the obligors remain unchanged and are used to define the threshold below which defaults occur according to $N(B_i(t)) = q_i(t)$ for obligor i. If more mass is placed into the left-tail of $P(X_i)$ then this implies that there will be more occurrences of default events.

In this section we introduce a number of models that have appeared in the literature for 'tuning' the shape of the loss distribution until the prices produced by the model reproduce those observed in the market. An important point to note is that these models attempt to fit simultaneously the prices of all the tranches by suitable choice of parameters. This is in contrast to the base correlation approach introduced in Chapter 9 where each tranche is calibrated sequentially (moving up the capital structure) via a bootstrapping procedure.

The variations of the basic model discussed here introduce more parameters into the model. More parameters (simplistically) imply greater freedom to calibrate the model simultaneously such that it reproduces the market observed prices of standardised tranches. In practice, however, it is not always the case that an enhanced parameter space facilitates simultaneous calibration to all the tranches; if the model is fundamentally poor no amount of free parameters will enable calibration. Worse still, if a poor model with a large number of adjustable parameters does by virtue of its flexibility somehow calibrate to the market, it may give misleading sensitivities. A good model should therefore permit calibration with the minimum number of adjustable parameters. A detailed comparative analysis of different copula models can be found in Burtschell *et al.* [2005] and Claudio [2006].

It is also important to note that not all valuation methods are appropriate for calibrating to market observed data. Typically, to calibrate the model to specific data, many tranche revaluations must be performed. This therefore precludes Monte Carlo methods, and means the semi-analytic methods are necessary. In addition to this, if we are using a stochastic optimisation algorithm, then it is necessary to feed it deterministically generated tranche valuations (deterministic given a particular set of parameters), also precluding Monte Carlo methods.

10.2.1 The mixed copula

A very simple extension of the model, while still remaining within the framework of the normal copula, is to replace the systemic factor in $X_i = \rho_i V + \sqrt{1-\rho_i^2}\varepsilon_i$ with a mixture of Gaussian distributions [Wang *et al.* 2006, 2007]

$$V = \sum_{k=1}^{m} p_k V_k$$

where $V_k \sim N(\mu_k, \sigma_k)$ is a normally distributed random variable and p_k is the weight associated with the kth Gaussian (and $\sum_k p_k = 1$). The idiosyncratic factor is still a standardised normal random variable.

By a suitable choice of m, μ_i, σ_i, p_k a very wide range of distributions can be obtained (see Alexander [2001] and Alexander and Narayanan [2001] for some examples). This implies that there is a large amount of flexibility available to 'mould' the distribution of X_i to fit the market implied shape. In practice it is found that this method is capable of calibrating to quite a wide range of market observed prices. However, the number of free parameters in the model (means, volatilities and mixing probabilities for the m different normal distributions) is quite large meaning that it is necessary to make some quite arbitrary choices in the calibration procedure.

10.2.2 The Student *t* and double *t* copula

If the problem is tail events, then an obvious extension to the model is to replace the Gaussian copula with a Student t dependence structure [Cherubini *et al.* 2004]. This, by construction, has a higher tail dependence than the Gaussian dependence structure.

In this set-up the asset value is given by

$$X_i = \left[\rho_i V + \sqrt{1-\rho_i^2}\varepsilon_i\right]\sqrt{W}$$

where the systemic and idiosyncratic factors are normally distributed, but W is an inverse gamma distribution with parameter equal to $\nu/2$. The conditional probability distribution now becomes

$$q_i(t|V, W) = N\left(\frac{t_\nu^{-1}(q_i(t)) - \rho_i V}{\sqrt{W}\sqrt{1-\rho_i^2}}\right).$$

However the Student t distribution is symmetric meaning that the mass in both left- and right-tails increases compared to the Gaussian distribution. To remedy this it has also been proposed [Hull and White 2004] to model the asset value according to

$$X_i = \sqrt{\frac{\nu-2}{\nu}}\rho_i V + \sqrt{\frac{\bar{\nu}-2}{\bar{\nu}}}\sqrt{1-\rho_i^2}\varepsilon_i$$

where both the systemic and idiosyncratic factors are Student t distributed (with ν and $\bar{\nu}$ degrees of freedom respectively – the double t model). Under these distributions the conditional probability distribution is given by

$$q_i(t|V) = t_{\bar{\nu}}\left(\sqrt{\frac{\bar{\nu}}{\bar{\nu}-2}}\frac{H_i^{-1}(q_i(t)) - \sqrt{\frac{\nu-2}{\nu}}\rho_i V}{\sqrt{1-\rho_i^2}}\right)$$

where $H_i(X_i) = p_i(t)$ is the cumulative distribution function of X_i. This has to be computed numerically since the Student t distribution is not stable under convolution.

10.2.3 The normal inverse Gaussian (NIG) model

Normal inverse Gaussian distributions have been applied to the problem of synthetic CDO skew modelling by Guegan and Houdain [2005] and Kalemanova *et al.* [2005]. Normal inverse Gaussian distributions are a very flexible family of distributions (they are also considered as a special case of generalised hyperbolic distributions by Eberlein *et al.* [2007]). Depending on the parameterisation of the distribution, a wide variety of distributions (with varying degrees of non-normality) can be obtained. The NIG distribution is characterised by four parameters $(\alpha, \beta, \mu, \delta)$ making the model quite parsimonious. These parameters correspond to

- steepness α,
- symmetry β,
- location μ,
- scale δ.

The NIG$(\alpha, \beta, \mu, \delta)$ probability density function is given by

$$\text{NIG}(x; \alpha, \beta, \mu, \delta) \sim a(\alpha, \beta, \mu, \delta) q\left(\frac{x-\mu}{\delta}\right)^{-1} K_1\left(\delta\alpha q\left(\frac{x-\mu}{\delta}\right)\right) e^{\beta x}$$

Table 10.1 *Ten different possible NIG distributions*

Distribution	α	β	μ	δ	Mean	Variance	Skewness	Kurtosis
1	17.32	0	0	17.32	0	1	0	3
2	1	0	0	1	0	1	0	6
3	0.5774	0	0	0.5774	0	1	0	12
4	0.4201	0	0	0.4201	0	1	0	20
5	1.6771	−0.75	0.6	1.2	0	1	−1	6
6	1.6771	0.75	−0.6	1.2	0	1	1	6
7	1.4214	−0.8751	0.5455	0.7216	0	1	−2	12
8	1.4214	0.8751	−0.5455	0.7216	0	1	2	12
9	1.9365	−1.5	0.6	0.4899	0	1	−3	20
10	1.9365	1.5	−0.6	0.4899	0	1	3	20

where $q(x) = \sqrt{1+x^2}$, $a(\alpha, \beta, \mu, \delta) = \dfrac{1}{\pi}\alpha e^{\delta\sqrt{\alpha^2-\beta^2}-\beta\mu}$ and

$$K_1(x) = x\int_1^\infty e^{-xt}\sqrt{t^2-1}\,dt$$

(is a modified Bessel function of the third kind with index 1). The necessary conditions for a non-degenerate density are $\alpha > 0$, $\delta > 0$ and $|\beta| < \alpha$.

Table 10.1 shows 10 different possible NIG distributions depending upon the parameters specified [Guegan and Houdain 2005]. Distributions 1–4 have zero skewness (i.e. symmetrical) but increasing kurtosis. Distributions 5, 6 and 7, 8 and 9, 10 have non-zero and opposite skewness and excess kurtosis. Figures 10.1 and 10.2 plot these distributions. Clearly the NIG distribution can display a wide variety of shapes. A very nice property of the NIG distribution in that it is stable under convolution. That is if $X_1 \sim \text{NIG}(\alpha, \beta, \mu_1, \delta_1)$ and $X_2 \sim \text{NIG}(\alpha, \beta, \mu_2, \delta_2)$ then

$$X_1 + X_2 \sim \text{NIG}(\alpha, \beta, \mu_1+\mu_2, \delta_1+\delta_2),$$

i.e. the NIG distribution is stable under convolution (the sum of two NIG distributed random variables is also NIG distributed).[1] The scaling property of the NIG distribution is also simple. If $X \sim \text{NIG}(\alpha, \beta, \mu, \delta)$ then $cX \sim \text{NIG}(\alpha/c, \beta/c, c\mu, c\delta)$. From the moment generating function of the NIG distribution, which is

$$M_X(u) = e^{u\mu+\delta\left(\sqrt{\alpha^2-\beta^2}-\sqrt{\alpha^2-(\beta+\mu)^2}\right)},$$

[1] Consider two random variables X and Y with distribution functions $f_X(x)$ and $f_Y(x)$ respectively. The distribution of the sum $X+Y$ is given by their convolution $f_{X+Y}(x) = f_X * f_Y(x) = \int_{-\infty}^{+\infty} f_X(z)f_Y(x-z)dx$ [Grimmett and Stirzaker 2004].

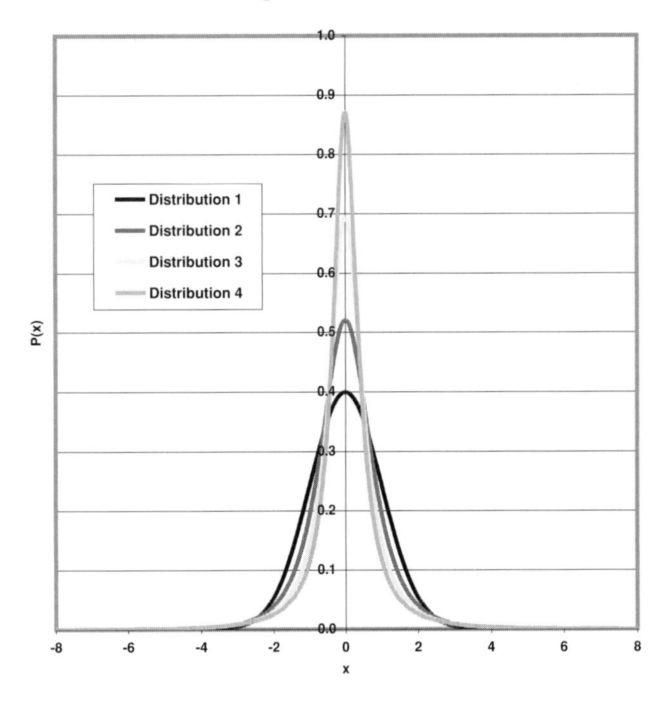

Figure 10.1 NIG distributions 1–4.

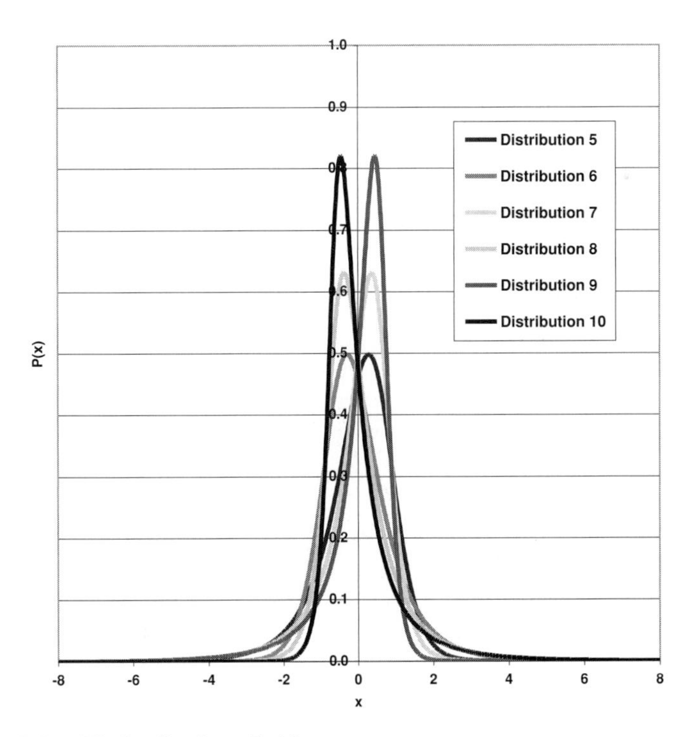

Figure 10.2 NIG distributions 5–10.

the first four moments of the distribution can be shown to be

$$E(X) = \mu + \frac{\delta\beta}{\gamma},$$

$$\sigma(X) = \delta\frac{\alpha^2}{\gamma^3},$$

$$s(X) = 3\frac{\beta}{\alpha}\frac{1}{\sqrt{\delta\gamma}},$$

$$\kappa(X) = 3\left[1 + 4\left(\frac{\beta}{\alpha}\right)^2\right]\frac{1}{\delta\gamma}$$

where $\gamma^2 = \alpha^2 - \beta^2$. These equations imply that $\kappa(X) > 3$ and

$$3 + \frac{5}{3}s(X)^2 < \kappa(X).$$

Consequently there exists a bound on the skewness relative to the kurtosis.

As $\alpha \to \infty$ we see that $\gamma^2 \sim \alpha^2$, hence $E(X) \sim \mu$, var$(X) \sim \delta$ (justifying the identification of μ and δ as location and scale respectively), $s(X) \sim 0$ and $\kappa(X) \sim 0$. Hence the normal distribution is a limiting case when $\alpha \to \infty$. The Cauchy distribution is obtained for $\alpha = 0$. When $\beta = 0$ the distribution is symmetric. $\beta \neq 0$ gives a distribution which has positive or negative skew.

If we are given the distributions moments $E(X), \sigma(X), s(X), \kappa(X)$ then the NIG distribution parameters are given by

$$\alpha = \sqrt{\frac{3\kappa - 4s^2 - 9}{\sigma\left(k - \frac{5}{3}s^2 - 3\right)^2}},$$

$$\beta = \frac{s}{\sqrt{\sigma}\left(k - \frac{5}{3}s^2 - 3\right)},$$

$$\mu = E - \frac{3s\sqrt{\sigma}}{3k - 4s^2 - 9},$$

$$\delta = \sqrt{3^3}\frac{\sqrt{\sigma\left(k - \frac{5}{3}s^2 - 3\right)}}{3k - 4s^2 - 9}.$$

Turning now to CDO valuation, following Kalemanova *et al.* [2005] we assume that the systemic and idiosyncratic factors have the following normal inverse

Gaussian distributions:

$$V \sim \text{NIG}\left(\alpha, \beta, -\frac{\alpha\beta}{\sqrt{\alpha^2-\beta^2}}, \alpha\right),$$

$$\varepsilon_i \sim \text{NIG}\left(\frac{\sqrt{1-\rho^2}}{\rho}\alpha, \frac{\sqrt{1-\rho^2}}{\rho}\beta, -\frac{\sqrt{1-\rho^2}}{\rho}\frac{\alpha\beta}{\sqrt{\alpha^2-\beta^2}}, \frac{\sqrt{1-\rho^2}}{\rho}\alpha\right).$$

Therefore the obligor return is also NIG distributed according to (using the scaling and convolution properties of the NIG distribution)

$$X_i \sim \text{NIG}\left(\frac{\alpha}{\rho}, \frac{\beta}{\rho}, -\frac{1}{\rho}\frac{\alpha\beta}{\sqrt{\alpha^2-\beta^2}}, \frac{\alpha}{\rho}\right).$$

The distribution of the latent variable and the systemic and idiosyncratic factors is therefore fully characterised by the parameter triple (α, β, ρ). Because the NIG distribution is stable under convolution we can determine the distribution of the latent variable analytically (without recourse to computing the convolution numerically). This is important from an implementation point of view given the need to perform multiple tranche revaluations to calibrate to the market spreads.

It is important to note that although the systemic and idiosyncratic factors are both NIG distributed random variables, they are in fact different parameterisations of the NIG distribution. Therefore the functional form could be quite different. It is also worth noting that it is possible to parameterise the two NIG distributions so as to introduce additional parameters. This should in principle allow greater flexibility in the calibration of the model if it proves to be necessary.

The default threshold for the obligor i is calculated according to $N_{X_i}(B_i(t)) = q_i(t)$ where

$$N_{X_i}(x) = \int_{-\infty}^{x} P(X_i')\mathrm{d}X_i'$$

is the cumulative distribution function (cdf) of the random variable X_i. The default threshold is computed from the inverse of this cdf. The conditional default probability is given by

$$P(X_i < B_i|V) = P\left(\varepsilon_i < \frac{B_i(t) - \rho_i V}{\sqrt{1-\rho_i^2}}|V\right),$$

i.e.

$$q_i(t|V) = N_{\varepsilon_i}\left(\frac{B_i(t) - \rho_i V}{\sqrt{1-\rho_i^2}}\right)$$

where

$$N_{\varepsilon_i}(x) = \int_{-\infty}^{x} P(\varepsilon_i')\mathrm{d}\varepsilon_i'$$

is the cdf of the random variable ε_i.

The portfolio conditional loss distribution is now obtained using the same recursive method outlined previously. The unconditional portfolio loss distribution is also obtained as before, where the integration is weighted by the probability density function of the systemic factor V (which has a different NIG distribution to X_i and ε_i). The marginal default probability can be recovered from the conditional default probability by

$$q_i(t) = \int_{-\infty}^{+\infty} q_i(t|V)P(V)\mathrm{d}V.$$

Note that this integration is weighted by the probability density function of the systemic factor V (which has a different NIG distribution to X_i and ε_i) and the integration scheme and limits of integration must be considered carefully as discussed previously.

In its current formulation of the double NIG model the distribution of the latent variable X_i is completely specified by the parameter triple (α, β, ρ). In terms of suitable values for these parameters we expect $0 < \rho < 1$, $|\beta| < \alpha$ and $0 < \alpha < \infty$. In practice we might further expect $\beta < 0$ (to give more weight to the left-tail of the distribution where the losses occur) and $0 < \alpha < 5$ (5 is chosen somewhat arbitrarily, but for this value of α the NIG distribution is starting to approach a normal distribution so anything larger than this and the model will start to become equivalent to the standard normal copula).

The simplest approach for determining suitable parameters is to perform a systematic search over this parameter space. Analysis of these results will identify the region(s) of parameter space where a simultaneous fit to the observed market spreads is obtained (if indeed such regions exist). The search can then be further refined in this region to close in more precisely on the best fit parameters. To quantify the fit more precisely, a 'cost' function can be specified. For example we could use

$$E\left(\alpha, \beta, \rho, \{s_k^{\text{Market}}\}\right) = \sum_{k=1}^{N_{\text{Tranches}}} \left(\frac{s_k^{\text{Model}} - s_k^{\text{Market}}}{s_k^{\text{Model}}}\right)^2$$

as a cost function. The objective of the calibration procedure is then to find the parameter triple (α, β, ρ) which minimises the cost function (simultaneously replicating the market observed prices of all the traded tranches). For the double NIG model there are only three independent parameters to vary to fit prices. For other models (such as the mixture of Gaussians model) there are more free parameters. In these cases determining suitable values for the parameters which simultaneously fit all the observed prices will require some form of optimisation procedure. As always, simulated annealing is a useful technique to apply (see Appendix B).

It is important to note that as $\alpha \to 0$ and for $\beta \neq 0$ the NIG distribution deviates significantly from the normal distribution. Care must be taken to ensure that numerical tasks such as the integration out of the systemic factor are performed correctly. Monitoring the price of the 0–100% tranche is an effective way of achieving this since the par spread of this tranche should be independent of the copula choice (because it is independent of the dependence structure between the obligors).

There are of course other copulas that can be used to characterise the dependence structure, for example the Archimedean and Marshall–Olkin copulas (see Cherubini *et al.* [2004] for more details). Each choice of copula has its own pros and cons in terms of range of dependence structures that can be generated (and ultimately the shapes of the portfolio loss distributions generated), the number of parameters that are used to characterise the copula and the ease with which the copula can be parameterised.

The overall objective of the choice of dependence structure is to facilitate calibration of the model to the market observed prices. As with any calibration problem (not just in finance but in all technical applications), there are certain desirable features of the process. Firstly the calibration parameters should be relatively stable over time. This is especially true in the current context of capturing market observed synthetic CDO prices. The calibration procedures for each of the models in this section all attempt to 'fine-tune' the shape of the distribution of the asset value (and by extension the shape of the loss distribution). Different parameters imply different shapes for the distribution. Calibrating the model to the market observed prices is therefore equivalent to uncovering the market's implied view of the portfolio loss distribution. It is difficult to believe that the market's view of the loss distribution would vary significantly on a day-to-day basis (shifting from uni- to bi-model and back again for example). The parameters of a calibrated model should therefore remain relatively stable over time, displaying only gradual evolution from their initial values. If the parameters are not stable, it is a likely indication that the model is not adequate. During periods of market stress (such as the correlation crisis of May 2005) models are pushed beyond their range of applicability and previously successful calibration procedures can break down. These periods of market turbulence can be very useful for driving the evolution of models (the same is true in natural sciences where considering the behaviour of models under extreme conditions – quantum mechanics and cosmology come to mind – leads to refinements of our understanding and the development of better models). Secondly the model itself should be relatively parsimonious. A model with a very large number of parameters (degrees of freedom) will very likely be able to calibrate to a wide range of market conditions. But its explanatory power will be limited. A far more desirable model is one which has a minimum number of free parameters to vary.

10.2.4 The functional copula

An alternative approach to those outlined previously is to invert the problem [Hull and White 2006a]. That is, rather than postulating a dependence structure and parameterising it so as to fit the observed market data, why not use the observed market data to infer what the distribution of hazard rate inputs into the model should be. This approach eliminates the need for making a parametric assumption regarding the dependence structure.

Starting from the single-factor model $X_i = \rho_i V + \sqrt{1 - \rho_i^2}\varepsilon_i$ we observe that there will be a large number of defaults when $X_i < B_i$ if $V \to -\infty$ (irrespective of the idiosyncratic factor) and a small number of defaults if $V \to +\infty$. The conditional probability of default in the one-factor model is of course given by

$$P(\tau_i \leq t|V) = 1 - N\left(\frac{B_i(t) - \rho_i V}{\sqrt{1 - \rho_i^2}}\right)$$

so the distribution of the systemic factor $P(V)$ acts as a proxy for the distribution of static default environments. If we assume that the term structure of hazard rates is flat then $P(\tau_i \leq t|V) = 1 - \mathrm{e}^{-h_i^V t}$ where the static pseudo conditional (on V) hazard rate is given by

$$h_i^V = -\frac{1}{t}\ln N\left(\frac{\rho_i V - B_i(t)}{\sqrt{1 - \rho_i^2}}\right).$$

The distribution of V therefore leads to a distribution of static, conditional pseudo hazard rates. In the current formulation of the model the systemic and idiosyncratic factors are normally distributed. There is in fact no requirement for this to be the case, and any distribution is possible. The key insight of the model is not to assume any parametric form for any of the variables.

The approach taken in [Hull and White 2006a] to determine the empirical distribution of hazard rates from market observed synthetic CDO data is as follows.

- Assume a series of possible default rates at the maturity of the instrument and extract the corresponding pseudo hazard rates.
- Revalue the CDO tranches for each pseudo hazard rate (these are tranche values conditional on the systemic factor).
- Calculate the unconditional tranche value as a weighted combination of the conditional values.

10.2.5 Stochastic correlation

The models introduced previously attempt to calibrate to the market by a somewhat arbitrary modification to the dependence structure. In all of these models the obligor default correlation is a flat, deterministic parameter. Stochastic correlation models aim to provide a somewhat more fundamental approach to actually capturing the features of the product and the market rather than simply fitting to the market [Gregory and Laurent 2003, 2004, Burtschell *et al.* 2007]. The fundamental insight in these models is that correlation amongst obligors is not constant through all time, and that it can in fact switch between different regimes depending upon the overall macro-environment. Stochastic correlation models therefore extend the existing framework to include correlation as an additional random variable. This introduces additional degrees of freedom into the model, permitting greater flexibility in the overall behaviour of the model (in turn implying greater freedom to calibrate to market observed prices).

In the vanilla one-factor model the correlation ρ_i is a deterministic parameter. Stochastic correlation models introduce a correlation which is stochastic, but independent of the factors (thereby being *state independent*). When the correlation is stochastic, and dependent on the factor, it is *state dependent*. The three state independent stochastic correlation model has an asset return of the form $X_i = \tilde{\rho}_i V + \sqrt{1 - \tilde{\rho}_i{}^2}\varepsilon_i$ where $\tilde{\rho}_i = (1 - B_S)(1 - B_i)\rho + B_S$. This implies that

$$X_i = [(1 - B_S)(1 - B_i)\rho + B_S]\,V + \lfloor (1 - B_S)(1 - B_i)\sqrt{1 - \rho^2} + B_i \rfloor \varepsilon_i$$

where the idiosyncratic ε_i and systemic V factors are standard normal random variables, B_S, B_i are Bernoulli random variables (independent of the systemic and idiosyncratic factors) and $0 \leq \rho \leq 1$. Writing $q = Q(B_i = 1)$ and $q_S = Q(B_S = 1)$ we can see that the marginal distribution of $\tilde{\rho}_i$ is discrete, taking a value 0 with probability $q(1 - q_S)$, ρ with probability $(1 - q)(1 - q_S)$ and 1 with probability q_S. The model therefore has three different correlation states: zero correlation, intermediate correlation and perfect correlation. The zero correlation state corresponds to a regime where idiosyncratic risk is the dominant factor (an important driving factor in the May 2005 correlation crisis), and perfect correlation to a regime where systemic factors are dominant (such as was observed in the summer of 2007).

Conditional on V and B_S the default times are independent. The conditional default probabilities can be written as

$$q_i(t|V, B_S) = (1 - B_S)\left[(1-q)N\left(\frac{B_i(t) - \rho V}{\sqrt{1-\rho^2}}\right) + q q_i(t)\right] + B_S \mathbf{1}_{V < B_i(t)}$$

where $N(B_i(t)) = q_i(t)$ is the marginal probability of default. Once the conditional default probabilities have been determined it is possible to compute the full loss distribution. We outline the details of this calculation following Burtschell *et al.* [2007] as it introduces another method for computing the conditional portfolio loss distribution. The accumulated portfolio losses at time t are given by

$$L(t) = \sum_{i=1}^{n} (1 - \delta_i)\mathbf{1}_{\tau_i < t}.$$

The probability generating function [Grimmett and Stirzaker 2004] for the losses is given by $\psi_{L(t)}(u) = \mathbf{E}[u^{L(t)}]$ which is

$$\psi_{L(t)}(u) = q_S \mathbf{E}\left[u^{L(t)}|B_S = 1\right] + (1 - q_S)\mathbf{E}\left[u^{L(t)}|B_S = 0\right].$$

Now

$$\mathbf{E}\left[u^{L(t)}|B_S = 0\right] = \int \mathbf{E}\left[u^{L(t)}|V = v, B_S = 0\right]P(V)\mathrm{d}V$$

(where $P(V) \sim \mathrm{e}^{V^2/2}$ is the Gaussian probability density up to a normalisation constant) and

$$\mathbf{E}\left[u^{L(t)}|V, B_S = 0\right] = \prod_{i=1}^{n}\left[(1 - q_i(t|V, B_S = 0)) + q_i(t|V, B_S = 0)u^{(1-\delta_i)}\right].$$

From these expressions the $\mathbf{E}[u^{L(t)}|B_S = 0]$ term can be evaluated by a numerical integration procedure. The $\mathbf{E}[u^{L(t)}|B_S = 1]$ term can also be evaluated (see the original paper of Burtschell and Stirzaker [2007] for details). Putting all of these components together allows the portfolio loss distribution to be computed.

The model has three free parameters to use to calibrate to the market: the idiosyncratic risk q, the systemic risk q_S and the correlation ρ (it is therefore rather parsimonious as a good model should be). Results reported by Burtschell and Stirzaker [2007] demonstrate that the model (characterised by the expression for the conditional obligor default probability) does indeed allow for a fairly good fit with both iTraxx and CDX market quotes. It is also demonstrated that the parameters are relatively stable over time (accounting for the credit market dislocations over the period around May 2005).

By introducing three discrete correlation states, stochastic correlation models start to remove the modelling assumption in the standard market model that default correlation between obligors is homogeneous. The practical consequence of homogeneous default correlation is that each element in the full correlation matrix $\mathbf{C}$ is approximated as $\rho_{ij} \to \rho$. There is evidence to suggest [Hager and Schobel 2006] that introducing more granularity into the correlation matrix can in fact allow the standard market model (without any modifications to the dependence structure or factor model formulation) to generate a correlation skew. This is an interesting possibility as it would indicate the correlation is the key mechanism driving skew.

10.2.6 Random factor loadings

In this extension of the one-factor model we write

$$X_i = m_i(V) + \sigma_i(V)\varepsilon_i$$

where $V \sim N(0, 1)$ and $\varepsilon_i \sim N(0, 1)$ for $i = 1, \ldots, n$. In the model of Andersen and Sidenius [2004] $\sigma_i(V)$ is a constant and in the local correlation model of Turc *et al.* [2005] $\sigma_i(V) = \sqrt{1 - m_i(V)^2}$. The simplest form of this model has

$$X_i = m + (l\mathbf{1}_{V<e} + h\mathbf{1}_{V\geq e})V + \nu(V)\varepsilon_i,$$

for some parameters l, h, e. The m and ν parameters are chosen to satisfy $\mathbf{E}[X_i] = 0$ and $\mathbf{E}[X_i^2] = 1$. This is equivalent to a random factor loading model since the pre-factor of the systemic factor is a state dependent random variable. It is therefore similar to the stochastic correlation approach.

As noted by Burtschell and Stirzaker [2007], stochastic and local correlation models do not 'sever the umbilical cord' with the Gaussian copula framework. They are a natural extension of it developed as an attempt to explain and fit the correlation skew. These models, like all Gaussian copula based approaches, suffer from consistency problems when attempting to calibrate simultaneously tranches across different maturities as well as from the more fundamental problems related to lack of credit spread dynamics and a replicating portfolio. Nevertheless, all the approaches introduced in this section succeed in further developing understanding of the weaknesses of the current modelling approach. For this reason alone they serve a very useful purpose.

10.2.7 Gamma process models

Some of the weaknesses of the Gaussian copula model (lack of credit spread dynamics, arbitrage, difficulties extending to other products) have been addressed recently by Baxter [2006, 2007]. The model in question is a structural model where default occurs when an obligor's value decreases below a certain threshold. The value of the obligor, in contrast to previous models, is driven by gamma (diffusion) processes. These processes essentially introduce instantaneous jumps into the obligor's value process; some of these jumps are small and some of the jumps are large corresponding to idiosyncratic and systemic effects. The obligor's value is given by

$$S_i(t) = S_i(t) \exp\left[-\Gamma_{\mathrm{g}}(t, \phi\gamma, \lambda) - \Gamma_{\mathrm{i}}(t, (1-\phi)\gamma, \lambda) + \gamma t \ln(1 + \lambda^{-1})\right]$$

where $\Gamma(t, \gamma, \lambda)$ is a pure-jump, increasing Levy process. The parameter γ is the rate at which jumps arrive and λ^{-1} is the jump size. The two gamma processes in the above equation correspond to global and idiosyncratic jumps respectively. ϕ represents the proportion of an obligor's value movements due to systemic events. Results reported by Baxter [2007] suggest that this model can fit market prices reasonably well (without a copula in sight). The model, however, has a significant advantage over the Gaussian copula in that it describes the dynamics of the correlated asset value processes in a consistent arbitrage-free manner. It remains to be seen how successful this approach will be in the long run. But its parsimonious nature and intuitive dynamics (the shocks to the obligor are driven by a cascade of local and global events) are certainly appealing.

10.2.8 Levy base correlation

The stochastic correlation models are an attempt to overcome the deficiencies of the vanilla Gaussian market model by introducing several correlation regimes. The NIG and other models modified the dependence structure. Levy base correlation, introduced and discussed by Albrecher *et al.* [2006], Garcia and Goossens [2007a, 2007b] and Garcia *et al.* [2007a, 2007b], is another variation on the dependence modification theme.

In this model [Garcia and Goossens 2007b, Garcia *et al.* 2007b] the latent variable representing the obligor asset value is constructed from two iid Levy processes modelling the systemic and idiosyncratic factors given by X_ρ and $X^i_{1-\rho}$ respectively (following the notation of the original papers). The asset value is $A_i = X_\rho + X^i_{1-\rho}$. The obligor is assumed to default if $A_i(t) = K_i(t)$ where $H(K_i(t)) = q_i(t)$ defines the asset barrier level and $H(x)$ is the cumulative distribution function of the Levy distributed random variables. From this the conditional default probability can be determined. We now construct the factors as $X_t = \sqrt{a}t - G_t$ where G_t is a gamma process (such that $G_0 = 0$ and the increments of the process are gamma distributed). The $\Gamma(a, b)$ distribution has a probability distribution function of the form $f(x, a, b) = b^a x^{a-1} \mathrm{e}^{-bx} / \Gamma(a)$. Having now specified the dynamics of the model, the recursive algorithm can be employed to construct the conditional and unconditional loss distribution. From the loss distribution the tranche fee and contingent legs can be computed.

Each of the Levy processes is represented as a shifted gamma process. This shifted gamma model has two free parameters, the drift a and the correlation ρ. The literature [Garcia and Goossens 2007b] reports that it is not possible to fit the full capital structure of tranche spreads simultaneously with just these two parameters. The authors therefore invoke the base correlation concept (hence the name of the method). There still remains the problem of how to specify the drift parameter (since it is not a base correlation related parameter per se, but a property of the shifted gamma process). The authors report results for the fitted base correlation values of the iTraxx for the 12–22% tranche (the rationale for this tranche being that it is the last in the chain to be bootstrapped and hence most sensitive to the previous calibrations). They find that, for a fixed tranche maturity, there is broadly good qualitative agreement between the results for the Levy and the normal copula models.

Fitting a base correlation surface across a range of tranche maturities demonstrates that at fixed detachment point, the Gaussian copula surface in the maturity dimension can display a smile structure. This makes interpolating prices for bespokes off this surface potentially difficult and possibly not arbitrage free. On the other hand, the surface calibrated with the Levy model displays no such smile (at a fixed attachment point). It is also observed that at long maturities (10 years) the base correlation curve (as the detachment point increases) displays a smile structure, with the 3–6% tranche having a smaller base correlation than the 0–3% tranche. It is postulated that the reason for this behaviour is that at such long maturities the relative benefits of subordination of the 3–6% tranche diminishes (as observed in Chapter 7).

10.2.9 Other modelling approaches

All of the different models described in this section are able to fit to the market data with varying degrees of success and ease. The choice of which model to use then becomes more subjective. The usual criteria of ease of calibration, number of free parameters used to calibrate, stability of calibration parameters over time and in different market environments, still apply. But in addition to this one of the key issues is the stability of the model sensitivities. The sensitivities determine the amount of hedging instruments purchased so they play a direct role in the overall P/L of the strategy.

In addition to the models outlined, there are a number of other different approaches that have been proposed to enable calibration to the market. These include the intensity gamma approach of Joshi and Stacey [2006], alpha-stable models [Prange and Scherer 2006], multi-factor models [Glasserman and Suchintabandid 2006], generalised Poisson processes [Brigo *et al.* 2007], normal gamma processes [Hooda 2006], a mixture of modelling the portfolio loss distribution as a function of three parameters and the hazard rates as stochastic [Longstaff and Rajan 2006], correlated variance gamma processes [Moosbrucker 2006], a structural model approach [Willemann 2005], extending the Gaussian copula to incorporate jumps [Xu 2006] and dynamic factor models [Inglis and Lipton 2007].

A very recent approach proposed by Hull and White [2007] takes a microscopic approach to modelling the behaviour of individual obligors and from this builds up the overall portfolio loss distribution. This is in contrast to portfolio loss models (to be discussed in the next section) which take a top-down approach to modelling, explicitly modelling the overall portfolio loss as the relevant state variable.

In this model the obligor survival probabilities follow deterministic drift processes with stochastic shocks in the form of jumps (assumed to be Poisson distributed)

$$-\mathrm{d}\ln S(0,t) = \mu(t)\mathrm{d}t + \mathrm{d}q(t)$$

where $\mathrm{d}q(t)$ is a jump process with intensity $\lambda(t)$ and magnitude H. In a short time interval Δt the probability of a jump such that $\mathrm{d}q = H$ is $\lambda(t)\Delta t$, and the probability of no jump is $1 - \lambda(t)\Delta t$. This formulation of the survival probability ensures that $S(0,t) \geq 0$ and $S(0,t) < S(0,t')$ for $t < t'$.

The probability of J jumps in the period $[0,t]$ is given by

$$P(J,t) = \Lambda(t)^J \frac{\mathrm{e}^{-\Lambda(t)}}{J!}$$

where

$$\Lambda(t) = \int_0^t \lambda(t')\mathrm{d}t'.$$

If there have been J jumps then the survival probability is given by

$$S(J,0,t) = \mathrm{e}^{-\int_0^t \mu(t')dt' - \sum_{j=1}^J H_j}.$$

Conditional on these J jumps, the probability of l defaults in the portfolio is given by

$$\Phi(l,t|J) = \frac{n!}{(n-l)!l!} q^l (1-q)^{n-l}$$

where $q(J,0,t) = 1 - S(J,0,t)$ is the default probability. The conditional (on the number of jumps) expected tranche loss is

$$\bar{L}(t|J) = \sum_{l=0}^{n} \Phi(l,t|J)L(l)$$

where $L(l)$ is the tranche payoff function. The unconditional expected tranche loss is

$$\bar{L}(t) = \sum_{J=0}^{n} P(J,t)\bar{L}(t|J)$$

and the tranche fee and contingent legs can be computed as normal. The model is clearly easy to implement and will be quick to perform tranche valuations.

To calibrate the model, several different variations can be considered. A formulation where the size of the jumps is modelled as $H_J = H_0 e^{\beta J}$ captures the empirical observations that defaults tend to occur more frequently in adverse market conditions [De Servigny and Renault 2002]. The authors report that the model is able to calibrate successfully the full term structure of index tranches. In addition to calibrating the model to the market, it is also shown how the model can be applied to forward starting CDOs, options on CDOs and leveraged super-senior tranches.

10.3 Dynamic portfolio loss models

Thus far we have considered modelling synthetic CDOs with no optionality. That is to say the protection seller and protection purchaser enter into a contract whereby they agree to swap cashflows from contract inception and the specification of the cashflows is fixed.

As the market for synthetic CDO products has evolved, the sophistication and subtlety of the views that market participants want to express has also evolved. A recent innovation in the marketplace has been the introduction of synthetic CDOs with embedded optionality, specifically forward starting synthetic CDOs and options on synthetic CDOs. To value an option on a tranche, for example, it is necessary to know the par spread of the tranche at some point in the future (and compare this with the strike to determine the option payoff). Unfortunately the standard market model based on the Gaussian (or its extensions) copula framework is incapable of capturing this information (although see Hull and White [2006b] who invoke an appropriate change of measure to cast the valuation problem into the Black option valuation model, Hull and White [2007] or Jackson and Zhang [2007], Totouom and Armstrong [2007]). This is because the copula framework is essentially a static one. The par CDS spreads of the individual obligors used to calibrate the marginal default curves are static, the generation of default times represents a single static default scenario, the correlation amongst the obligors is assumed constant. Once all these parameters are fixed, the portfolio loss distribution is also fixed and hence known with certainty at time zero.

Portfolio loss models represent the current state-of-the-art in synthetic CDO modelling. They are a move away from the Gaussian copula approach. The fundamental approach of these models is to model the dynamics of portfolio losses directly as the underlying stochastic variable (hence enabling products with loss dependent payoffs, such as options, to be valued). In addition to this, these models are designed to calibrate to the market by construction. The calibration is also to the full term structure of credit spreads (across the whole capital structure for differing tranche maturities to yield a calibration surface). The first such models were introduced by Sidenius *et al.* [2005] (the SPA model), Bennani [2005, 2006], Walker [2006, 2007 and Schonbucher 2006]. These models utilise technology introduced in the interest rate world (specifically HJM/BGM type formulations [Rebonato 1998, 2002, Brigo and Mercurio 2001]).

Portfolio loss models have a number of features which make them very attractive for portfolio credit derivative modelling. At the time of writing it is not clear how easy to use the models will be from a practical (calibration, implementation and performance) point of view. The reader is encouraged to consult the original research papers on this subject and perhaps attempt to implement the models to formulate their own opinion.

10.4 Chapter review

In this chapter we have considered a number of extensions to the standard market model that have been proposed in the literature. The motivation for these models has been to facilitate better calibration of synthetic CDO pricing models to market observed prices (in general it is observed that the standard model underprices the equity and senior tranches and overprices the mezz). Distinct from the base correlation approach, these models attempt to provide a more fundamental approach to explaining and reproducing the correlation skew observed in the market and also attempt to calibrate simultaneously to all the market observed prices (rather than bootstrapping one value at a time as in the base correlation framework).

The model extensions considered generally fall into one of three categories: extensions to the standard model by modification of the underlying dependence structure (Student t, NIG etc.), the introduction of more complex factor models (stochastic correlation or random factor loadings) or portfolio loss models. Each type of approach has its own strengths and weaknesses and it is difficult to conclude with any certainty which will eventually (if any) turn out to be the superior modelling approach. The reader is encouraged to consult the original research papers, implement the models and test them on real market data in order to arrive at their own view on the relative merits of each particular model.

In the next chapter we will turn our attention away from synthetic CDOs and consider some more exotic variations on the basic technology.

11
Exotic CDOs

11.1 Introduction

The majority of the text up until now has been concerned with plain vanilla synthetic CDOs (however the sobriquet 'plain vanilla' does not really do justice to the complexity of the product). In this chapter we will briefly consider a number of variations on the basic synthetic CDO/securitisation theme.

Section 11.2 will look at CDO squared and higher (cubed etc.) trades. These trades are a natural extension of the basic pooling and tranching concept where the assets in the underlying pool are themselves CDO assets. However, as we will see, the introduction of an additional layer of tranching introduces additional complexities into the modelling methodology and the risk management. Section 11.3 relaxes the constraint that the underlying collateral is CDSs and considers (typically cashflow) CDOs of general asset backed security (ABS) collateral. The marketplace for ABSCDOs is enormously diverse and a thorough treatment of this part of the structured finance world warrants a full textbook on its own. The discussion presented here will be necessarily brief. ABSCDOs have also received an enormous amount of coverage in popular media due to the credit crunch of 2007 and the role that securitisation of risky assets played in it.

Section 11.4 introduces CDSs where the underlying reference entity is not a corporate or sovereign entity, but is instead an ABS. This apparently simple extension of the basic CDS concept introduces a whole host of additional modelling complexities. The natural extension of ABCDSs is then to construct CDOs referencing pools of ABCDSs, so called synthetic ABSCDOs. This is the subject of Section 11.5 where we introduce the ABX indices and tranches defined on this index. Finally Section 11.6 reviews what has been introduced in the chapter.

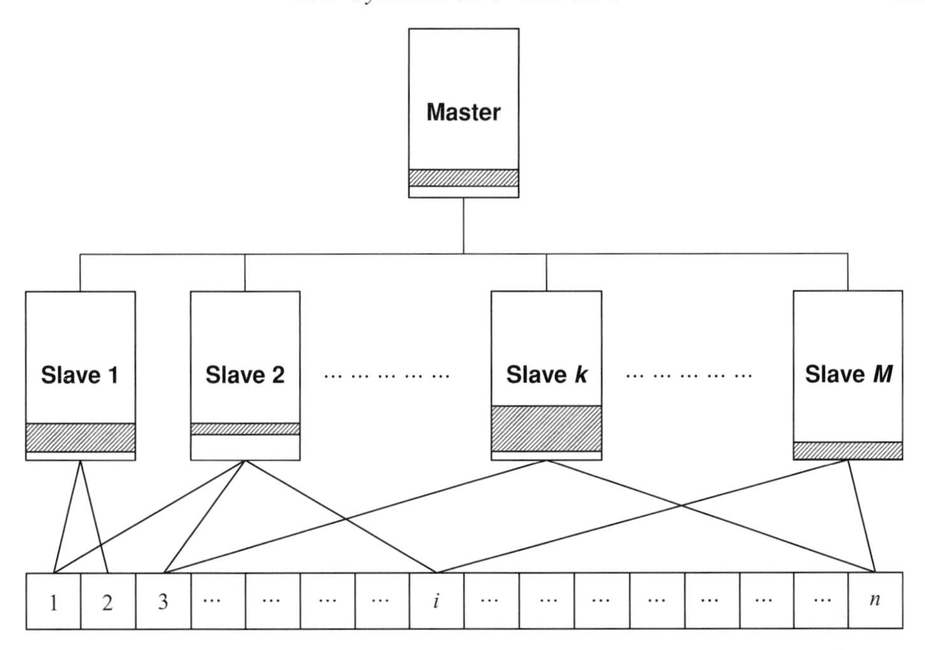

Figure 11.1 Schematic diagram of a synthetic CDO squared trade. At the lowest level there is an underlying universe of obligors $i = 1, \ldots, n$. These obligors are wired (connected) into a set of slave synthetic CDO tranches (each with their own amounts of subordination and tranche thickness). An individual obligor may appear in multiple slave tranches. The slave tranches are connected to the master synthetic CDO tranche. Losses in the underlying universe propagate upwards through the slave tranches and are aggregated together to impact the master tranche.

11.2 Synthetic CDO^2 and CDO^n

A natural extension of a synthetic CDO from a financial engineering perspective is a CDO squared. A plain vanilla synthetic CDO has a collateral pool which is composed of CDSs referencing individual obligors. A synthetic CDO squared (or CDO^2) is a (master) CDO where each of the assets in the underlying collateral pool is itself a (slave) synthetic CDO (see Rajan *et al.* [2007] for further details of the mechanics and investor rationales for trading these structures). Figure 11.1 shows schematically how a CDO squared is structured. At the lowest level we have a total pool (universe) of corporate obligors. At the next level is the pool of synthetic CDOs. Finally at the highest level is the master CDO itself. Each of the slave CDOs (also referred to as 'inner' or 'baby' CDOs) has their own attachment/detachment points. The master CDO also has a specified attachment and detachment point. When defaults occur amongst the obligors in the underlying universe of obligors, the associated losses are registered in the relevant tranches. The losses to the slave tranche are of course zero if the total cumulative losses over time do not exceed the attachment point of the slave tranche. The losses from the slave tranches are aggregated together to form an overall loss number. When this overall loss exceeds the attachment point of the master tranche the protection seller must compensate the protection purchaser. The notional amount of the master tranche is also reduced by the amount of losses suffered by the master tranche. The slave tranches therefore act as loss filters, buffering the master tranche from individual defaults (for example, an equity master tranche will not be impacted by defaults – in contrast to a vanilla CDO – if the slave tranches still have some subordination remaining to absorb the losses).

Synthetic CDO squareds are structurally identical to vanilla CDOs in terms of their cashflow mechanics. The only difference between a vanilla and a higher order synthetic CDO is the presence of additional layers of subordination. It is not difficult to conceptualise that higher orders of subordination (e.g. CDO cubed, quartic etc.) could be constructed, although their practical usefulness is not clear. Indeed with a number of asset backed CDO structures there are additional layers of subordination separating the master tranche from the underlying collateral. Typically the slave tranches are not actual physically traded assets, but only serve as a means to define the subordination structure of the master tranche. The additional layer of subordination results in additional leverage (amplification of spreads) compared to similarly rated vanilla synthetic CDO tranches.

Other extensions to the CDO squared structure allow for mixtures of different types of collateral in the pools (utilising a combination of STCDO and cashflow CDO technologies). This can include individual corporate entities, asset backed securities, ABSCDOs etc. It is easy to see that virtually any risk/return profile could be structured by appropriate choice of collateral. It is also easy to see that a CDO squared with a cocktail of ABSCDOs, synthetic CDOs and other types of collateral underpinning it could quickly become very opaque when trying to determine precisely what the fundamental exposure of the trade was to, for example, US sub-prime assets. However, in principle this type of CDO squared structure can increase asset class diversification with respect to an individual synthetic CDO.

CDO squareds are typically less sensitive to idiosyncratic (individual) default events than normal synthetic CDOs because of the additional layer of subordination. However, they are more susceptible to 'cliff risk' (the risk of suddenly falling over the edge of a precipice) because the default of a single highly connected obligor can have a significant impact upon the master tranche.[1]

[1] This is reflected in a probability of default for the master tranche which is 'small' but a loss-given-default which is 'large' (meaning that the tranche expected loss is approximately the same as a vanilla CDO referencing all the obligors).

11.2.1 Valuation using Monte Carlo simulation

The simplest and most transparent method for valuation of a synthetic CDO squared is to extend the Monte Carlo methods utilised for vanilla synthetic CDOs [Gilkes and Dressler 2003, Neugebauer *et al.* 2004a, 2004b]. For example, we would undertake the following steps to value a synthetic CDO squared tranche.

1. Generate correlated default times for all the obligors in the universe.
2. Determine the cumulative losses to the slave tranches over time.
3. Determine the cumulative loss to the master tranche due to this particular default scenario.
4. Compute the master tranche fee and contingent legs.

We assume that there is an underlying universe of $i = 1, \ldots, n$ separate obligors which can be referenced in the different slave CDOs. There is no restriction on the same obligor appearing in multiple slave CDOs (in fact it is very likely that this will be the case given that there is only a finite universe of obligors for which there are synthetic exposures). To characterise the connections between an obligor and the different slave CDOs we introduce a connectivity matrix $\mathbf{J}$ where the element $J_{i\gamma} = 0$ if obligor i is not referenced in (slave) CDO pool γ or $J_{i\gamma} = 1$ if it is.[2] Each obligor has a recovery rate given by δ_i and the survival curve of obligor i is denoted by $S_i(t, T)$. The slave or inner CDOs are denoted by $\gamma = 1, \ldots, m$ ($m \sim$ 5–15 is a typical value for the number of inner tranches) and the notional of obligor i in an inner CDO γ is N_i^γ. The inner CDOs have a set of attachment/detachment points $\{(K_{\mathrm{L}}^\gamma, K_{\mathrm{U}}^\gamma) : \gamma = 1, \ldots, m\}$ and the master CDO has attachment/detachment points $(K_{\mathrm{L}}^{\mathrm{M}}, K_{\mathrm{U}}^{\mathrm{M}})$. Obligor i's default time on simulation path α is denoted by $\tau_i^\alpha \in [0, \infty)$. The default times are computed from $\tau_i^\alpha = S_i^{-1}(N(\phi_i^\alpha))$ where $\vec{\phi}^\alpha = \mathbf{A}\vec{\varepsilon}^\alpha$ is a vector of correlated normal deviates, $\vec{\varepsilon}^\alpha \sim N(\mathbf{0}, \mathbf{I})$ is a vector of uncorrelated normal deviates and $\mathbf{C} = \mathbf{A}\mathbf{A}^{\mathrm{T}}$ is used to determine the Cholesky decomposition $\mathbf{A}$ of the correlation matrix $\mathbf{C}$.

For a particular slave tranche γ, the loss to the pool at time T_j which that tranche references is given by

$$L_\gamma^\alpha(T_j) = \sum_{i=1}^{n} N_i^\gamma (1 - \delta_i) \mathbf{1}_{\tau_i^\alpha < T_j} J_{i\gamma}$$

and the cumulative pool loss up to time T_j is $\tilde{L}_\gamma^\alpha(T_j) = \sum_{r=1}^{j} L_\gamma^\alpha(T_r)$. The tranche loss is therefore

$$\hat{L}_\gamma^\alpha(T_j) = \max\left[\min\left(\tilde{L}_\gamma^\alpha(T_j), K_{\mathrm{U}}^\gamma\right) - K_{\mathrm{L}}^\gamma, 0\right].$$

[2] We will look in more detail at the connectivity matrix $\mathbf{J}$ in Chapter 13.

The total loss that is propagated up from the slave tranches to the master tranche is $L_{\mathrm{M}}^{\alpha}(T_j) = \sum_{\gamma=1}^{m} \hat{L}_{\gamma}^{\alpha}(T_{\gamma})$ and the master CDO tranche loss is given by

$$\hat{L}_{\mathrm{M}}^{\alpha}(T_j) = \max\left\lfloor \min\left(L_{\mathrm{M}}^{\alpha}(T_j), K_{\mathrm{U}}^{\mathrm{M}}\right) - K_{\mathrm{L}}^{\mathrm{M}}, 0\right\rfloor.$$

This procedure is repeated a large number of times in order to compute the expected master tranche loss at each coupon date. We can then compute the fee and contingent legs of the master tranche using the usual (single-name CDS) expressions. It is also straightforward to compute other quantities for the master tranche such as the probability of default, tranche expected loss, loss-given-default etc.

This valuation method is simple to implement (although care needs to be taken to ensure that all the 'plumbing' between the individual obligors and the slave CDO tranches is implemented correctly). It is also flexible enough to deal with all sorts of additional bells and whistles that a particular trade may have, for example, if the CDO squared trade has cross-subordination. Cross-subordination means that the master tranche will not begin to register losses until the subordination of all the slave tranches has been wiped out. In an ordinary CDO squared, when the loss of a slave tranche exceeds the attachment point, the losses are propagated up to the master tranche. When cross-subordination is employed losses are only propagated up to the master tranche once the total losses across all the slave tranches exceed the total subordination. Cross-subordination reduces idiosyncratic risk (since a single obligor defaulting is unlikely to wipe out all of the subordination), but this is at the expense of increasing systemic risk. On the downside, the calculation of sensitivities is even more problematic than for a vanilla synthetic CDO. This is because there are now two layers of subordination that losses must eat through before the master CDO is impacted. To register non-zero losses at the master tranche level therefore requires more simulations than if there was only a single layer of subordination.

In terms of risk management, CDO squareds will have the same sorts of risk sensitivities as vanilla synthetic CDOs, namely spread sensitivities, correlation sensitivities and default sensitivity. The risk measures introduced in Chapter 8 can be used to quantify these risks. In addition to these risks we also have to quantify the impact of overlap between the pool constituents of the slave tranches. An obligor which is referenced in a single slave tranche will have less of an impact upon the risk than an obligor that appears in multiple tranches. This is an aspect of risk that we will consider more carefully in the context of risk managing portfolios of synthetic CDO tranches in Chapter 13.

11.2.2 Valuation using semi-analytic methods

The Monte Carlo valuation model outlined above is simple to understand and implement (it is particularly amenable to a clever object-oriented implementation), and will work. On the other hand, because of the additional layer of subordination the calculation of sensitivities will be very problematic, requiring large numbers of simulations to obtain stable estimates for the Greeks. It would be nice to be able to apply the semi-analytic methodologies developed for vanilla synthetic CDOs to this problem.

Unfortunately it is not straightforward to adapt semi-analytic methods directly to the valuation of CDO squareds and above. The reason for this is the possibility that the same obligor may appear in multiple slave tranches. Because of this the loss distributions of the slave tranches need not be conditionally independent, even if the defaults of the individual obligors are. Despite this difficulty, some researchers have extended the normal proxy [Shelton 2004] and recursive methods [Baheti *et al.* 2004, 2005] or developed alternative methods [Li and Liang 2005] to cope with multiple underlying tranches with encouraging results. The reader is encouraged to look at the original papers by these authors for more details.

11.3 Cashflow CDOs

11.3.1 Description of cashflow CDOs

Much of this book has focused on synthetic CDOs. Synthetic CDOs are not backed by cashflows of actual assets, but rather are linked synthetically to the underlying collateral by credit default swaps. The originator of the synthetic CDO does not have to purchase the assets physically since the exposure can be gained synthetically via the derivative security. This has been a key determinant in the popularity of the synthetic CDO form since not having to purchase the assets in the collateral pool greatly simplifies and speeds up the trading process.

Cashflow CDOs[3] on the other hand are backed by the real cashflows of 'true' assets such as loans (a collateralised loan obligation, CLO) or bonds (a collateralised bond obligation, CBO). Good discussions of different types of cashflow CDOs can be found in De Servigny and Jobst [2007], Rajan *et al.* [2007] and Lucas *et al.* [2007]. Since the liabilities are backed (meaning that the liability payments are met by the collateral on the asset side of the structure) by real cashflows, the originator of the cashflow CDO must physically purchase the assets. This leads to the need to ramp up the asset portfolio over time as well as warehouse the assets that have previously been purchased. Cashflow CDOs are either static where the underlying collateral pool does not change over time, or managed where the collateral pool composition can be actively evolved by a portfolio manager (who will charge a fee for their expertise). In a cashflow CDO the proceeds from investors (who purchase

[3] Including cashflow CDOs in a chapter on exotics is a bit disingenuous – they are not really exotic, and in fact the basic structure constitutes the majority of securitisations in the ABS world.

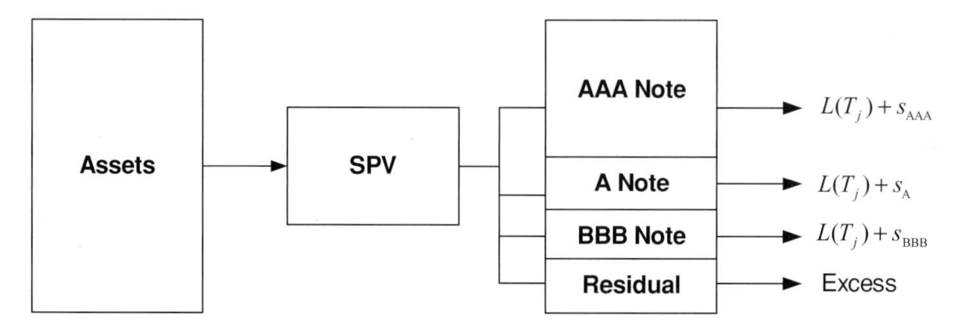

Figure 11.2 A typical cashflow CDO. On the asset side are the physical assets (e.g. loans) that make periodic coupon payments as well as repayment of principal at maturity. All of the cashflows are typically handled by an SPV which is a bankruptcy remote vehicle. On the liability side are the issued notes (of differing levels of seniority). At each payment (distribution) date T_j the aggregate cashflows from the asset side are fed down through a cashflow waterfall on the liability side, paying off the liabilities due at this time. Each of the notes pays a coupon of LIBOR plus a spread (based on the current remaining notional of the tranche).

the liabilities) are used to purchase the collateral (bonds, loans etc.) that makes up the asset side of the transaction.

Assets produce a stream of coupons as well as principal repayment (when the asset matures or prepays or re-finances etc.) or contingent recovery payments (when the asset defaults). The asset cashflows are used to meet the issued liabilities. Cashflow CDOs differ from synthetic CDOs in the manner in which the liabilities are met. Cashflow CDOs typically have complex waterfall structures where the income generated on the asset side is fed into the top of the waterfall and propagated down through it, sequentially paying off the senior liabilities before the junior liabilities (as well as paying management fees and hedging costs). What monies that are left remaining after all the liabilities have been met goes to the equity (residual) piece. The waterfall feature requires a match between the cashflows on the asset and liability sides of the structure. Figure 11.2 shows a schematic view of a typical cashflow CDO structure. The principal risks of a cashflow CDO to the investor in the CDO (the note holders who invest capital in the structure in return for periodic coupons and return of principal) are clearly interruptions to the stream of payments from the collateral portfolio on the asset side. This can occur for the following reasons.

- The assets defaulting (e.g. an obligor defaulting on a loan).
- The assets terminating (e.g. a corporate entity takes advantage of low interest rates to re-finance a corporate loan).
- The assets maturing. When the assets mature – without defaulting – the principal is returned. This represents returned cash that the CDO manager must reinvest. The risk

is that the new assets that are purchased may not perform as well as the asset that has matured and that the prevailing economic environment may not be favourable
- Assets gradually prepaying over time (e.g. individuals repaying their mortgage early).

In all these cases, if the cashflows from the collateral portfolio diminish, then the income to be distributed amongst the investors on the liability side will also diminish.

To protect senior investors, the senior notes in a cashflow CDO are protected by subordination and over-collateralisation. Senior notes are further protected by the presence of coverage tests (typically over-collateralisation and interest coverage tests). Coverage tests are designed to test whether the principal and interest liabilities of a note can be met. If there is insufficient cash coming in from the asset side to meet the liabilities for a senior note then cash will be diverted from paying off more junior tranches into paying down the notional of the senior tranche. At the bottom of the pile are equity (or residual) note holders. Once all the other liabilities have been met, whatever is remaining goes to the equity piece. If the assets on the asset side do not default or prepay then the return for equity note holders can be substantial. Conversely, if the assets under-perform the returns can be zero.

In a synthetic CDO the underlying assets are typically actively traded. There is therefore efficient price discovery in the marketplace for these assets. For a synthetic CDO the underlying obligor par CDSs are observed in the marketplace and used to construct their survival curves (thereby ensuring a lack of arbitrage for the underlying – individual – assets). In a cashflow CDO the assets in the pool are typically not so liquidly traded. Price indications for these assets typically come through knowledge of their ratings (and rating movements) or market implied probabilities of default (for example from a Merton-type model).

11.3.2 Example – a collateralised loan obligation

Just as synthetic CDOs are very homogeneous and consistent to model, cashflow CDOs are precisely the opposite. Each deal is extremely bespoke in terms of the composition of the collateral pool (including the mix of different asset classes) and also in terms of how the cash from the asset side is propagated down through the waterfalls on the liability side. It is difficult therefore to provide a general prescription for modelling a CLO, CBO or other ABSCDO. Instead we will focus on a specific example of a sample trade and use that as a template for how to approach similar trades.

A CLO is a bankruptcy-remote special purpose vehicle (SPV) incorporated for the only purpose to purchase and securitise assets (in the case of CLOs the collateral assets are generally senior secured bank syndicated loans). The underlying collateral is typically leveraged corporate loans. These loans receive periodic coupon payments as well as return of principal. The coupon and principal payments can be interrupted by early redemption of the loan (with the loan potentially being re-financed at a more favourable rate). Additionally the loan cashflows can be interrupted by an outright default of the obligor. An outright default has the impact of curtailing any further coupon payments and replacing the return of principal with a recovery payment.

The corporate loans typically pay floating coupons of 3M LIBOR plus a floating spread. Occasionally they pay monthly, i.e. 1M LIBOR plus a spread. The CLO pays quarterly the interest due on the liability tranches issued and all expenses for running the SPV, such as admin fees, trustee fees, management fees, etc. The interest cashflows from the collateral loans are paid into the interest collections account of the CLO. Any principal cashflows from the loans (generally principal repayments upon maturity and from re-financing and recovery amounts upon default), are paid into the principal collections account of the CLO. These accounts are held with the trustee.

CLOs typically have a three to five year reinvestment (replenishment) period. During this period, all principal proceeds held in the principal collections account are used by the CLO manager to buy additional collateral subject to the eligibility criteria stated in the indenture (or the offering memorandum for the CLO). All interest proceeds are used to pay the fees and expenses and the interest due on the CLO liabilities. The remainder goes to the junior most CLO tranche, the equity (or residual) tranche. The CLO liabilities pay a coupon of 3M LIBOR plus a floating spread, which is paid quarterly on each payment date, and is calculated as ACT/360.

The indenture of each CLO specifies cashflow diversion tests (over-collateralisation tests and interest coverage tests). If any of these tests are breached on a payment date, they have to be remedied by a mechanism which diverts cashflows from their normal path through the waterfall. Generally, interest proceeds are used first to cure an over-collateralisation or interest coverage breach by diverting all cash away from the junior tranches below the test breached towards principal redemption of the outstanding notional on the senior CLO tranches until the test is back in compliance. If interest proceeds are not sufficient, principal proceeds are used.

The interest and principal cashflows are propagated down through the cashflow waterfall for each coupon date. The end result of this is a temporal profile of the cashflows received by the different note investors on the liability side (e.g. the equity residual investors). From this set of cashflows quantities such as the internal rate of return can be determined and the sensitivity of this to variations in the input parameters and modelling assumptions (such as the likely default rate of the assets on the asset side) quantified.

11.3.2.1 Modelling defaults

The simplest approach to modelling cashflow CDOs makes a specific assumption about the occurrence of defaults in the asset pool (a scenario). This is usually in the form of a temporal profile, e.g. 1% of the pool will default in the first year, rising to 3% in year 2, falling to 2% in year 3 etc. This vector of default rates is used to amortise the total notional of the assets and the resulting cashflows to the liabilities are computed. This approach may seem crude and somewhat arbitrary but it is in fact reasonable given the complexity of the underlying collateral on the asset side. If there are hundreds or thousands of corporate loans in the asset pool, determining detailed information about each and every loan may be completely impractical. Aggregating all of the likely default behaviour together into a single vector of default rates simplifies the problem significantly and at least allows the modeller to explore the overall impact of different default assumptions (e.g. corporate defaults will increase next year due to the rising cost of credit) upon the CLO cashflows.

However, the problem with this scenario based approach is that the modeller is expressing a single view on a particular state of the world based on their subjective (but often highly experienced and insightful) view of the market. To understand better the likely future performance of the CLO, the modeller must look at different scenarios and try to assess the probabilities of these scenarios occurring to come up with an assessment of the most likely outcome (and base their investment decisions on this accordingly). Quantities such as expected internal rates of return used to market CDOs to potential investors will therefore have significant exposure to the modelling assumptions and also to the assumptions used to parameterise the model.

A simulation based approach would help to mitigate this situation by generating many different scenarios. In the spirit of the synthetic CDO modelling methodologies introduced in previous chapters we postulate the following simple approach for modelling corporate default times (an alternative, more sophisticated approach based on rating transition mechanics is postulated in Picone [2005] and De Servigny and Jobst [2007]). Default times for individual obligors are calibrated to fit the marginal default probabilities implied from market data for each obligor (or alternatively as input by the user). The marginal distributions are 'stitched' together using a Gaussian copula (although other copulas could be used if required) to form the multivariate joint default time distribution. This is the market standard approach for valuation of synthetic CDOs.

This approach is reasonable for CLOs where the underlying collateral is corporate loans. For corporate entities it is likely that there exists enough available information to make an informed guess about cumulative default probabilities. The same cannot be said for other types of ABS collateral such as AAA rated MBS bonds. In this case default is not likely to be a simple and clear-cut event, and instead the security will 'bleed' notional over an extended period of time (which can in fact be of the order of years). The simple Gaussian copula approach in this case will in all likelihood not yield useful information.

We assume that the obligor default correlation matrix $\mathbf{C}$ (of dimension $n \times n$) is available as an input. It is noted that determining this input is problematic – there is no liquid market for these instruments which can be used to imply a correlation from the observed price which is the case for synthetic CDOs. Furthermore it is assumed that the full correlation matrix of $n(n-1)/2$ distinct entries can be approximated by a single homogeneous correlation number ρ. This is a standard modelling assumption for synthetics. The Cholesky decomposition of the correlation matrix is computed from $\mathbf{C} = \mathbf{A}\mathbf{A}^{\mathrm{T}}$ and the matrix $\mathbf{A}$ is used to generate a vector of correlated random numbers $\vec{\phi}$ according to $\vec{\phi} = \mathbf{A}\vec{\varepsilon}$ (where $\vec{\varepsilon} \sim N(\mathbf{0}, \mathbf{I})$ is a vector of uncorrelated normal deviates). Default times $\{\tau_i : i = 1, \ldots, n\}$ are computed according to $S(0, \tau_i) = N(\phi_i)$, i.e. $\tau_i = S_i^{-1}(N(\phi_i))$ (inverting the survival curve) where $N(x)$ is the cumulative normal distribution function. Following this standard procedure allows us to generate default time scenarios for the assets on the asset side.

11.3.2.2 Modelling prepayments

Similar to the modelling of defaults, the simplest approach to modelling prepayments is to make an assumption as to the rate of prepayment of an asset, for example it pays down 10% of notional per year. This amortisation schedule is then applied to the asset's cashflows during the course of the simulated default scenario. Of course if the asset defaults before maturity the cashflows are halted and the recovery amount determined (based on the remaining notional at that time).

11.3.2.3 The asset side

We now describe the composition of the assets that make up the underlying collateral and how the cashflows from the assets are treated. We assume that there are $i = 1, \ldots, n$ separate corporate loans in the underlying asset pool. Let the notional amount of the collateral for obligor i be N_i, the coupon rate be c_i and the fraction of the notional amount recovered in the event of a default be δ_i. This implies that the loss given default for obligor i is $L_i = (1 - \delta_i)N_i$ and the recovery amount given default is $R_i = N_i \delta_i$. It is assumed that the recovery fraction for each obligor is an input. We assume that all of the asset coupons are of equal tenor (in practice this of course is not the case). Let the maturity of the assets be $\{T_i : i = 1, \ldots, n\}$ (the maximum asset maturity is given by $T_{\text{Max}} = \max(\{T_i : i = 1, \ldots, n\}$ which represents the maximum number of coupon dates). Let the survival probability curve for obligor i be given by $S_i(t, T)$ which represents the probability of survival as seen at time t until the time T. It is assumed that the survival curve for each obligor is available as an input.

The total notional amount of the collateral pool at time T_j is given by

$$N_{\text{Total}}^{\alpha}(T_j) = \sum_{i=1}^{n} N_i \mathbf{1}_{T_j < T_i} - N_i \delta_i \mathbf{1}_{\tau_i^{\alpha} < T_j}.$$

This formula reflects the reduction in notional of the asset pool due to principal redemptions when the assets mature (the first term) and also the reduction in notional due to assets defaulting (the second term). Once the default times $\{\tau_i^{\alpha} : i = 1, \ldots, n\}$ for a particular default simulation α are determined the asset cashflows can be computed at each coupon date T_j. Since the defaults are simulation specific, so too is the amortisation of the total asset pool notional (hence the time dependence in the above formula). We assume that the day-count fraction between coupon payment dates T_{j-1} and T_j is $\Delta_j = T_j - T_{j-1}$. We also assume that all the coupon payment dates are uniformly distributed. At coupon date T_j the interest, principal and recovery cashflows for an obligor i are given by

$$I_i^{\alpha}(T_j) = Z(0, T_j)[L(T_j) + c_i] N_i \Delta_j \mathbf{1}_{T_j < \tau_i^{\alpha}},$$
$$P_i^{\alpha}(T_{N_i}) = Z(0, T_{N_i}) N_i \delta(T_j - T_{N_i}) \mathbf{1}_{T_{N_i} < \tau_i^{\alpha}},$$
$$R_i^{\alpha}(\tau_i^{\alpha}) = Z(0, \tau_i^{\alpha}) N_i \delta_i \mathbf{1}_{\tau_i^{\alpha} < T_{N_i}},$$

respectively. The interest payments are based on a rate equal to LIBOR plus the coupon of the asset and are paid for as long as the asset has not defaulted. The principal repayment is made at the maturity of the asset (again assuming that the asset has not previously defaulted). If the asset does default then the recovery amount is collected (instead of the principal repayment) and further coupon payments are not made. Note that the notation here is slightly confusing. N_i where it appears multiplied by a discount factor $Z(0, t)$ is the notional amount of obligor i. N_i where it appears attached to a time, as in T_{N_i}, refers to the terminal cashflow (maturity) of the asset attached to obligor i.

At each coupon date T_j the cashflows (principal repayments, interest payments and recovery amounts) of all the assets in the collateral pool are aggregated according to

$$I^{\alpha}_{\text{Total}}(T_j) = \sum_{i=1}^{n} I^{\alpha}_i(T_j),$$

$$P^{\alpha}_{\text{Total}}(T_j) = \sum_{i=1}^{n} P^{\alpha}_i(T_j),$$

$$R^{\alpha}_{\text{Total}}(T_j) = \sum_{i=1}^{n} R^{\alpha}_i(T_j),$$

i.e. summing the cashflows from each asset on the asset side.

The total interest proceeds from the assets at each coupon date are placed into the interest collection account. The sum of the proceeds from principal repayment and recovery are placed into the principal collection account.

Monte Carlo simulation is used to simulate many possible default scenarios in order to estimate possible future cashflows on the asset side. For each default scenario and at each coupon date the proceeds from the asset side are used to meet the liabilities. In the current formulation of the model we are assuming that the tenor of all the assets and liabilities are identical. Of course in practice this will not be the case and will introduce additional difficulties for matching cashflows on asset and liability sides. From the ensemble of scenarios generated, expectation values for various quantities can be computed. For example, the expected IRR of the residual tranche as a function of the correlation assumption can be computed.

11.3.2.4 The liability side

The cashflows generated by the collateral on the asset side are used to meet the liabilities. Figure 11.3 shows a typical liability structure for a cashflow CDO. The interest cashflow $I^{\alpha}_{\text{Total}}(T_j)$ at each coupon date T_j is fed into the interest collections account at the top of the waterfall and propagated down through the rest of the waterfall. In this example we have assumed that there are two sets of fees: senior fees which are paid before any other liabilities and include payments to administrators (to cover for example running costs), trustees etc., and junior fees which are paid before the residual note holders. It is assumed that there are several different classes of note: AAA, AA, A, BBB and residual notes. The AAA notes are the most senior and the coupon payments due on this note $V^{\alpha}_{\text{AAA}}(T_j)$ are the first to be paid from the interest collection account.

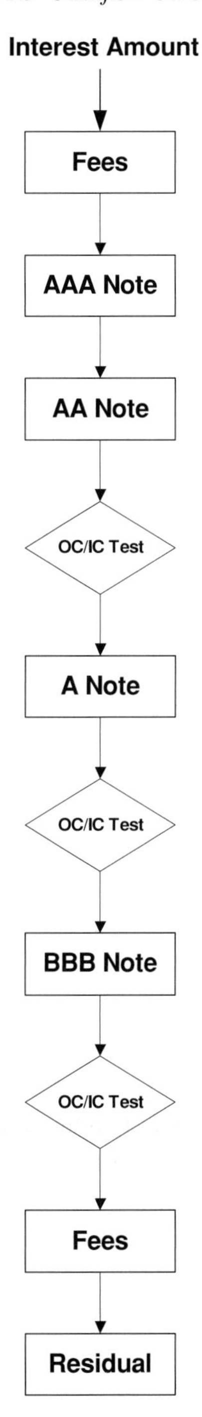

Figure 11.3 Schematic diagram of a cashflow CDO waterfall. Interest proceeds enter at the top of the waterfall and propagate down through the various fee and note payments and coverage tests.

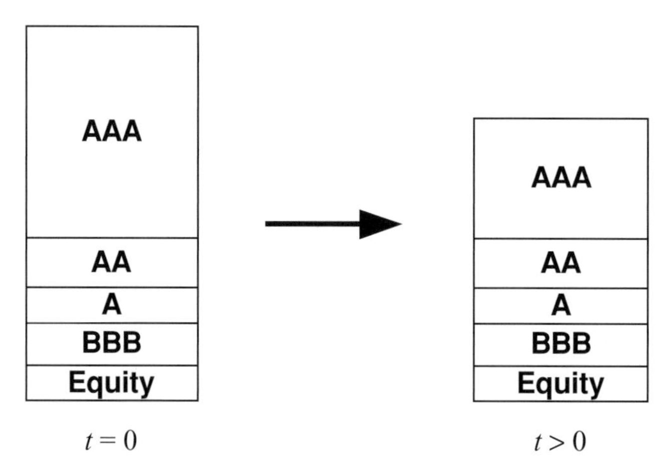

Figure 11.4 The capital structure on the left is the initial structure. As time progresses the notional amount of the most senior notes is written down in order to cure the OC and IC tests.

It is important to note that the capital structure, characterised by the notional amounts of the different notes, changes over time, Figure 11.4. For example, the initial capital structure of the notes may be of the form

$$
\begin{aligned}
N^{\alpha}_{\text{AAA}}(0) &= w_{\text{AAA}} N^{\alpha}_{\text{Total}}(0) \\
N^{\alpha}_{\text{AA}}(0) &= w_{\text{AA}} N^{\alpha}_{\text{Total}}(0) \\
N^{\alpha}_{\text{A}}(0) &= w_{\text{A}} N^{\alpha}_{\text{Total}}(0) \\
N^{\alpha}_{\text{BBB}}(0) &= w_{\text{BBB}} N^{\alpha}_{\text{Total}}(0) \\
N^{\alpha}_{\text{Equity}}(0) &= w_{\text{Equity}} N^{\alpha}_{\text{Total}}(0)
\end{aligned}
$$

where $w_{\text{AAA}} + w_{\text{AA}} + w_{\text{A}} + w_{\text{BBB}} + w_{\text{Equity}} = 1$ are the fractional sizes of the notes. The notional amounts of the different notes are reset to these values at the beginning of each simulation α. For a particular simulation α we have a certain amount of interest notional $I^{\alpha}_{\text{Total}}(T_j)$ generated by the assets on the asset side that is available at each coupon date to meet the liabilities. If at some stage of the propagation of this cash down through the waterfall one of the coverage tests (to be outlined below) fails, then cash must be freed up to bring the test back into compliance. The mechanism for doing this is to pay down some of the current (at time T_j) notional amount outstanding on the most senior notes, and working down the notes in order of seniority in order to cure the test. For example, the AAA/AA test in Figure 11.3 may fail. If we require \$10 m to bring the test back into compliance the first thing to do is to write down the remaining notional amount of the AAA note. If the remaining notional amount of this note is \$7 m, then this amount is written down (taking the notional of this note to be \$0) and we check whether we can write down some of the remaining notional of the AA note to cure the test. This results in a note notional that can reduce over time (and also vary from simulation to simulation depending upon when the assets default). The most senior notes are the first to be paid down (ensuring that the senior note holders receive their principal investment back) reflecting their relatively higher quality compared to the junior notes.

Upon the payment of a liability (fees or note coupon), the remaining amount in the interest collection account is updated. It is important to note that this amount is of course floored at zero (i.e. can never pay out more than what has been received).

11.3.2.5 Coverage tests

In a cashflow CDO senior notes are protected by subordination and over-collateralisation. Senior notes are further protected by coverage tests. Coverage tests are designed to ensure that there are sufficient resources (i.e. cash) to meet the note liabilities. If there is not enough cash, then cash has to be redirected through the waterfall (which is also referred to as the priority of payments) in order to bring the test back into compliance.

An over-collateralisation (OC) test quantifies the total remaining principal of the collateral securities on the asset side compared to the amount required to pay back the notional of the note being tested and for all the notes senior to it. The OC test for a particular note is essentially the ratio of the outstanding amount of cash left in the asset pool to the amount of liability notional outstanding on the current note and all notes senior to it. For example, consider an OC test for a note of rating R (e.g. an OC test for the AA note in the structure outlined in Figure 11.3). The OC test for this note (which will depend on the particular simulation and the coupon timenode) is given by

$$\mathrm{OC}_R^\alpha(T_j) = \frac{N_{\mathrm{Total}}^\alpha(T_j)}{\sum_{R' \geq R} N_{R'}^\alpha(T_j)}.$$

The sum in the denominator is over all the notes with a rating in excess of the note being tested (in the example given this would be the AAA and AA notes). In the contract indenture it will be specified what the critical values of the OC ratio should be. For a note of rating R we represent this as $\mathrm{OC}_R^{\mathrm{Critical}}$. It is usually the case that $\mathrm{OC}_R^{\mathrm{Critical}} > 100\ \%$ (e.g. 105%) to ensure that there is sufficient collateral in the pool to pay off the liabilities rated R and above.

If an OC test fails for a specific simulation at any particular coupon distribution date, i.e. $\mathrm{OC}_R^\alpha(T_j) < \mathrm{OC}_R^{\mathrm{Critical}}$ then cash must be diverted from the liabilities rated below R to cure the test. This is achieved by paying down the remaining notional of the liabilities rated R and above. To cure the test we need to pay down an amount $\Delta N^\alpha(T_j)$. This is calculated from

$$\mathrm{OC}_R^{\mathrm{Critical}} = \frac{N_{\mathrm{Total}}^\alpha(T_j)}{-\Delta N^\alpha(T_j) + \sum_{R' \geq R} N_{R'}^\alpha(T_j)}.$$

Once $\Delta N^\alpha(T_j)$ has been calculated the notional amounts of the rated notes are paid down. The first note to be paid down is the most senior, in this example the AAA note. The amount paid down from the AAA note is $\min[N_{\mathrm{AAA}}^\alpha(T_j), \Delta N^\alpha(T_j)]$, i.e. pay down as much from the AAA note as possible. After this note has been paid down to cure the OC test, the next most senior note is paid down. This process is continued until all of $\Delta N^\alpha(T_j)$ has been paid down, or all the note notional amounts have been repaid.

An interest coverage (IC) test proceeds similarly to the OC test and quantifies whether the interest proceeds from the assets in the collateral pool are sufficient to pay the fees and coupons of issued notes. If the test fails then notional amounts are again written down starting from the most senior notes. Writing down the notional amounts reduces the future interest liabilities, thereby improving the IC test ratio. Both the OC and IC tests have to exceed user-defined ratios (e.g. 105%) in order to pass. If the tests are not passed then the notional amounts of the notes are written down (starting with the most senior notes) as outlined earlier until the test is back in compliance.

It is worth stressing that the cashflow CDO introduced in this section probably represents the most vanilla structure trading in the marketplace. As with synthetic CDO modelling, it is recommended that the reader implement their own cashflow CDO model. This can be done very effectively for simple models in a spreadsheet (for example see Allman [2007], Tick [2007]). Once a simple model has been implemented more sophistication can gradually be incorporated. A word of warning is appropriate, however. Real-life cashflow CDOs can have a large amount of bespoke customisation and have many separate components to model (representing a lot of moving parts to keep track of). Spreadsheets are a good place to start, but soon the complexity of the model can make spreadsheet implementations difficult to maintain and susceptible to implementation errors (managing a $1 bn cashflow CDO from a spreadsheet is close to the wire). A code based solution is highly recommended. In particular the use of custom written macros [Roman 1999] for simple tasks is recommended. For more complex tasks DLLs or add-ins [Dalton 2005] are a very good means for incorporating the power of object oriented, compiled languages such as C++, with the user-interface and ease of use of a spreadsheet.

11.4 Asset backed CDS (ABCDS)

In this section we briefly digress from dealing with collateralised instruments to consider a variation of the basic CDS structure. The variation in question is asset backed CDSs [Davies *et al.* 2005, Flanagan *et al.* 2005, Lucas *et al.* 2007, Rajan *et al.* 2007, Schultz *et al.* 2007]. As was the case for a vanilla CDS, there are three parties to the ABCDS trade: a protection seller, a protection purchaser and a reference entity. In the ABCDS case the reference obligation that the contract refers to is no longer a simple bond of the reference entity but is an asset backed security (ABS). This seemingly simple change has enormous implications from a modelling and risk management perspective primarily given the potentially vast variety of collateral that can be securitised into an asset backed bond, and the different characteristics that this collateral can exhibit.

The ABCDS market grew rapidly with the introduction of standardised ISDA documentation in mid 2005. Standardising documentation and trades enables potential basis risk due to dealing with different counterparties (with different terms and conditions) to be eliminated. ABCDSs have a number of features that distinguish them from simple CDSs, specifically relating to the definition of a credit event, principal writedown (which can occur due to prepayment and default) and to the settlement mechanics. These features have previously hindered the development of the ABCDS market. The development of standardised ISDA documentation (that is acceptable to both parties in the transaction) has greatly enhanced the functioning of the ABCDS market and facilitated liquidity (although liquidity is still nowhere near the levels observed in the corporate CDS market). In developments that paralleled the evolution of the single-name corporate CDS market, standardised indices referencing a pool of ABCDSs were introduced in early 2006 (the ABX indices) and trading in standardised tranches referencing these indices began in 2007 (the TABX tranches to be discussed in the following section).

The reference entity in an ABCDS is an asset backed security. Asset backed securities are themselves, of course, securitisations of multiple individual assets. For example, the rights to the future cashflows of several thousand (illiquid) mortgages or home equity loans or credit card receivables may be packaged (securitised) up into notes of varying seniority (rating) and sold on to other investors. Much like a cashflow CDO, the cashflows from the assets in the underlying collateral pool (e.g. the repayments on a home equity loan) are used to service the liabilities (notes) that have been issued. The primary risks to the cashflows of an ABS therefore are events that will interrupt this stream of cashflows. The standard ISDA pay-as-you-go (PAUG) template for ABCDS recognises the following [Lucas *et al.* 2007]:

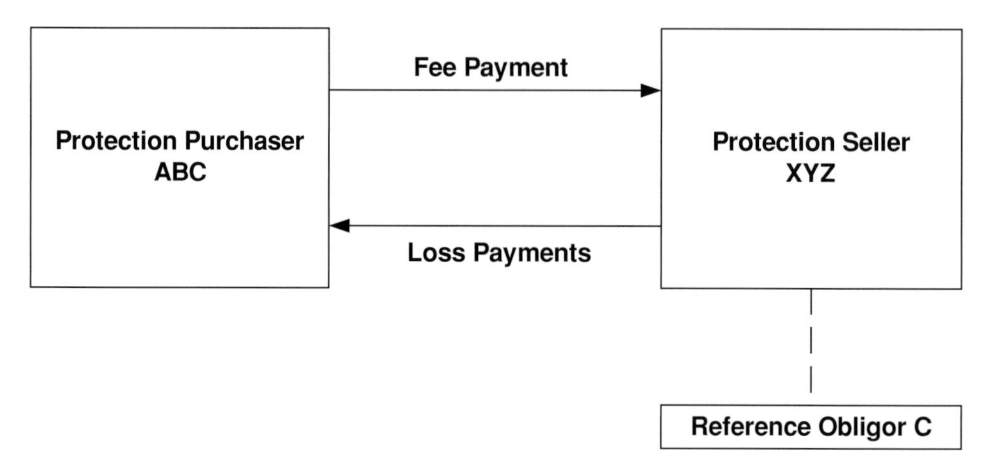

Figure 11.5 Schematic diagram of an ABCDS contract.

- failure to pay (the ABS fails to make a scheduled interest or principal payment);
- writedown (the principal of the ABS is written down and deemed irrecoverable);
- distressed ratings downgrade.

ABS securities typically have less room for manoeuvre in the event of credit events or writedowns than ordinary corporate CDSs. This is because a corporate can more readily restructure debt to avoid, for example, a missed payment. But an ABS relies upon a stream of cashflows from many individual debt securities. Asset backed securities may also suffer principal writedowns when the prepayments or losses to the security exceed the available subordination. Uncertainty in the timing of prepayments therefore also impacts the ABS. Because the ABS can prepay, 'failure-to-pay' is not appropriate to use to identify a credit event. Prepayment and default also mean that the overall notional amount of the ABCDS declines over time. Finally, the maturity of an ABCDS is typically assumed to be that of the longest maturity asset in the underlying ABS pool. For some assets, such as mortgages, this can be very long indeed (of the order of decades).

Figure 11.5 shows a schematic diagram of the cashflows in a typical ABCDS contract. The protection purchaser pays the protection seller a periodic fee (coupon) payment. In return for this protection, the protection purchaser receives payments from the protection seller covering cashflow shortfalls resulting from failure to pay or writedowns of the underlying reference ABS. If a writedown is eventually reversed the protection purchaser must return the payment. The protection (loss) payments are made on an ongoing basis (PAUG) as-and-when cashflow shortfalls occur. There can therefore be multiple contingent payments. This can be converted to physical delivery (the protection purchaser delivers the ABS to the protection seller in return for par). This settlement mechanism should be contrasted with that for a corporate CDS where there is only a single contingent payment upon a credit event. Cash settlement is also possible where upon the default event the protection seller delivers par minus the recovery rate to the protection buyer (the recovery rate is determined by dealer poll). The difficulty with cash settlement is that in a distressed market it may be difficult to obtain reliable, transparent or accurate marks for the value of the ABSs (the problem of mark-to-market versus mark-to-make-believe).

PAUG settlement differs from cash and physical settlement in that a single payment (cash) or exchange of bond for par (physical) may not take place. Under PAUG, protection sellers make contingent cash payments equivalent to any writedowns on the bond as the writedowns on the bond occur. Part of the rationale for PAUG is that the protection purchaser does not want to wait until the final legal maturity of the ABCDS (which could be many years) in order to receive their contingent payment. It is also the case that writedowns and default events in the ABS world happen on much longer timescales than in the corporate credit world. The emergence of CDSs referencing corporate assets enabled holders of the corporate bonds (or debt) to hedge their exposure. In a similar fashion ABCDSs enable investors to hedge their ABS exposure or short the market or take on risk synthetically (without owning the ABS asset) by being a seller of protection.

In terms of valuation of ABCDSs, a lot of the technology from the corporate CDS world is directly transplanted. For example the MtM of an ABCDS would be determined from $\text{MtM}(t) = [s(t) - s(0)]\text{dV01}$. The spreads are observable in the market. But the real difficulty is in determining the dV01. For a corporate CDS this is simply the survival probability weighted coupon payments. The same will apply to the ABCDS, but now it is necessary to model the ABS payments. This is not an easy task. It would in principle be possible to apply synthetic CDO techniques, as outlined in the section on cashflow CDOs (using Monte Carlo simulation to model different default and prepayment scenarios – clearly the timing of cashflow events impacting the ABS asset impacts the overall valuation), but the main problem with this approach is how to parameterise the survival curves of the underlying collateral and how to model the correlation between default timings (not to mention the computational difficulties associated with modelling the behaviour of very large pools of collateral).

11.5 ABX indices and tranched ABX (TABX) exposures

In this section we briefly introduce the ABX indices of ABSs and the tranched exposures to these indices (TABX).

11.5.1 ABX index and sub-indices

The ABX index comprises 20 ABCDSs referencing sub-prime home equity loans. The index began trading at the beginning of 2006. The ABX index has five separate sub-indices comprising pools of ABS CDS securities referencing home equity loans. Each of the different sub-indices references a different rating class of the underlying ABX collateral pool and they are called ABX.HE.AAA, ABX.HE.AA, ABX.HE.A, ABX.HE.BBB and ABX.HE.BBB– respectively. The BBB and BBB– sub-indices are the most actively traded.

Analogous to the CDS indices, the ABX indices are rolled every six months (on 19 January and 19 July each year). The series of ABX indices are referred to by their year and issue within that year, for example 06–1, 06–2 and 07–1. There is no overlap across each index in terms of the underlying collateral. Specific (transparent) eligibility criteria are used to determine which reference obligations may be included in the indices (for more details of the rules for composing the indices and their trading mechanics see the index administrator's website www.markit.com). ABX uses fixed cap PAUG ABCDS documentation with cash-only settlement.

The ABX indices enable investors to go short ABS exposure, diversify their exposure and also to express rapidly macro views on the ABS market (just like the iTraxx and CDX indices for the corporate world). It also facilitates other relative value trades for example capital structure, vintage, single names versus the index etc. Each ABX sub-index trades on price like a bond (compared to corporate CDS indices which trade on spreads). They trade on price in order to incorporate credit and prepayment assumptions as well as durations. Running payments are made monthly and the price is quoted as a percentage of par with any premium/discount representing the amount to be exchanged upfront (similar to CDS indices). Similar to CDS indices, the coupon/spread of each index is fixed over the life of the contract. Buying the index corresponds to selling protection (and receiving the coupon payments) and is referred to as being long risk. Selling the index is being short risk. The spread of the index is the weighted spread of the individual index constituents (the 20 ABCDS securities). Each of the market makers submits their valuations for the ABCDS securities. Each market maker has their own specific assumptions about home price appreciation, constant prepayment rates and idiosyncratic characteristics of the underlying mortgage pools in each home equity loan deal. In principle, therefore, the ABX index should represent the market's aggregate view about prepayment rates etc., which as we saw in Section 11.3 on cashflow CDOs is one of the significant uncertainties in ABS valuation.

Figure 11.6 shows the ABX cashflow mechanics. At contract inception there is an exchange of upfront payments. After this the protection buyer makes the periodic coupon payments to the protection seller. The payments are equal to the

Table 11.1 *Set coupons (in bps) for the ABX series 07–1 indices*

ABX.HE.07–1.AAA	9
ABX.HE.07–1.AA	15
ABX.HE.07–1.A	64
ABX.HE.07–1.BBB	224
ABX.HE.07–1.BBB–	389

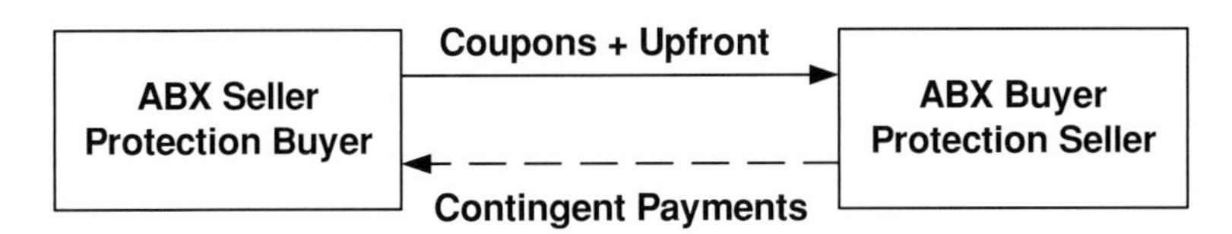

Figure 11.6 Schematic diagram of the cashflows in an ABX position.

set coupon times the remaining index notional. Contingent payments can occur due to principal losses (writedowns), prepayments and amortisation and interest shortfalls. Writedowns trigger a contingent payment and reduce the ABX index notional. Loss and writedown contingent payments can be reversed (paid back) if the loss or writedown event is subsequently reversed. In contrast, prepayments and amortisation on the underlying home equity loan deals do not trigger contingent payments, but do reduce the notional of the ABX index (and hence the coupon payments). The ABX indices use a fixed cap structure under which the protection seller's exposure to interest shortfalls in the underlying pool is capped at the ABX set coupon.

Finally Table 11.1 shows the set coupon levels for the ABX series 07–1 indices. As expected, the sub-indices with high ratings have a small coupon and the coupon increases as the rating decreases in quality.

11.5.2 Tranched ABX (TABX)

TABX are the standardised tranched exposures to the ABX indices [Flanagan *et al.* 2007]. Only the BBB and BBB– ABX sub-indices have thus far been tranched. These tranches are based on standardised parameters:

- standardised attachment/detachment points;
- static pool of reference obligations (hence the trades are not managed in comparison to most mezz ABSCDOs);
- same maturity as the index.

Table 11.2 *Attachment/detachment points (in per cent) for the BBB and BBB– TABX tranches*

BBB	BBB–
0–3	0–5
3–7	5–10
7–12	10–15
12–20	15–25
20–35	25–40
35–100	40–100

When the underlying ABX index rolls, two new standardised tranched products are introduced. The attachment/detachment points for the two different rated portfolios are shown in Table 11.2. TABX has some similarities with other ABSCDOs. However, ABX indices are based on home equity securitisations alone; ABSCDOs can reference a whole variety of underlying collateral including MBS, RMBS, CMBS, commercial real estate CDOs, credit card receivables etc. ABSCDOs also have bespoke cashflow waterfalls that determine how the liabilities are met from the assets in the collateral pool. ABX tranches (like corporate synthetic tranches) have a linear writedown structure. As principal in the collateral pool is written down it requires a contingent payment to be made. This reduces the subordination of the tranches. Finally TABX trades have collateral which is static. ABSCDO deals tend to be managed.

As in the corporate world, one of the motivations for the introduction of tranched exposures is leverage. A position in the equity tranche provides the investor with enhanced leverage compared to the index itself. TABX also offer the ability to express relative value trades (for example investing in different parts of the capital structure) as well as potentially acting as a hedge for other bespoke synthetic ABSCDOs. If the market for TABX eventually becomes liquid enough, the standardised tranches could also act as a calibration target (benchmark) for valuation models (opening up the possibility of applying some of the technology from synthetic CDOs such as base correlation concepts to TABX trades). This would allow bespoke tranches to be priced by an appropriate 'mapping' procedure. Finally the introduction of the standardised indices may itself prompt further liquidity in the ABCDS market (as was observed in the corporate CDS world).

Standardised ABX tranches trade under ISDA PAUG documentation with principal writedown and principal shortfall as credit events. Like ABCDS credit events are reversible and any reimbursements are also reflected in standardised tranches.

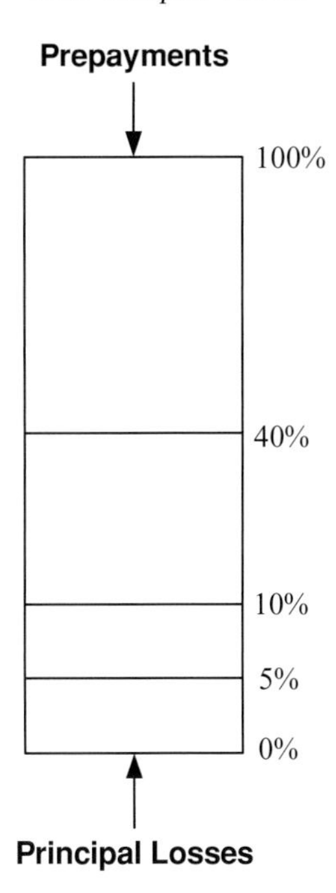

Figure 11.7 Schematic diagram showing the impact of losses and prepayments on the TABX tranches. Losses eat away at the equity tranches. Prepayments reduce the notional of the most senior tranche.

Each TABX will have a final legal maturity that matches the longest maturity of any reference obligation in the pool.

TABX tranches make contingent payments due to the occurrence of either writedown (loss) of principal or repayment of principal. Writedowns impact the equity and junior tranches first. Repayment of principal and amortisation impacts the senior tranches first (i.e. amortises the pool notional from the top down). Once the most senior tranche is fully redeemed, the next most senior tranche is impacted by further principal repayments. Writedowns reduce the amount of subordination of the equity tranche; repayment reduces the detachment point of the senior tranche. Interest shortfalls are not a credit event in TABX. The loss and prepayment mechanics of TABX are shown schematically in Figure 11.7.

11.6 Chapter review

In this chapter we have briefly introduced some of the variations on the basic securitisation and tranching concept. A natural extension of synthetic CDOs is to consider the case where the underlying collateral is itself composed of synthetic CDOs. In terms of modelling it is a simple matter to apply the standard Monte Carlo methods to valuation and calculation of risk sensitivities. However, the additional layer of subordination means that many more simulations are necessary to obtain reliable estimators. Because of the possibility of the same obligor appearing in multiple slave tranches it is not possible to apply semi-analytic methods easily to mitigate these problems. However, some promising approaches have been developed.

Cashflow CDOs were then introduced. Where synthetic CDOs are very homogeneous in their composition and cashflow mechanics, cashflow CDOs display a large degree of diversity. Firstly the underlying collateral can be composed from a whole range of different types of debt security (bonds, loans, CDSs, ABSs or any combination of the previous). Secondly the cashflows generated from the assets in the collateral pool are used to meet the issued liabilities by propagating the cashflows down through a bespoke cashflow waterfall. The waterfall structures traded in the marketplace tend to have a lot of features in common (coverage tests etc.), but the precise details of each waterfall will vary from trade to trade. The only commonly adopted valuation methodology in the marketplace is to make specific assumptions regarding the default, recovery and prepayment rates of the underlying asset pool (at a coarse grained level) and apply these assumptions to a deterministic cashflow model representing the waterfall. It would be desirable to apply simulation based techniques to assess many different scenarios, but practical problems remain in terms of parameterising and calibrating suitable models.

Finally we introduced CDSs where the underlying reference entities are asset backed securities (ABCDSs). A parallel evolution of this marketplace to the corporate world has been the introduction of standardised indices of ABSs (ABX) and tranching of these indices (TABX). The market for these products is nascent and it remains to be seen how the products and technology used to value them will evolve over time.

The topic of ABS collateral and modelling warrants a whole book of its own accord and this chapter has only briefly touched on the subject. For the interested reader there are a number of other resources available which describe the economics and business rationale of asset backed securities in great detail [Kothari 2006, Rajan *et al.* 2007].

12

Correlation trading of synthetic CDO tranches

The market can stay irrational longer than you can stay solvent.
John Maynard Keynes

12.1 Introduction

The previous chapters have presented models for the valuation, analysis and calibration of synthetic CDOs. In this chapter we analyse some of the more common trading strategies involving synthetic CDOs that have been employed in the marketplace and introduce the concept of correlation as an asset class to trade.

Originally credit derivatives were developed in order to isolate and hedge specific credit risks, essentially removing that risk from the balance sheet. They were illiquid instruments. Once off the balance sheet the protection seller no longer needed to hold regulatory capital against the potential loss of the underlying capital (thereby freeing up this capital for more profitable investments). The positions were typically buy-and-hold until maturity and very bespoke to the requirements of particular clients. With the advent of liquid markets trading standardised products this changed.

Probably the single biggest impact of liquid markets in portfolio products has been the emergence of a new asset class to trade: correlation. Valuing a synthetic CDO requires single obligor credit spreads and recovery rates as inputs. These are typically observable in the market and the values are (more or less) agreed upon by all market participants. The final fundamental input is the default correlation. As we have seen in Chapter 9, market participants strip from observed prices an implied correlation number. Trading tranches therefore is analogous to trading a view on default correlation; entering into a tranche position is the same as expressing a view as to what the default correlation value is.

Correlation trading is the concept. The reality is that positions also need to be hedged. This is where the difficulties start to arise. As we have seen previously, the standard market model does not allow perfect replication of the synthetic CDO contingent claim. Copula models provide prices for synthetic CDO tranches without really considering in detail the dynamics of the underlying obligors (and in particular the dynamics of their credit spreads). This means that there is no perfect hedge that simultaneously immunises a synthetic CDO position against market and credit risk and therefore sellers of protection are exposed to market and credit risks that are difficult to hedge against. This chapter will introduce some structures and the empirical hedging strategies employed to isolate and trade particular risks as well as expressing a particular view on future market movements.

The composition of this chapter is as follows. Section 12.2 defines some of the terminology and concepts of correlation trading. Section 12.3 will describe the general approach to hedging synthetic CDO positions and present results on delta hedging of various tranches. Following this in Section 12.4 we will describe and analyse a number of different single-tranche/index/CDS trading strategies that have developed as the sophistication of the single-tranche market has increased. Section 12.5 discusses recent credit market dislocations, and finally in Section 12.6 we will review what has been discussed in the chapter, and highlight the key points.

12.2 An overview of correlation trading

Correlation trading refers to investments or trades that arise from long or short exposure to correlation sensitive products (such as default baskets and synthetic CDOs) as well as positions in associated hedging instruments such as single-name CDSs referencing individual obligors or credit indices (iTraxx and CDX) [Kakodkar *et al.* 2003, Rajan *et al.* 2007]. Correlation investors typically consist of fundamental credit investors looking to invest in synthetic assets in an effort to enhance yield by acquiring cheap convexity, volatility or correlation. They are typically long-term investors meaning that positions are not actively traded on an intra-day basis, but that positions are entered or exited on the basis of fundamental views of credit quality of obligors or the market. Because credit derivative positions are exposed to market and credit risk factors, correlation traders are concerned not only with the credit (default) risk of a position, but also with the daily mark-to-market fluctuations. These MtM gains or losses arise due to fluctuations in the parameters used as inputs to value the tranche (linear and non-linear spread sensitivity and correlation sensitivity). On a daily basis P/L is crystallised as positions are entered and exited. Even buy-and-hold investors are sensitive to market fluctuations if their positions need to be marked-to-market (for accounting and reporting purposes).

It was principally the development of single-tranche synthetic CDOs (STCDOs) referencing the standardized credit indices (iTraxx and CDX) that facilitated the emergence of correlation as a tradable asset class. In cashflow (i.e. non-synthetic) CDOs the originating institution sells off the entire capital structure to investors (although they will typically retain the first-loss piece). This assumes that investors can be found for all the different portions of risk. It is typically never the case that all the desired assets in the underlying collateral pool can be purchased immediately, or that investors for all the issued liabilities can be found simultaneously. Consequently there is usually a ramp-up period when the asset portfolio is put into place and also a prolonged period when the different slices of risk are sold off. In the meantime the originating bank keeps large slices of risk on their own balance sheet. Secondary market liquidity is also typically low since the underlying collateral tends to be bespoke to the requirements of the particular trade. During the time it takes to ramp up the portfolio and find investors, a lot can happen in the market and the initial attraction of the debt to investors can change. On the positive side, however, because the whole capital structure is being sold off, correlation between defaults in the underlying collateral pool does not really matter.

The introduction of STCDOs allowed for very rapid trading of tranched exposures to pools of obligors (since there are only two parties to the trade). STCDOs allow for a separation of views on default risk and market/liquidity risk, thereby facilitating the expression of complex views on obligor default. If the entire capital structure of the index were traded then correlation as an asset class would not exist. However, because an STCDO only trades a particular slice of the portfolio loss distribution, the correlation becomes a key element of the trade. This is because the portfolio loss distribution (and therefore the tranche loss), as we have seen on numerous occasions, is strongly dependent on the value of the correlation input. The seller of protection assumes the credit (default) risk of the tranche (and provides compensation for losses in the reference portfolio); the buyer of protection assumes the market risk (which can be hedged with CDS products). Differences in supply and demand for different parts of the correlation structure are resolved through differences in the levels of implied default correlation. Dealers take on correlation risk when selling STCDO transactions to investors (such as hedge funds) because they are only placing a portion of the overall portfolio loss. To mitigate this risk the dealer can hedge the exposure using single-name CDSs, CDS indices or other credit products.

Correlation traders will invest in combinations of tranches, indices, CDSs etc. in order to create a position that isolates particular risks. The payoff can then be manipulated by appropriate choice of contract parameters. In order to achieve this it is necessary to understand clearly all the parameter sensitivities of the instruments (linear and non-linear risks in particular). The (empirically calculated) Greeks provide an estimate of the sensitivity of a position to fluctuations in the market parameters. For vanilla, linear derivatives (such as futures or forwards) the Greeks are very simple and intuitive. As demonstrated in Chapter 8, sensitivities of synthetic CDOs are not trivial to characterise or control. In terms of hedging of tranche positions there are generally two types of hedge that can be applied.

- Macro hedges – for example buying or selling index protection expresses a macro view on the constituents in the underlying portfolio.
- Micro hedges – exposures to individual obligors (for which there may be particular concerns about their survival probability) can be hedged by taking out an offsetting position in a CDS. Alternatively exposure to the default of small subsets of the underlying portfolio can also be hedged by positions in default baskets. An important consideration when micro hedging is the impact on the overall carry of the position (too many discrete hedges can erode the carry too much).

12.2.1 Definition of long/short correlation

In order to avoid confusion it is a good idea to spell out explicitly the definition of long and short correlation. The correlation is an indication of market sentiment about the distribution of risk across the capital structure. As correlation increases, the probability of large losses in the pool increases (Chapter 7) because the propensity for simultaneous default amongst the obligors is increased. Hence the likelihood of losses impacting the mezz and senior tranches will increase. An investor selling protection on these tranches will therefore demand a larger coupon (spread) for providing protection. Conversely, as more of the mass of the probability loss distribution is 'squeezed' from low to high losses as correlation increases, the probability of losses impacting the equity tranche will decrease leading to a reduction in the spread demanded by an equity tranche protection seller. As the correlation decreases, idiosyncratic risk dominates over systemic risk (and vice versa).

It is important to remember that the total risk of the portfolio is characterised by the portfolio loss distribution $p(l, \{s_i\}, \rho)$ (making explicit the parameter dependence of the loss distribution). Since

$$\int_0^{+\infty} p(l, \{s_i\}\rho)\mathrm{d}l = 1$$

then the total risk of the portfolio at some time t for fixed parameters is conserved. Tranches artificially slice up the risk of the portfolio by setting fixed attachment/detachment points. These points never move (unless defaults occur), but the loss distribution does 'shift' around these points as parameters like the correlation are varied (as was discussed in detail in Chapter 7). Therefore as correlation and spreads fluctuate, the proportion of the loss distribution falling between a given pair of attachment/detachment points will vary. But if probability mass moves from, for example, the equity tranche (due to the correlation increasing), by the conservation of probability mass, risk must be shifted up the capital structure. Market supply/demand dynamics will determine the distribution of risks amongst the tranches, but the overall amount of risk is constant. This means that there is a relationship between the movements of the spreads of the different tranches. As equity spreads go down, mezz spreads must go up and so on. At 100% correlation the risk of all the tranches is identical (since if one obligor defaults, they all default impacting all tranches). This means that at 100% correlation the par spreads of all the tranches should be identical. This is indeed what is observed in the graphs presented in Chapter 7.

The strategy of selling protection (making the contingent payments and receiving the fee payments) on the equity tranche benefits from high correlation, or increases in the correlation from the level at which the position was entered. This is because the probability of zero or low losses increases as correlation increases. Lower losses imply that the seller of default protection will make fewer contingent payments. The opposite is true for the case of selling protection on senior tranches (since the probability of large losses increases with correlation).

In analogy with the conventions in equity markets, selling protection on low-lying tranches (such as the equity tranche) is termed being 'long correlation'. Selling protection on senior tranches is termed being 'short correlation'.

12.2.2 Tranche leverage

Tranches implicitly provide leveraged exposure to losses compared to the index. Figure 12.1 shows the percentage losses to the standardised iTraxx tranches and the 0–100% (index) tranche as the number of defaulting obligors is increased. The leverage is given by the relative slopes of the tranche loss with respect to the index loss.

Because of the tranched nature of the exposure to the underling portfolio that STCDOs introduce, the relative speed with which the tranches are written down relative to the index is greatly enhanced ('The light that burns twice as bright, burns twice as fast'). However this enhanced exposure means that someone long tranche risk will receive an enhanced yield compared to the index. This can be quantified by defining a spread leverage measure $\lambda_s = s_{\text{Tranche}}/s_{\text{Index}}$. Tranches are also leveraged with respect to obligor spreads. This is characterised by the tranche's delta value (credit spread sensitivity).

12.2.3 Single-tranche trading mechanics

Investors have various choices to make when trading an STCDO. These include the following.

- Pool composition (can be bespoke or based on a standardised index or a mixture of both) – the choice of pool can be used to provide natural offsetting hedges for certain worrisome obligors or sectors.

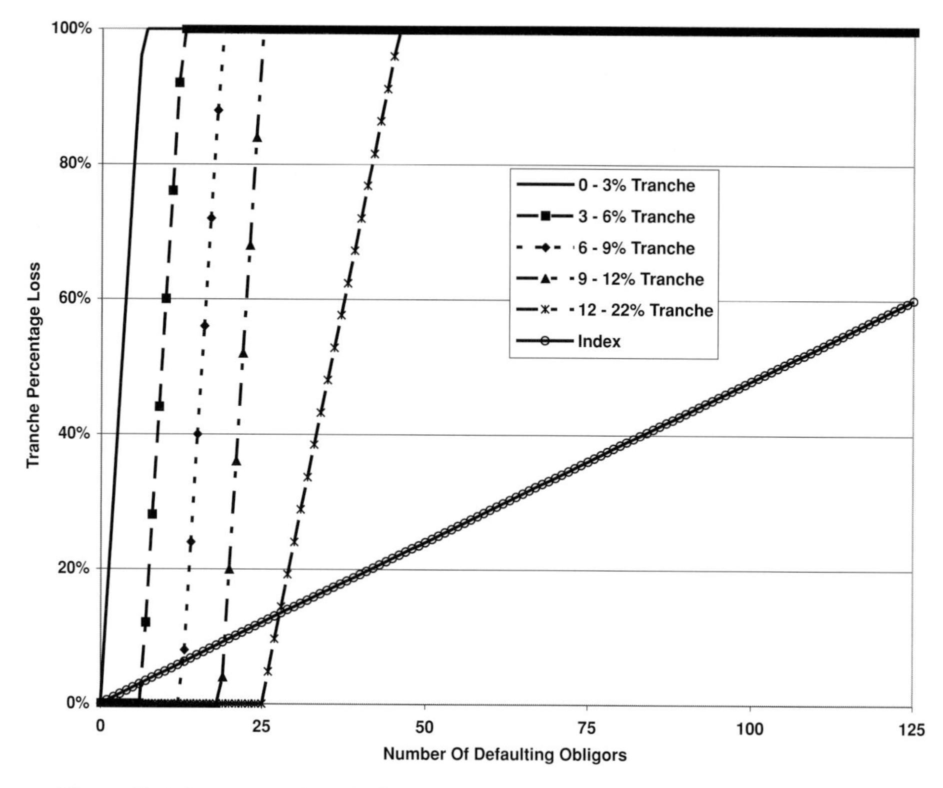

Figure 12.1 Leverage of the index tranches relative to the index.

- Tranche maturity (reflecting the period for which protection is either sought or provided).
- Tranche subordination and thickness (allowing the investor to choose the slice of portfolio loss they are exposed to, hence determining the tranche leverage).
- How the coupon payments are made (all upfront, homogeneously distributed over time etc.).
- Whether the portfolio is managed (usually by an external manager) or static. An important consideration in managed trades is the P/L that is crystallised when names are substituted in and out of the pool.
- Whether the position is intended to be buy-and-hold (in which case only defaults are really of concern) or whether it is to be actively traded (meaning daily MtM fluctuations are important).
- Funded or unfunded (swap variety – the most common variant).

In addition to these considerations, the investor also has to decide whether the tranche position is to be delta hedged or not. Delta hedging (described in the next section) has to be done on a dynamic basis and can in fact be very expensive to maintain. Delta hedging is facilitated by requesting BWICs (bids wanted in competition) from dealers on lists of obligors when taking long positions and OWICs (offers wanted in competition) when taking short positions. These bid lists can have a significant impact upon traded CDS spreads in the market. Also, delta hedging has the potential to delay deals because different parties to a trade will have different views of the hedge ratios (because they are using proprietary models, calibrations etc. to compute the delta values). The market solution to this is to quote delta values as 'traded' deltas and state clearly that the quoted index tranche spreads are delta exchange spreads. Delta exchange is the process whereby index tranches and their (index) hedges are traded simultaneously (delta exchange is essentially the spread hedge for a tranche). A good example of the delta exchange mechanism can be found in Bluhm and Overbeck [2007].

12.2.4 Strategies to express specific views

The popularity of STCDOs has been due in large part to the flexibility the product allows for expressing a multitude of views. We now describe some of the more common market views that can be expressed with tranche positions.

12.2.4.1 Utilising tranche leverage with respect to indices

Taking a position in a credit index (long or short) exposes the investor to the default risk of the entire portfolio of obligors (for which a protection seller may typically receive a spread of 30–40 bps). Taking a position in a tranche defined on the index exposes the investor only to a slice of the overall portfolio risk, with a significantly enhanced spread that the protection seller receives (e.g. 200 bps for a junior mezz tranche). The tranched position therefore allows the investor to gain a targeted exposure to the overall portfolio default risk. A typical leverage trade will be to sell protection on a mezz tranche. This has a positive carry (because we are selling protection) and the spread earned on this is typically greater than that earned on the index. A greater carry could be obtained by selling protection on the equity tranche, but this entails significant default risk. Leverage trades like this therefore are suitable for investors who are reasonably bearish on default rates (expecting some default activity). Leverage strategies are typically buy-to-hold. However, it is to be noted that negative carry effects can be important. If the spread on the 3–6% tranche is 150 bps, the spread on the index 30 bps and the hedge ratio of mezz spread sensitivity to index spread sensitivity is 5, then a delta hedged mezz tranche will receive 150 bps (from selling mezz protection) but have to pay 5×30 bps to hedge the position against the index, resulting in a net carry of zero. The views that tranches allow to be expressed can have very attractive carry, but this can quickly be wiped out by the requirements of delta hedging positions.

12.2.4.2 Relative value trades

Simple relative value trades include the following:

- long/short the on-the-run tranche versus the off-the-run tranche, index versus the tranche;
- long/short the X year tranche versus the Y year tranche;
- synthetically create an equity tranche (long x–100%, short delta hedged amount of index leaves a synthetic 0–x% tranche) and compare this with a straight 0–x% tranche to estimate the relative value).

12.2.4.3 Micro/macro hedging strategies

As an example consider shorting an equity tranche (purchasing protection) referencing a bespoke pool of obligors. The bespoke nature of the pool allows the investor to customise the names they are hedged against. Delta values are usually exchanged on the single-name CDSs as well. Purchasing protection in the form of an equity tranche is cheaper than purchasing protection (in the form of single-name CDSs) on the individual obligors. The downside of course is that we only receive protection on the first layer of losses.

12.2.4.4 Expressing correlation views

STCDOs allow investors to express views on the direction of implied correlation of a tranche. As was seen in Chapter 9, implied correlation is driven (at least partially) by the supply/demand dynamics of tranches observed in the market. In order to isolate correlation exposure, correlation traders typically will delta hedge their positions so as to be spread neutral. Going long/short a particular tranche enables the trader to express a view about the directionality of correlation. In order to understand how correlation views can be expressed, the key quantity to understand is the correlation sensitivity of specific tranches. This has been quantified and explained in detail in Chapter 7. In summary, the equity tranche par spread is monotonically decreasing in correlation and exhibits strong correlation dependence. The senior tranche par spreads are monotonically increasing but display less correlation sensitivity. Finally the mezz tranches are 'confused' displaying little correlation sensitivity (compared to the equity tranche).

A long correlation position (selling protection) will benefit from an increase in correlation. Selling equity protection and buying senior protection both benefit from increases in correlation. Conversely, a short correlation position will benefit from a decrease in correlation. Buying equity protection or selling senior protection allows an investor to express the view that correlation will decrease. If an investor does not have a strong view as to the direction of correlation, this can be crystallised, at least approximately, by investing in a mezz tranche, for example the 3–6% tranche. Alternatively the investor could form the portfolio

$\Pi = V_A(K_L^A, K_U^A) + \Delta^{\text{Hedge}} V_B(K_L^B, K_U^B)$, i.e. invest in two mezz tranches A and B with differing attachment/detachment points but referencing identical pools (the position is also delta hedged to maintain approximate spread neutrality). The objective is to achieve a correlation neutral view, i.e. $\partial\Pi/\partial\rho \approx 0$. This can be done by choosing suitable values for the attachment and detachment points of the two tranches. It is left as an exercise for the reader to implement a suitable optimisation scheme (once again simulated annealing would be a good choice although in this case the number of independent degrees of freedom is relatively small) to determine the best choice of attachment/detachment points which gives the flattest correlation sensitivity profile.

It is worth remembering that although the portfolio may be correlation insensitive at trade inception, due to time decay it will not remain correlation neutral (even with all other parameters such as credit spreads remaining constant). This is because of the shifting risk profile of the tranches as time passes.

12.2.4.5 Expressing market views – long equity

Going long equity means selling protection. This is suitable therefore for an investor who is bullish on default risk (and does not believe they will ever have to compensate the protection purchaser for losses). However, it is necessary to manage spread risk via appropriate delta hedging. Default risk to specific obligors can also be hedged by a discrete number of CDSs (purchasing protection). The trade should still have a positive carry. A more sophisticated form of strategy is to sell protection on managed synthetic CDOs. In this case the default risk can be actively managed by swapping names in and out of the portfolio (at the prevailing MtM profit/loss).

12.2.4.6 Expressing market views – short senior tranche

In this position protection is purchased. The investor can use this position to express a bearish view of systematic spread widening (if the spreads widen the contingent payments will increase). Typically this will be cheaper than purchasing index protection (e.g. index protection can cost 60 bps, senior protection 20 bps).

12.2.4.7 All-upfront basis and zero-coupon equity tranches

The standard way to trade equity tranches is with an upfront payment to the protection seller followed by a running spread payment of 500 bps. It is also possible to trade the equity tranche in different ways. For example, investors can buy or sell equity protection on an all-upfront basis. In this case all of the running spread payments are bundled up into the one-off upfront payment which is paid to the protection seller. The protection seller still has to compensate the protection purchaser for losses to the tranche. The rationale for this trading mechanism for the protection seller is that they receive the total payments up front (which can be reinvested), and their total liabilities are capped at the tranche notional. The value of the upfront payment is of course given by setting $s_{\text{Running}} = 0$ in $V_{\text{Cont}} = u + s_{\text{Running}} V_{\text{Fee}}$ leading to $u = V_{\text{Cont}} = \mathbf{E}[L_{\text{Tranche}}]$ as the upfront payment. In addition to this the delta value of the tranche is reduced when trading on an all-upfront basis.

Another trading mechanism is to treat the equity tranche as a zero-coupon bond. In this case investors pay a premium at contract inception to receive at maturity 100% less the notional amount of any tranche losses (at maturity). The value of the zero-coupon equity tranche is given by $V_{\text{ZCE}}(T) = (1 - \bar{L}_{\text{Equity}}(T))e^{-rT}$ where T is the maturity of the tranche.

12.3 Delta hedging of synthetic CDO tranches

In Chapter 8 we introduced measures to quantify the risk sensitivities of synthetic CDO positions. Now we will consider how it is possible (empirically if not theoretically) to hedge CDO positions. To hedge a synthetic CDO position the simplest hedging portfolio to construct is composed of positions in CDSs referencing the obligors in the underlying pool of the synthetic CDO. This hedging strategy will immunise the CDO position to individual defaults and small fluctuations in the credit spreads of the underlying obligors in the pool. However, due to the strong convexity of the tranche PVs (as shown in Chapter 8) but weak CDS convexity as a function of the fluctuations in spread, large fluctuations in the credit spreads will lead to a P/L leakage in the overall hedged position (since the response of the tranche and the hedge is not balanced). Basis between the CDO and CDS spreads will also lead to P/L 'bleed'.

An additional complication is that the hedge will require constant rebalancing, i.e. to remain delta neutral the hedging strategy must be dynamic. This is because the obligor par spreads fluctuate on a daily basis and the magnitude of the fluctuations changes also. This implies that the response of the tranches and the underlying CDSs changes daily leading to new delta values. In principle the delta hedge should be rebalanced as often as possible, ideally on a daily basis. In practice market illiquidity (it may not be possible to transact the model specified amount of a CDS on a particular obligor), transaction costs and bid-offer spreads render daily rebalancing impractical.

Finally, the delta hedge strategy immunises the position against small fluctuations in the credit spreads. But as we have seen there is also outright default risk to contend with. In principle a separate CDS could be purchased for each obligor in the pool (with the exact same notional as that of the obligor in the CDO tranche pool). This would mitigate the default risk of each and every obligor, but would be prohibitively expensive. In practice CDS protection is purchased for only a small subset of obligors for which there may be particular concerns about their survival prospects. Alternatively protection can be purchased on small baskets of obligors in the form of FtDs. In the analysis to follow we only consider how to hedge a tranche position to fluctuations in the credit spreads of the underlying obligors.

12.3.1 The principles of delta hedging

Consider the case of delta hedging the credit spread fluctuations of a synthetic CDO tranche. Assume that the tranche position is hedged with individual CDSs referencing each obligor in the underlying pool. The value of the hedging portfolio is

$$V^{\text{Hedge}}(t) = \sum_{i=1}^{n} \Delta_i^{\text{Hedge}}(t) V_i^{\text{CDS}}(t) = \Delta_i^{\text{Hedge}}(t) V_i^{\text{CDS}}(t) + \sum_{\substack{j=1 \\ j \neq i}}^{n} \Delta_j^{\text{Hedge}}(t) V_j^{\text{CDS}}(t)$$

where $\Delta_i^{\text{Hedge}}(t)$ is the amount of CDS of obligor i that must be purchased at time t to immunise the portfolio to small fluctuations in the credit spread of obligor i (remembering that we have assumed the term structure of credit spreads is flat, but inhomogeneous, for all obligors). The value of the overall instrument portfolio is given by

$$\Pi(t) = V^{\text{CDO}}(t) + V^{\text{Hedge}}(t)$$

and the portfolio cashflows are illustrated schematically in Figure 12.2. If the spread of the obligor fluctuates according to $s_i(t) \to s_i(t) + \delta s_i(t)$ (and hence $h_i(t) \to h_i(t) + \delta h_i(t)$ is the change in the hazard rate induced by the change in the spread) – which only affects obligor i – then the value of the hedging portfolio and the CDO will be impacted where

$$\Delta V^{\text{Hedge}}(t) = V^{\text{Hedge}}(s_i + \delta s_i, t) - V^{\text{Hedge}}(s_i, t)$$

is the change in value of the hedging (CDS) portfolio and

$$\Delta V^{\text{CDO}}(t) = V^{\text{CDO}}(s_i + \delta s_i, t) - V^{\text{CDO}}(s_i, t)$$

is the change in value of the synthetic CDO. Requiring that $\partial \Pi(t)/\partial s_i \equiv 0$, i.e. the overall portfolio position is insensitive to fluctuations in the credit spreads of obligor i implies that

$$\Delta_i^{\text{Hedge}}(t) = -\frac{\partial V^{\text{CDO}}(s_i, t)/\partial s_i}{\partial V_i^{\text{CDS}}(s_i, t)/\partial s_i} = -\frac{\Delta^{\text{CDO}}(t)}{\Delta_i^{\text{CDS}}(t)}$$

is the amount of CDS of obligor i that must be purchased to immunise the hedged portfolio against fluctuations in the credit spread of obligor i (in the limit of infinitesimal fluctuations in the underlying spreads).

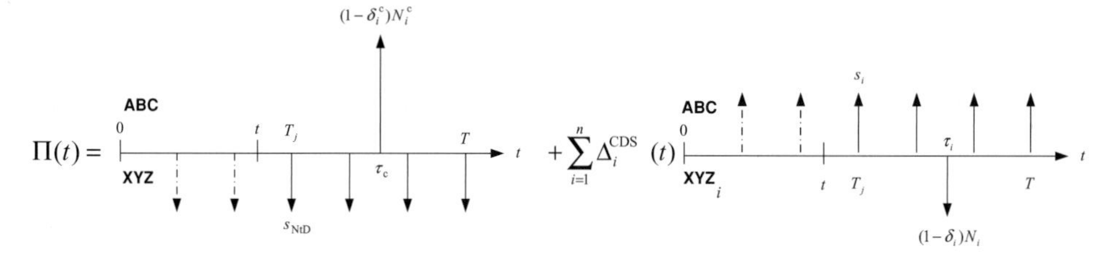

Figure 12.2 Schematic diagram of the cashflow vectors for the delta hedged synthetic CDO (hedged with a portfolio of CDSs). Obligor ABC is purchasing protection on the tranche from obligor XYZ. To hedge this exposure obligor ABC sells protection on a portfolio of individual CDSs (transacted with a different set of obligors XYZ $_i$).

In the more general case when each instrument depends upon more than one credit spread, the portfolio value is (making explicit the dependence of each of the terms on the obligor credit spreads)

$$\Pi(t) = V^{\text{CDO}}(\{s_i(t)\}t, T) + \sum_{i=1}^{n} \Delta_i^{\text{Hedge}}(t) V_i^{\text{CDS}}(s_i(t), t, T).$$

The requirement of delta hedging is that $\nabla_s \Pi(t) = \mathbf{0}$ for all t (where we have slightly abused the gradient operator notation such that $\nabla_s = \partial/\partial s_i$ for $i = 1, \ldots, n$). Writing this out in full generality it becomes

$$\begin{pmatrix} \frac{\partial}{\partial s_1} V^{\text{CDO}} \\ \cdots \\ \cdots \\ \cdots \\ \frac{\partial}{\partial s_n} V^{\text{CDO}} \end{pmatrix} + \begin{pmatrix} \frac{\partial}{\partial s_1} V_1^{\text{CDS}} & \cdots & \cdots & \cdots & \frac{\partial}{\partial s_1} V_n^{\text{CDS}} \\ \cdots & \cdots & \cdots & \cdots & \cdots \\ \cdots & \cdots & \cdots & \cdots & \cdots \\ \cdots & \cdots & \cdots & \cdots & \cdots \\ \frac{\partial}{\partial s_n} V_1^{\text{CDS}} & \cdots & \cdots & \cdots & \frac{\partial}{\partial s_n} V_n^{\text{CDS}} \end{pmatrix} \begin{pmatrix} \Delta_1^{\text{Hedge}} \\ \cdots \\ \cdots \\ \cdots \\ \Delta_n^{\text{Hedge}} \end{pmatrix} = 0.$$

In the simplified case where $\partial V_i^{\text{CDS}}/\partial s_j \equiv 0$ for $i \neq j$, the off-diagonal terms vanish. Writing the general form of the hedging problem in matrix form we have $\vec{\Delta}^{\text{Hedge}} = -\mathbf{H}^{-1}\vec{\Delta}^{\text{CDO}}$ (for each time t). The hedging calculation therefore requires a (potentially large) matrix inversion to be computed at each time (in addition to the instrument revaluations and calculations of sensitivities).

Delta hedging using individual CDSs may not be possible due to the illiquidity of the market in individual CDSs or due to the transaction costs involved in buying and selling so many single-name contracts. Certainly attempting simultaneously to transact 125 separate CDS positions with different counterparties is operationally impractical (not to say dangerous). A more popular and practical alternative is to hedge using an index such as iTraxx or CDX. In this case the value of the hedging portfolio is given by $V^{\text{Hedge}}(t) = \Delta^{\text{Hedge}}(t)V^{\text{Index}}(t)$ where $\Delta^{\text{Hedge}}(t) = -\Delta^{\text{CDO}}(t)/\Delta^{\text{Index}}(t)$ is the hedge ratio which quantifies the sensitivity of the tranche PV to fluctuations in the index PV, and $V^{\text{Index}}(t)$ is the PV of the index. The hedged position is given by the sum of the tranche PV and hedging portfolio PV.

The index trades at a spread that is the (dV01) weighted average of all the obligors in the underlying portfolio. The delta value of a tranche depends on the CDS spread of the individual obligors. If an obligor has a CDS spread greater than the average spread of the pool, hedging using the index will underweight the hedging portfolio relative to the risk of this obligor. The opposite is also true if the spread of the obligor is less than the average spread of the index.

12.3.2 Delta hedging tranche positions with the index

In the examples to be discussed below we assume the perspective of an investor who is purchasing protection on the tranche (making protection payments and receiving contingent payments when the cumulative pool losses impact the tranche), and hedging this position by selling index protection (receiving the protection payments, but making the contingent payments). Note that this example is purely for illustrative purposes and is not meant to be representative of a real trading strategy (examples of these will be presented later). The value of the total portfolio position is given by

$$\Pi(\{s_i\}) = V^{\text{CDO}}(\{s_i\}) + \Delta^{\text{Hedge}}(\{s_i\})V^{\text{Index}}(\{s_i\})$$

where the hedge ratio is calculated according to $\Delta^{\text{Hedge}}(\{s_i\}) = -\Delta^{\text{CDO}}(\{s_i\})/\Delta^{\text{Index}}(\{s_i\})$. In the results reported here the delta values are calculated by simultaneously bumping all of the obligor spreads by 1 bp. The dependence on the spread of the delta values is explicitly included (the valuations used to estimate the delta values clearly depend on the current level of the spread) because it is this dependence that leads to the observed behaviour shown in Figure 12.3. Also plotted on this figure are the spread sensitivities of the tranche and index positions. The hedge ratios are calculated according to the prescription above for an initial (homogeneous) spread of 60 bps for the obligors. These hedge ratios are then kept fixed as the homogeneous obligor spread is varied in the range [−60 bps, 120 bps].

It is because the hedge ratios are kept fixed that there is P/L leakage from the hedging strategy. Indeed, for very small bumps of ± 10 bps, the P/L leakage is approximately $\pm 1\%$ of the equity tranche notional; but at large positive spread bumps the leakage can be almost 100% of the tranche notional since at this magnitude of spread bump the equity tranche PV is approaching its saturation point (the contingent payment is approaching its maximal value and the fee payment is approaching zero), but the index tranche PV is still linearly increasing.

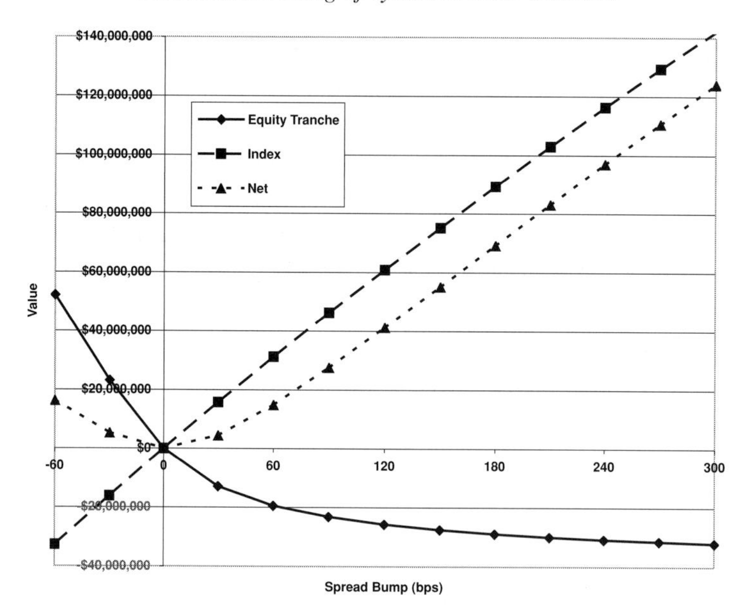

Figure 12.3 Delta hedging an equity tranche.

It is easy to explain qualitatively the form of this spread sensitivity profile. The value of the index is a linearly increasing function of the magnitude of the spread bump. To explain this, consider the index as the value of the 0–100% tranche. Because the attachment point is 0% of the portfolio notional, bumps to the spread impact the value immediately (like an equity tranche). Similarly, because there is no detachment point, as the spreads increase, the risk of the position increases in direct proportion to the spread; because there is no exhaustion point for the tranche (other than total 100% loss to the underlying portfolio) there are no saturation effects on the contingent payments payable. We saw for default baskets and synthetic CDOs that it was the saturation of the contingent payments when the tranche notional became exhausted that led to the spread convexity observed in Figure 12.3 for the bare equity tranche position. Consequently the index PV increases linearly for all magnitudes of spread bumps considered here. The equity tranche PV as a function of the spread bump has the same form as reported and explained in Chapter 8. For the 0–3% tranche we see that the portfolio is not well immunised against large fluctuations in the par spreads away from the spread value at which the hedge ratios were calculated (60 bps). The important point to note, however, is that the strategy has positive convexity, having a positive P/L for both spread widening and contractions. Because of this it is a very popular strategy for carry hungry investors such as hedge funds who are prepared to live with the default risk in return for the enhanced return.

The behaviour of the 3–6% tranche delta hedged with the index is as explained previously. In this case the overall behaviour of the hedged portfolio is a bit more complex than for the equity case. For negative spread bumps the mezzanine tranche is barely impacted because of its layer of subordination and so the hedged position is a linearly decreasing function of the spread bump. Once the tranche begins to be impacted by the spread bumps, the hedged position begins to increase again. Eventually, however, the tranche begins to saturate and so its PV starts to level off; the index tranche is still increasing in value. Therefore the hedged position will eventually start to lose money as the spread increases still further. On the other hand the 6–9% tranche appears to be well hedged by the index position. However, this is deceptive; the hedge ratio for this tranche is very close to zero since it has 6% of subordination (therefore the portfolio value is approximately equal to tranche value). This is intuitively as expected since the 6–9% tranche is more insensitive to fluctuations of the spreads than the 0–3% and 3–6% tranches.

These examples clearly demonstrate why delta hedging needs to be done on a dynamic basis. The hedged portfolio leaks P/L when the spread moves from the value used to calculate the hedge ratios. Because the spreads evolve dynamically on a day-to-day basis, the efficacy of the hedge will also fluctuate. To maintain the delta neutrality the hedges must be recalculated as the spreads move. Unfortunately this is impractical to do on a day-to-day basis, principally owing to liquidity issues, timing issues (it may not be possible to execute the exact amount of the index trade required, when required – although the standardised indices tend to trade very liquidly) and transaction costs. Delta hedging therefore produces only a partially effective hedge as time passes. Hedging the risk of synthetic CDOs is akin to trying to grab hold of a jelly – just when you think you have managed to pin it down, it moves away again.

Finally it is worth stressing that delta hedging of spread fluctuations does not immunise the position against default risk. Consider a default in the underlying pool where we have hedged a long position in an equity tranche with the index (the 0–100% tranche). The percentage loss to the tranche due to the default of obligor i (expressed as a fraction of the tranche notional) is given by $N_i(1-\delta_i)/N_{\text{Tranche}}$ and the percentage loss to the index is $N_i(1-\delta_i)/N_{\text{Index}}$. The change in the hedge value is therefore $\Delta^{\text{Hedge}} N_i(1-\delta_i)/N_{\text{Index}}$ which is not equal to the percentage loss of the tranche. The impact of the default therefore will be P/L leakage.

12.3.3 Crystallising P/L from delta hedging

A tranche position that has positive spread convexity will make MtM gains from volatility in the underlying (in this case the volatility of the underlying credit spreads – hence as the volatility increases, so too does the attractiveness of the strategy). Convexity in tranches (quantified by the gamma value) is computed by calculating the changes in the delta value of the tranche according to $\Gamma_i = \partial \Delta / \partial s_i$ (for obligor i). Frequent re-hedging of a delta hedged tranche position will allow an investor to crystallise the P/L gains from the convexity.

Consider the delta hedged equity tranche strategy that we have previously introduced. The portfolio is to sell protection on the equity tranche and purchase protection on the index (in an amount Δ). As we saw this strategy has positive convexity. To illustrate how P/L can be crystallised from this strategy we consider a simple example. Assume that an investor sells \$10 m of 0–3% equity tranche protection (receiving the upfront payment and the running spread) and at this time the index spread was 40 bps and the delta value was 13. The value of the hedge portfolio is therefore $13 \times \$10$ m which is the amount of index protection the investor would need to purchase (and the coupon of 40 bps $\times$ \$130 m would be paid quarterly for the protection).

Consider what happens if at some time t after the trade is entered (which is at time 0) the spreads have widened (if the spread widening is small the delta hedge would lead to only a marginal change in the P/L). As the spreads widen the equity tranche becomes riskier and the upfront payment will increase to reflect this increased risk. If the initial upfront payment was 35% (of the \$10 m tranche notional) the seller of protection would receive \$3.5 m at trade inception. If the spreads move wider and the upfront payment increases to 50% this equates to an upfront payment of \$5 m. To monetise the P/L the trade must be unwound. This means the investor sells the equity tranche position to another investor (leading to a net upfront payment of +\$3.5 m – \$5 m to the investor).

As the spreads move wider the delta value of the hedged position decreases. This means that less protection (than the original \$130 m) is now needed to hedge the tranche position. The investor can therefore unwind the index position and monetise the P/L gain due to the change in MtM of the index position. The change in value of the index position is given by $\text{MtM}_{\text{Index}} = (s_t - s_0)\text{dV01} > 0$ since the spreads have widened. The hedge is now re-established by selling protection on the equity tranche and purchasing protection on the index (using the current delta value). As spreads tighten the delta value of the hedged portfolio increases and the investor can realise the profit by buying additional index protection at the tighter spread (hence lower cost). By following this procedure dynamically (the standard delta hedging procedure) the hedged position can be periodically re-delta-hedged, monetising the P/L in the process.

12.4 Analysis of common correlation trading strategies

The previous section explained how to delta hedge tranche positions. In this section we will examine in more detail how positions in STCDOs can be used to express specific market views. To understand and tailor a strategy it is necessary to understand its

- sensitivity to credit spread fluctuations (both idiosyncratic and systemic),
- sensitivity to correlation fluctuations, and
- sensitivity to obligor defaults.

In this section we outline some common trading strategies used to express different market views or expectations. The discussion in this section draws heavily on some excellent sources [Douglas 2007, De Servigny and Jobst 2007, Rajan *et al.* 2007]. The strategies to be assessed all have a positive carry (when including the upfront payments), and are spot delta neutral (i.e. are hedged against small movements in spreads from their instantaneous values). However, they are not hedged against fluctuations in correlation or outright default. They are therefore very attractive to trading desks since they appear to be generating a positive income for no risk. The reality of course is very different.

12.4.1 Delta hedged equity tranche

This is perhaps the most common of the delta hedged correlation strategies. The strategy description is as follows:

- sell protection on the 0–3% equity index tranche (receive periodic coupons, make contingent payments);
- purchase protection on the index (make periodic coupon payments, receive contingent payments).

Because we are selling protection on the equity tranche this strategy will typically have a positive carry (the spread earned from the equity tranche will be greater than the spread paid on the index). The amount of index protection purchased is chosen so as to make the overall portfolio spread neutral. That is, the portfolio value is given by $\Pi = V^{\text{Equity}} + \Delta^{\text{Hedge}} V^{\text{Index}}$ and the hedge ratio to achieve $\partial\Pi/\partial s = 0$ is $\Delta^{\text{Hedge}} = -\Delta^{\text{Equity}}/\Delta^{\text{Index}}$. This hedging strategy immunises the overall portfolio to 'small' movements in the spreads. The cashflows in this strategy are represented schematically in Figure 12.2. The hedge ratio is calculated by revaluing the tranche and index positions under a 1 bp widening of all credit spreads simultaneously.

Note that for all the strategies analysed in this chapter we are ignoring the upfront payments. At execution, the premium received as a result of selling credit protection exceeds the premium paid to immunise the position against small scale spread moves. Upfront payments can be amortised over the tranche duration and added to the running premium to provide an estimate of the coupon received for selling equity tranche protection. Subtracting from this the coupon paid for receiving index protection provides an estimate of the net carry of the trade. At execution the net

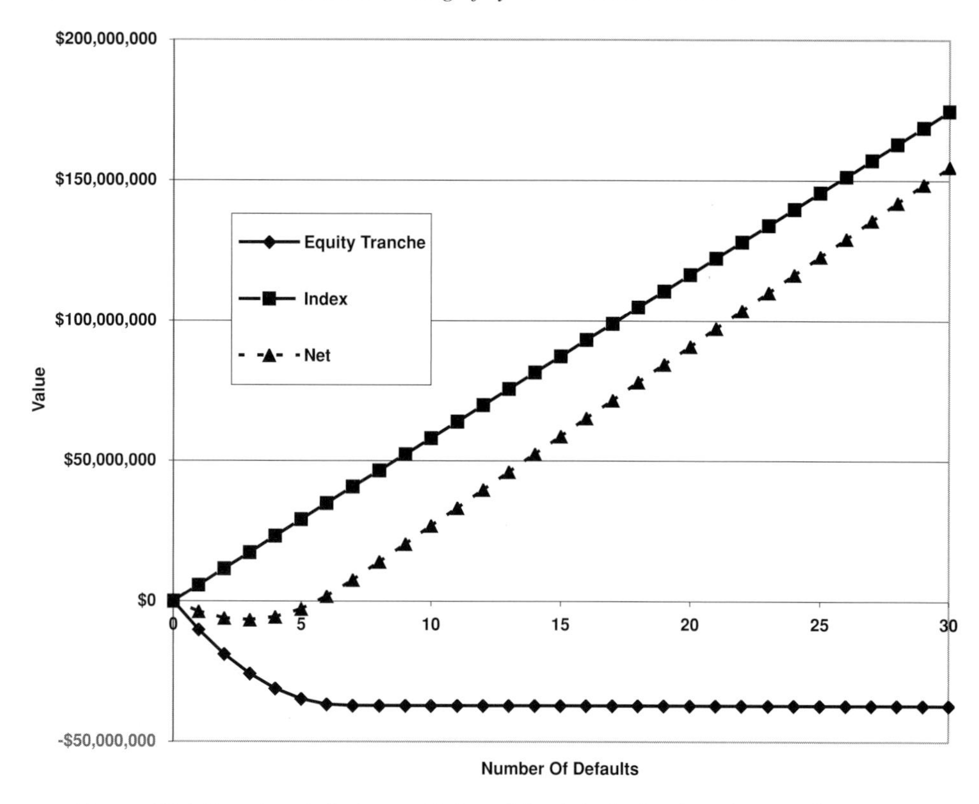

Figure 12.4 Running default sensitivity of the delta hedged equity tranche.

carry is approximately equal to a no-default scenario cashflow (which is close to the value of the upfront payment amortised over the risk-free duration of the tranche).

Figure 12.3 plots the value of the portfolio as the obligor credit spreads are bumped from their initial value (for simplicity all of the obligors are assumed to have a homogeneous par spread). All of the obligor spreads are bumped simultaneously by the same amount. The strategy has positive convexity under both spread widening and contracting scenarios. The form of this spread sensitivity was explained in Section 12.3 and will not be repeated here. We will add, however, that it is apparent from this Figure that once the spreads have widened to such a degree as to render the equity tranche irrelevant (when the equity tranche P/L saturates), the subsequent spread response of the strategy is the same as that for a plain index position. Although this strategy is (approximately) delta hedged, there is still significant default risk. This can be seen from Figure 12.4, which plots the strategy P/L as a function of the number of obligors instantaneously defaulted. As the number of defaults increases from zero, P/L is immediately lost from selling protection on the equity tranche. This is only partially compensated for by the payments received from the protection purchased on the index because the hedge ratio

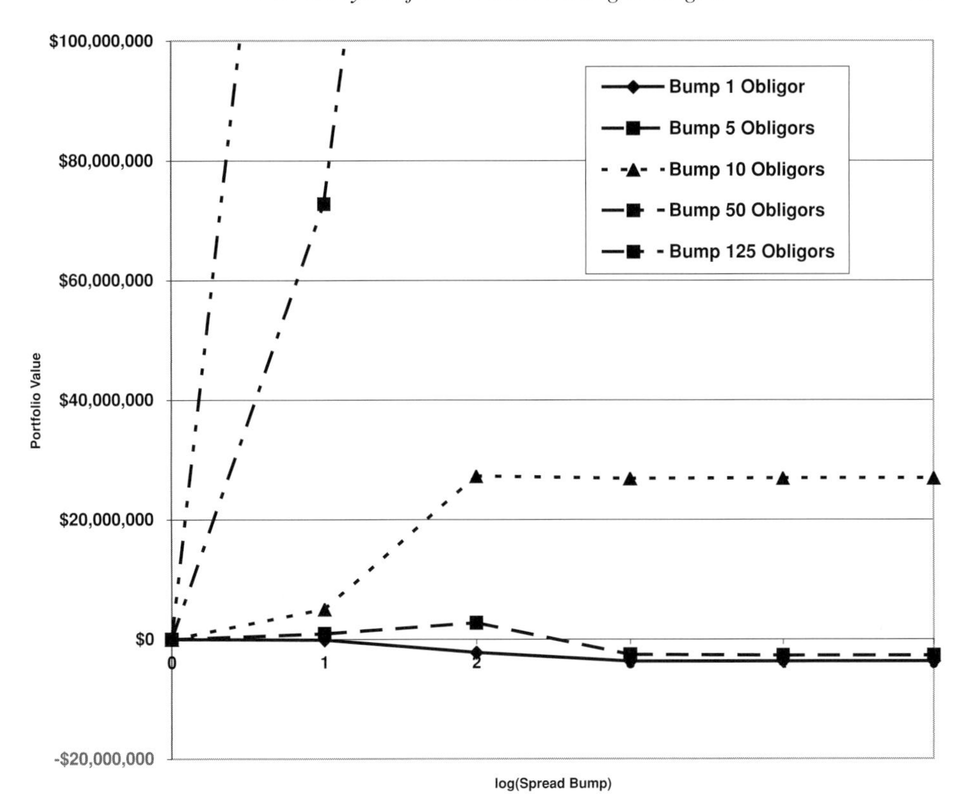

Figure 12.5 The impact of spread dispersion upon the P/L for the delta hedged equity tranche.

is greater than unity (the index protection payments are also received immediately upon the occurrence of the first defaults). As the number of defaults increases further, the protection payments from the index continue to be received indefinitely. But eventually the equity tranche is burnt out (once the total portfolio losses exceed the detachment point) meaning there are no more contingent payments to make. The P/L therefore increases linearly with the number of defaults past a certain breakeven point (approximately five defaults in this example). This strategy therefore benefits from large numbers of defaults that result in exhaustion of the equity tranche. Conversely, it suffers for small numbers of defaults. It therefore provides good protection against systemic default risk, but poor protection against idiosyncratic defaults.

Figure 12.3 plotted the spread sensitivity for a homogeneous spread bump applied to all the obligors simultaneously (and the magnitude of the spread bump was in fact relatively small). In practice, however, not all the obligor spreads will move in synchronisation. An important issue to consider therefore is the degree to which spread movements are dispersed. Figure 12.5 plots the P/L for the strategy as subsets of obligors have their spreads bumped by increasing amounts. We observe that for a small number of obligors widening (although the spread widening has to be quite significant) the P/L of this positive carry strategy is negatively impacted, i.e. it loses money. However, as the number of obligors whose spreads widen increases, the P/L impact inverts and the strategy has a positive P/L. For an intermediate number of spreads widening the P/L of the strategy appears to saturate. As large numbers of spreads widen the P/L increases unboundedly. This behaviour is consistent with the results obtained for the running default sensitivity. The conclusion that can be reached from this figure is that correlation amongst the spread movements of the individual obligors is another important factor to consider when designing tranche trading strategies. This figure shows that there is a clear dependence of the strategy P/L on the degree to which obligor spread moves are synchronised.

In terms of correlation sensitivity, the index position is correlation insensitive (since it is a 0–100% tranche). The correlation sensitivity of this strategy is therefore simply that of the long equity tranche (which has been reported elsewhere and so will not be repeated here). Overall this strategy performs poorly for small numbers of defaults or idiosyncratic spread widening.

12.4.2 Delta hedged mezz tranche

The strategy description is as follows:

- sell protection on the index (receive periodic coupons, make contingent payments);
- purchase protection on the 6–9% tranche (make periodic coupon payments, receive contingent payments).

The amount of index protection purchased is chosen so as to make the overall portfolio spread neutral. That is, the portfolio value is given by $\Pi = V^{\text{Mezz}} + \Delta^{\text{Hedge}} V^{\text{Index}}$ and the hedge ratio is $\Delta^{\text{Hedge}} = -\Delta^{\text{Mezz}} / \Delta^{\text{Index}}$. This hedging strategy immunises the overall portfolio to 'small' movements in the spreads.

The strategy has positive convexity under both spread widening and contracting scenarios. Although this strategy is delta hedged, there is still significant default risk. As defaults occur, P/L is lost due to payout on the index CDS. However, as more defaults occur the senior mezz tranche kicks in and compensates for the losses due to the short position in the CDS. Eventually though, the senior mezz tranche reaches its exhaustion point and ceases to provide an offset to the losses from the index position (hence the P/L decreases again). This means that as the number of defaults increases beyond the exhaustion point of the mezz tranche, the strategy losses are unbounded (except insofar as the losses are absolutely capped at the total loss incurred if all the obligors in the index default). This strategy therefore expresses a view that defaults will exceed the mezz tranche attachment point, but not exceed the detachment point. Because one of the trades in the strategy is an index position (correlation insensitive), the overall correlation sensitivity of the strategy is that of the mezz tranche position.

12.4.3 Long equity–short mezz straddle (bull-bear trade)

The strategy description is as follows:

- sell protection on the 0–3% equity tranche (receive periodic coupons, make contingent payments);
- purchase protection on the 3–6% tranche (make periodic coupon payments, receive contingent payments).

The amount of mezz protection purchased is chosen so as to make the overall portfolio spread neutral. That is, the portfolio value is given by $\Pi = V^{\text{Equity}} + \Delta^{\text{Hedge}} V^{\text{Mezz}}$ and the hedge ratio is $\Delta^{\text{Hedge}} = -\Delta^{\text{Equity}}/\Delta^{\text{Mezz}}$. This hedging strategy immunises the overall portfolio to 'small' movements in the spreads, although it is to be noted that the degree of insensitivity to spread movements is not as good as when the equity tranche is hedged with an index position. This strategy expresses the view that default rates will be low (hence the long equity position), but that widespread spread widening may be possible (hence the protection that the junior mezz tranche provides).

The strategy has positive convexity under both spread widening and contracting scenarios. The larger the movement the larger the gains. Because both components in the strategy (the 0–3% and 3–6% tranches) have finite notional amounts, the tranche spread sensitivities will eventually saturate (because at large spread bumps the contingent payments saturate and the fee payments approach zero). This means that the strategy payoff will be capped at a certain maximum amount and will not increase indefinitely as spreads widen indefinitely. This is in contrast to hedging the equity tranche with the index. Although this strategy is delta hedged, there is still significant default risk. The default picture is similar to that for the delta hedged equity tranche (hedged with the index); initially P/L is lost due to defaults eating into the equity tranche. As defaults occur the subordination of the 3–6% tranche is reduced which impacts its value (resulting in a positive P/L for the purchaser of protection). Eventually the mezz tranche payouts start to compensate the losses to the equity tranche and the equity tranche also burns out, meaning that the 3–6% tranche continues to contribute a positive P/L once the equity tranche has been wiped out. Eventually the number of defaults is sufficient to have exhausted both the equity and mezz tranches and the P/L saturates. Therefore it makes money as the number of defaults increases. In this strategy more defaults are required to break even than in the delta hedged equity tranche strategy. This is because the mezz protection does not start to kick in until a finite number of defaults has occurred. This contrasts with hedging the equity tranche with the index; in this case the index protection against defaults kicks in immediately.

12.4.4 Equity curve flattener

The strategy description is as follows:

- sell protection on the 0–3% equity tranche of maturity, for example, 10 years (receive periodic coupons, make contingent payments);
- purchase protection on the 0–3% equity tranche of maturity, for example, 5 years (make periodic coupon payments, receive contingent payments).

Both tranches are delta hedged. This strategy has positive gamma value (so profits if spreads move either wider or tighter) and it is short correlation (therefore profits if correlation decreases). The strategy is also short default risk, profiting if there is a default. This strategy was potentially applicable to the market conditions observed in summer 2007. In this environment equity tranche correlations were at historic highs (and thought to be likely to decrease), but default risk was perceived to be high.

12.5 Credit market dislocations

The liquid market for correlation products such as STCDOs is relatively young, being only a few years old. Although young it has had an interesting and eventful life thus far. The history of the iTraxx index over the period September 2004–November 2007 was shown in Figure 4.1. It is observed that there have been two clearly identifiable 'events' in this period: around May 2005 and the summer of 2007. In this section we will provide some brief commentary and views on these events, their catalysts and their consequences.

12.5.1 The May 2005 correlation 'crisis'

The correlation crisis refers to the events centred around May 2005 [Crider 2005, Patel 2005, De Servigny and Jobst 2007]. During this period, credit spreads spiked to previously unobserved levels and equity tranche implied correlation decreased sharply. It is postulated that this was because investors became aware of an unpriced cost of hedging idiosyncratic spread jumps (which are not included in the standard model) [Petrelli *et al.* 2006]. This led to a re-pricing of risk, i.e. protection sellers realised they should be charging more for providing protection, leading to a jump in implied correlation. The commonly accepted view is that it was the downgrading of the debt of Ford and General Motors by S&P to non-investment grade that triggered this event. The market had anticipated the event, but the timing of the downgrades was not anticipated. The downgrades led to fears that these obligors would rapidly move into default. In response to the perceived increase in default risk of these obligors their par CDS spreads blew out. Because both Ford and General Motors were pernicious throughout a large number of STCDOs, both bespoke and standardised, the MtM impact of the spread widening on sellers of (delta hedged) equity protection was significant.

In terms of the market, at the time the majority of dealers (brokers) were long equity, short mezz and long super-senior. Hedge funds were also long equity and short mezz (with little super-senior exposure). Mezz pieces typically were in the hands of buy-and-hold investors (pension funds etc.) because they provide an efficient means of earning a greater spread than the index but with a degree of subordination to protect against idiosyncratic defaults.

The sell-off of General Motors and Ford led to some hedge funds hitting risk limits and trying to close positions to make margin calls. The hedge funds were significantly exposed to the MtM fluctuations due to idiosyncratic risk. Dealers were also long equity and therefore had little appetite for buying the hedge fund positions. This led to equity prices plummeting (because there was no demand) and correlation dropping sharply. As the idiosyncratic risk of General Motors and Ford increased, more risk was transferred from mezz tranches into the equity piece leading to the mezz trades outperforming. Those dealers with short mezz positions tried to buy them back (because these positions were now under-performing). But mezz pieces were in the hands of long-term investors who did not want to sell. Mezz spreads therefore dropped sharply. Mezz tightening of spreads implies a drop in correlation. As correlation varies, the shape of the loss distribution of course also shifts (as seen in Chapter 7). This has the effect of redistributing the losses amongst the different tranches. In this case as mezz spreads dropped, more risk was transferred to the equity tranche until it eventually became saturated (because it has a finite total exposure). Some risk therefore spilled over into the senior tranches from the mezz tranches, leading to a doubling of super-senior spreads. These events sparked the growth of the leveraged super-senior tranche.

One of the consequences of this event was that hedge funds realised that investors holding similar directional positions would withdraw liquidity in times of crisis. In order to mitigate against this in the future, hedge funds devised investment strategies whereby investors were required to lock-in their investments for a period of time, or where the initial investment principal is protected (for example CPPI structures). Another consequence of the crisis was that it helped to reveal that there were shortcomings with the standard market model.

12.5.2 The credit crunch, 2007

The events of May 2005 were mostly contained within the credit markets (the response of broader equity indices was relatively muted) and were also relatively short lived. The correlation crisis was sparked by idiosyncratic risk in the automobile sector. The credit market's quickly recovered and the overall equilibrium of trades was shifted to reflect the new realities of the market's understanding of risk. Fundamentally there was not a great deal wrong with individual corporates and the strength of their balance sheets. Summer 2007 was an altogether far more serious and wide ranging event (a very concise summary of the timeline of significant events is provided in Jarrow *et al.* [2007]).

The most significant empirical observation (from a credit derivative perspective) was that in the period around July 2007 credit spreads on all the main CDS indices blew out very quickly to the widest levels yet seen (corresponding to a systematic re-pricing of risk by market participants). The levels observed were iTraxx main $\sim$60 bps (see Figure 4.1), iTraxx Xover $\sim$450 bps, CDX main $\sim$80 bps, CDX Xover $\sim$350 bps. Prior to this period spreads had gradually been moving downwards to close to their lowest ever levels, representing the market's benign view of the risks to the credit environment. It is worth noting in passing that there was (in hindsight) a precursor to these events observed in February/March of 2007. It was at this time that the problems in the US sub-prime mortgage market began to become apparent. Mortgage servicers and brokers in the USA began to report large losses due to foreclosures and delinquencies of sub-prime mortgage borrowers and to go out of business. The spectre of a US recession was also first raised. The credit spreads of banks that were believed to be exposed to the sub-prime business widened significantly. Some of these banks were included in the credit indices, and this is a major contributing factor to the widening of the indices observed in this period.

The precise catalyst for the sharp spike in July is a matter of debate. As always, hindsight will prove to be a powerful analytical tool and it is likely that no one single event will ever be identified to be the root cause. However, during the period between March and summer 2007 the sentiment amongst market participants began to deteriorate. The word 'recession' was used more frequently and there were almost daily reports of small hedge funds facing difficulties (investor redemptions and mark-to-market losses on their sub-prime related assets). It could be argued that it was the demise in June of a number of large hedge funds with significant exposures to ABSCDOs that proved to be the straw that broke the camel's back. In addition to this the major rating agencies abruptly conducted a blanket review of the rating quality of large swathes of structured finance vehicles, mostly those heavily invested in home equity securitisations. These events acted as a massive jolt to the functioning of the asset backed commercial paper (ABCP) markets, forcing market participants to scrutinise in great detail what exactly was in their portfolios and the portfolios of paper that they had purchased. The growing feeling amongst financial institutions was that it was unclear exactly who was sitting on what exposures and potential losses to sub-prime related investments. The big fear in the market was that what had started out as problems in a relatively small segment of the mortgage market (sub-prime mortgages account for less than 20% of all mortgages in the USA), would spread to other areas of the economy. If consumers were having problems repaying their mortgages (and were unable to re-finance or sell their homes due to the falling prices or the fact they had 'spent' their home through home equity withdrawal schemes), then this would probably curtail their ability to spend on other items such as automobiles or white goods leading to a reduction in overall consumer spending (which is, arguably, the major driver of growth in the US economy). If consumer spending reduced, growth would slow and a recession might ensue. The key issue for the markets therefore was to what degree would contagion from the sub-prime sector spread to other sectors of the US and global economies?

How could problems in the US mortgage market affect the wider world economy? A possible mechanism for this is the following. Mortgage brokers in the USA arrange mortgages to clients offering them very attractive initial interest rates. These 'teaser' rates were offered in an economic environment where interest rates were already near historical lows, making previously unaffordable properties affordable (and fuelling house price inflation in the process). They were also offered to borrowers with little in the way of collateral or assets to secure the loan against. Low headline interest rates also encouraged individual consumers and corporate entities alike to take on record levels of debt to fund other purchases. These mortgages are funded by banks who securitise them and sell on the rights to the future cashflows to other investors. The income from these notes allows the banks to fund more business. These notes are in the form of plain ABSs, MBSs, ABSCDOs etc. and are purchased by investors all around the world. For example, municipal councils as far away as Australia bought financial products backed by these assets. These investors now have direct exposure to the US sub-prime mortgage market. As interest rates move higher and initial teaser periods come to an end, mortgage rates reset to higher levels. Many sub-prime borrowers can no longer afford the new payments and become delinquent on their mortgages, eventually leading to foreclosure. Defaults on individual mortgages themselves have almost no impact on the global economy. Large numbers of defaults might have an impact on the indigenous economy without any linkages to the outside world. But delinquency and defaults of a large number of mortgages which have been repackaged and sold on to other investors globally will spread the sub-prime infection far and wide.

Because almost all global banks were involved in one form or another in the securitisation of home equity loans, it was clear that banks would have exposure to these losses and potential mark-to-market writedowns of asset backed securities. What was not clear was who was sitting on what exposures and who would suffer funding problems going forward. For this reason banks became very cautious about who they lent to in the short-term interbank markets. The ABCP market also began to dry up as investors became very wary about purchasing commercial paper that might be backed by sub-prime assets (leading to the demise of several SIVs as described in Chapter 1). This led to short-term LIBOR rates moving out significantly wider from the official central bank rates, making it harder for banks who needed short-term funding to get it. As a consequence, liquidity in the financial system began to dry up. Liquidity to a financial system is like oil to a car engine – without it, it will cease to function. For this reason central banks around the world began to pump massive amounts of cash into the short-term money markets in order get the gears moving again.

The problem for retail banks is that they rely not only on the flow of deposits from individual investors, but also on access to the money markets (this is particularly true in the UK) to fund activities such as lending to individual investors for purchases such as mortgages, cars etc. They borrow at LIBOR+ spread and lend at a higher rate than this. As LIBOR rates increase they have to increase their rates to customers to maintain profit margins, thereby making the cost of credit for individuals higher (and making re-financing ongoing commitments such as mortgages more expensive). As the ABCP markets dried up some banks were unable to fund their ongoing business. As a consequence, retail banks became much more cautious about whom they would lend to, imposing much tighter restrictions on their lending practices (for example making 'buy-to-let' mortgages in the UK much more expensive to obtain and service). All of these factors led to a 'crunch' in the global credit markets. Even with central banks pumping in liquidity (to reduce short-term lending rates and bring them more into line with their headline rates), banks were still reluctant to lend to one another, instead preferring to build up liquidity. At the end of the day investor confidence is what drives financial markets, not Brownian motions.

What did all this do to the credit derivative markets? There was a systemic re-pricing of risk (across all sectors) leading to the iTraxx and CDX indices widening significantly. This was driven principally by market concerns regarding sub-prime contagion to other areas of the economy (if consumers stop spending, company profits are reduced, in turn impacting their future survival prospects – evidence of this began to emerge in the UK retail sector at the end of 2007). In addition, concerns regarding the availability of cheap money in the future for financing LBOs as well as investor appetite for taking on risky debt associated with LBOs led to spreads moving wider (overall about 45 corporate debt offerings were cancelled or delayed due to lack of investor appetite for taking on the risk, meaning that these exposures remained on the books of the originating banks). Compared to the first half of the year when LBO activity was high (and funded by the availability of cheap credit), the number of LBOs dropped off significantly. Despite these market concerns, corporate results reported during the summer were generally positive and exceeding expectations (although corporate results tend to be a backward looking indicator of performance). It remains to be seen how strong corporate earnings will remain in the future if sub-prime losses increase and act as a drag on the real economy through diminished consumer spending.

In addition to the systemic spread widening, it was observed that some spread curves 'inverted' during this period. Under normal circumstances the par CDS spread of an obligor should increase as the protection term increases. That is to say, the cost of 10 year protection on an obligor should be greater than the cost of 5 year protection. During this period it was observed that this relationship was inverted; the cost of short-term protection was greater than the cost of long-term protection. This reflected the nature of the credit crunch – short-term liquidity and funding (re-financing) concerns played havoc with the normal relationships between financial variables.

The re-pricing of risk led to a fundamental shift of risk from the equity to mezz tranches within synthetic CDO capital structures. Equity tranche spreads widened to reflect the wider spreads, but on a relative basis the widening observed in the lower mezz tranches was greater. This led to a relative out-performance of the equity relative to mezz tranches As equity tranche spreads widen the equity tranche becomes more attractive since more of the portfolio loss distribution is 'squeezed' towards higher losses. It was observed that the base correlation of the 0–3% iTraxx equity tranche touched historical highs of approximately 30% (for a long time previous to this the correlation was range bound around approximately 20%). This was observed in conjunction with the aforementioned systemic index widening, which represented a shift from an idiosyncratic risk environment (LBOs etc.) to a systemic risk environment (implying a rise in default correlation). Correlation would rise as index spreads widened. This is because the index widening increased the volatility of credit spreads. As seen in Section 12.3, increased volatility in spreads is good for delta hedged equity tranche positions (which are also long correlation) which have positive convexity. Therefore as indices widened, the increased volatility represented more opportunity to crystallise P/L from positive gamma trades, thereby increasing demand for selling protection on delta hedged equity tranches (which drives up correlation). As spreads tightened, expected volatility would decrease and long gamma investors would take profits driving correlation back down. It was also observed during this period that the correlation exposure of delta hedged equity tranches did not change much, but of course spreads widened. The gamma value of the strategy is clearly correlation dependent. So although the rise in correlation eroded some of the value of the trade, the increase in volatility more than made up for this. The equity tranche has the largest positive convexity, making it very attractive to risk-takers who are willing to take on the default risk. However, as spreads widen, the expected portfolio loss moves up the capital structure. Hence mezz tranches start to exhibit greater convexity for sold protection positions. However, this was not enough to provide a more attractive yield than the equity tranche. For this reason the equity curve flattener (discussed in Section 12.4) was also a popular trade at this time.

12.6 Chapter review

Correlation trading is the term used to describe investment strategies that exploit the sensitivities of synthetic CDOs to input parameters such as spreads and correlation in order to express a particular view on the likely behaviour of the market. In this chapter we have introduced the basic concepts of correlation trading (for example empirical hedging of positions against spread fluctuations) and spelt out in general terms how synthetic CDOs can be used to express particular market views. We also analysed in more detail some of the most common trading strategies (delta hedged equity tranche etc.). Finally in this chapter we provided some commentary and views on the two major credit market events that have occurred since the advent of liquid trading in STCDOs.

In the next chapter we will extend ideas introduced in this chapter and consider how to characterise and risk manage a portfolio of synthetic CDO tranches. As we will see, understanding a portfolio of tranches will bring together a large amount of the technology that has been introduced in previous chapters.

13

Risk management of a portfolio of synthetic CDOs

13.1 Introduction

In previous chapters we have analysed the valuation, calibration and risk management of synthetic CDOs as individual entities (trades). Although it is clearly of fundamental importance to understand this, in practice a financial institution (investment bank, hedge fund etc.) is not going to have just a single-tranche position on its books and will in all likelihood have many hundreds of tranches as well as many thousands of other flow instruments such as single-name CDSs and credit index positions (for both speculation and hedging purposes). All of these tranches will have different attachment/detachment points, maturities, portfolio compositions etc. and there will be a combination of long/short positions. Individual traders, portfolio managers and originators will typically only have an interest in the positions that they are actively responsible for managing on their desk and so may not be overly concerned with the overall risk picture of the entire institution.[1] However, a risk management function will want to understand what the overall exposure of the institution is to, for example, spread or correlation movements, be able to identify concentrations of obligor risk, and understand what the impact of proposed remedial/corrective action will be upon the overall exposure.

In this chapter we will consider the aggregate behaviour of a hypothetical portfolio of CDO tranches of varying composition. Of particular interest will be the following questions.

- What is the P/L profile of the portfolio under certain market scenarios (such as systemic or idiosyncratic spread widening and simultaneous collapse in base correlations).
- How can the portfolio be immunised against particular market scenarios (such as the default of an obligor to which there is a particularly large exposure or a large blow-out of

[1] This is not to say that the desk is not interested in risk, quite the opposite. All traders require timely and comprehensive risk information to enable them to make informed investment decisions based on the anticipated risk/return profile.

spreads in a particular sector or a systemic spread widening scenario as occurred during summer 2007).
- What is the impact of a new trade upon the exposure profile of the portfolio. How can the new trade be designed to best optimise the performance of the portfolio in order to achieve a specified risk/return goal (for example to minimise the spread sensitivity of the portfolio).
- What is the worst case loss scenario for a given portfolio.

The structure of this chapter is as follows. Firstly in Section 13.2 we will specify what the problem is that we want to address and then formulate the problem mathematically and algorithmically. We will then outline the important risk factors driving the sensitivity of the portfolio value and provide measures for quantifying this in Section 13.3. These ideas will be put into practice by constructing a sample portfolio and analysing its properties (in Sections 13.4 and 13.5). The objective of this analysis is not to analyse a wholly realistic problem (consisting of many hundreds of tranche positions and thousands of CDS positions), but rather to construct a simplified problem with the objective of developing a better understanding of how the portfolio behaves and how it can be analysed. Having quantified the risk characteristics of the baseline portfolio we will, in Sections 13.6 and 13.7, consider how we can make modifications to the portfolio and how the impact of modifications to the portfolio can be monitored and controlled so as to achieve specified risk/return objectives. Finally in Section 13.8 we summarise the important points that have been introduced in this chapter.

13.2 Set-up of the problem

A typical credit derivative book run by a large investment bank will comprise many thousands of positions in a variety of different instruments. The totality of credit exposures may also span across different desks and different parts of the business. Each desk will monitor their own aggregate exposures, but they will typically have little interest in the exposures of other desks and potential concentrations of risks across desks. Risk management on the other hand is responsible for understanding the risks at an overall institutional level, and not just at an individual desk level.

The types of credit related instrument on the books will typically include the following (in no particular order).

- Synthetic index tranches (based on iTraxx or CDX indices, i.e. STCDOs).
- Bespoke synthetic tranches (bespoke attachment/detachment points, bespoke pools, bespoke maturities) – this is often the most important and difficult to hedge exposure on the book).
- CLOs and CBOs. These typically fall under the auspices of cashflow CDOs. These will almost always have bespoke cashflow waterfalls including interest coverage tests,

over-collateralisation tests and other cashflow diversion features, which require Monte Carlo simulation methods for valuation.
- CDO squareds (synthetic).
- ABSCDOs (these can be cashflow CDOs or synthetic CDOs where the underlying collateral pool is synthetic exposure to the tranched ABX indices – TABX).
- Default baskets (typically first-to-default).
- Index CDS positions.
- Individual single-name CDSs (referencing corporate and sovereign entities).
- Credit default swaptions (referencing corporate and sovereign entities and indices).
- Simple bond and cash positions.
- Various FX and interest rate swaps to hedge currency and interest rate exposures (typically FX and interest rate exposures are hedged at the overall book level and not at the individual trade level).

Usually the cash and synthetic CDO exposures are maintained in separate books. Even ignoring the non-synthetic exposures, risk managing a book this complex is a hugely involved problem from both a modelling and a computational perspective. This is mainly due to the diversity of underlying instrument (and not forgetting the computational burden). Vanilla, synthetic CDOs can be valued using any of the techniques introduced in previous chapters. However, an ABSCDO, for example, may reference a pool of mortgage backed residential loans or home equity loans or, more likely, a whole range of types of collateral. The modelling of these types of collateral is very different from the modelling of vanilla CDSs (and much more complex).

For the sake of simplicity and clarity of exposition, the analysis in this chapter will focus on a greatly simplified subset of the above list of potential exposures. In particular we will consider an example portfolio composed solely of synthetic CDO exposures. As we will see, even this simplified problem raises a significant number of modelling and risk management issues.

Typically the sorts of things we will be interested in quantifying will be the portfolio's sensitivity to obligor spreads, correlation, defaults etc. But in addition to quantifying the sensitivity of an existing portfolio of tranches to fluctuations in market variables, there are also a number of other questions that are important to address.

- In the role of CDO originator/structurer (structuring from scratch a CDO which meets the client's desired risk/return characteristics), how can we construct an appropriate pool which best satisfies the investment criteria set out by the client with the best return for the originator. For example, what is the optimal pool constructed from the universe of possible pool constituents such that the average rating of the obligors is BBB and the geographical concentration is no more than 50% in any one region while achieving the optimal spread return?

- In the role of portfolio manager responsible for a managed synthetic CDO (where the names in the pool can be substituted on a regular basis), what are the optimal portfolio substitutions to make to maintain certain characteristics of the overall portfolio (e.g. keeping the exposure concentration to an individual obligor below a set limit)?

13.2.1 Definition of terminology

Previously when valuing a single synthetic CDO tranche it has been assumed (quite correctly) that all the obligors in the universe of obligors are contained within a single pool. Now however we are considering the situation where we have a total universe of obligors who are divided up into different pools by CDO originators and investors for the purpose of slicing up the pool exposures to achieve a certain risk/return characteristic. The standard indices are a simple example of this. The main iTraxx index is constructed from a subset of size 125 from the total universe of obligors (which is of the order of 1500 in total); we could in principle construct another pool composed of obligors some of which are part of the iTraxx portfolio, and some of which are not.

In order to be clear, it is worthwhile defining carefully the nomenclature used in this chapter. The *universe of obligors* Ω is defined as the total collection of debt-issuing obligors that a CDO manager/structurer can choose from to place into a pool. An individual manager will in all likelihood have constraints placed upon them regarding the type of collateral they can include (for example pension funds can only invest in high-quality collateral and there are usually set concentration limits for individual obligors). A *pool* p is defined as a subset of obligors chosen from the universe of obligors, i.e. $p \subseteq \Omega$; the risk/return profile of the pool is sliced up by defining tranches on the particular pool. A *tranche* is defined as an exposure to a particular portion of the loss distribution of a specified pool. Figure 13.1, demonstrates the interdependencies described above and Figure 13.2 shows a schematic view of the portfolio organization. Clearly there are a large number of possible pools that can be defined from the universe of obligors. For each pool we can define multiple tranches (e.g. 0–4%, 7–9% etc.) referencing that pool (and also different maturities).

13.2.2 Mathematical formulation

For clarity we will formulate the problem in terms of a Monte Carlo simulation (where there are $\alpha = 1, \ldots, N_{\text{Sims}}$ simulations). We assume that there are $i = 1, \ldots, n$ obligors in the overall universe. A typical value for the size of the universe is $n \sim 1500$. Each obligor has a recovery rate given by δ_i and a par CDS spread of s_i. The recovery rate and par CDS spreads are properties of individual obligors (and not related to a particular pool or tranche, hence there is only a

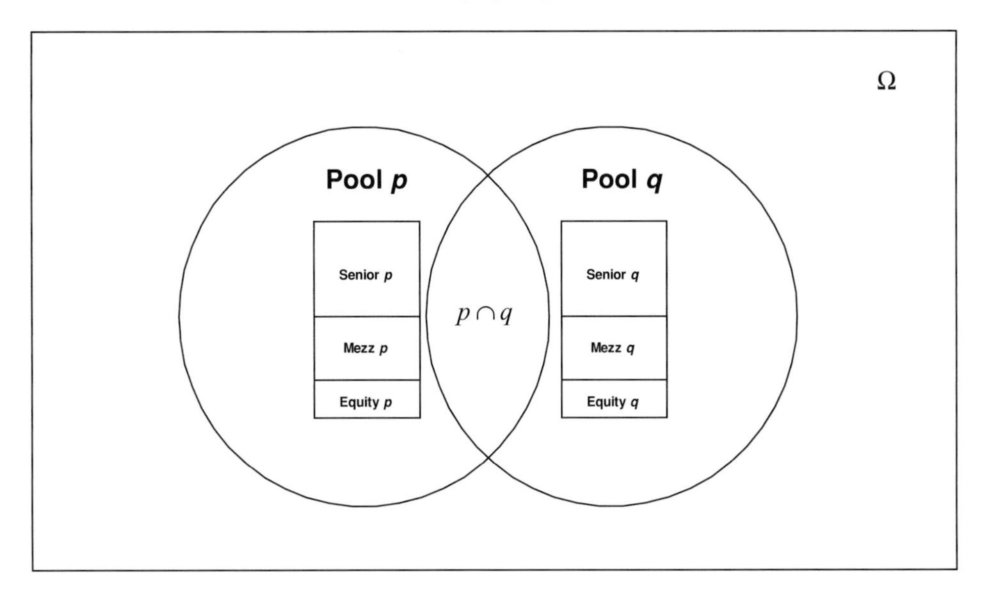

Figure 13.1 Venn diagram showing the structure and interrelationships of the universe of obligors and the pools and tranches.

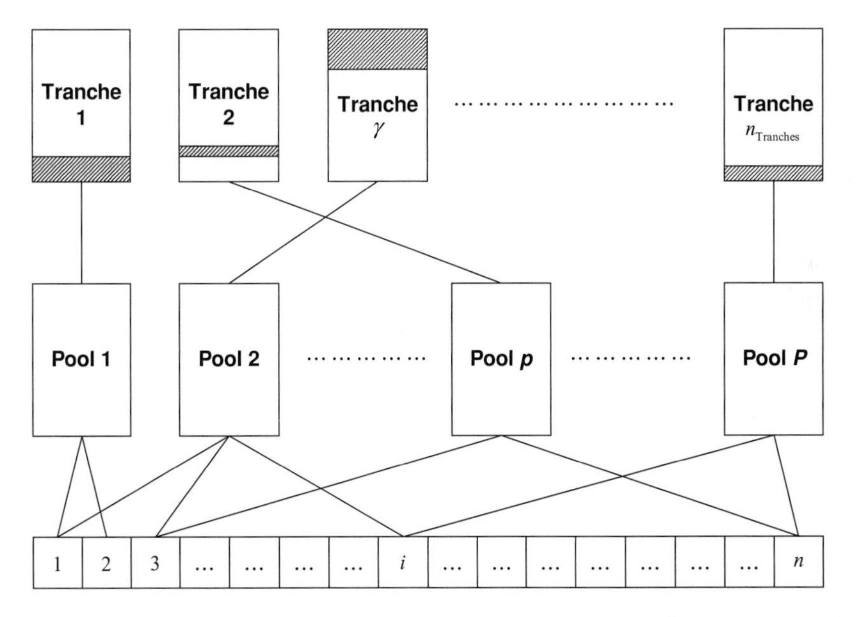

Figure 13.2 Schematic diagram showing how the individual obligors are grouped into pools and then into tranches. In this diagram, at the lowest level, we have the universe of obligors. Groups of obligors are combined into pools and finally tranches are defined which reference these pools and slice up their loss characteristics to achieve a desired risk/return profile. As we can see from this figure an individual obligor can appear in multiple pools (and by extension multiple tranches) but there is only a single linkage between a pool and a tranche.

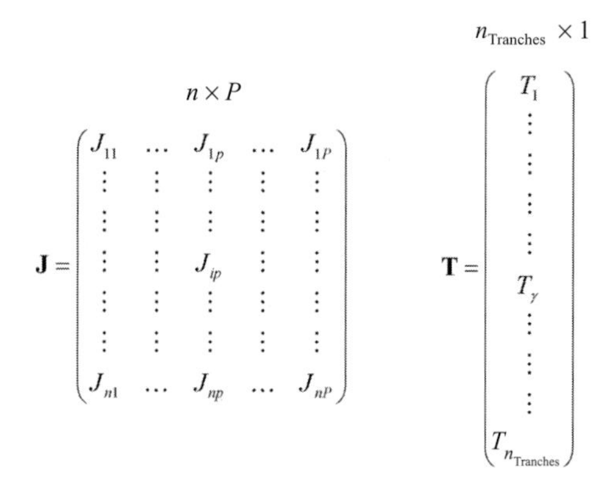

Figure 13.3 Examples of the obligor/pool connectivity matrix and tranche/pool vector.

single subscript i). Let there be $p = 1, \ldots, n_{\text{Pools}}$ separate pools into which these obligors are organised/referenced. Clearly an individual obligor may be referenced within several different pools. Tag a tranche by the index γ and assume there are $\gamma = 1, \ldots, n_{\text{Tranches}}$ separate tranches. We should have that $n_{\text{Pools}} < n_{\text{Tranches}}$ (otherwise we would have constructed some pools but then not defined any tranches which reference them). Each tranche has a maturity T^γ (for simplicity we will assume that all the tranches in the sample portfolio have the same maturity).

To characterise the connectivity of obligors and pools we define a matrix $\mathbf{J}$ where the element $J_{ip} = +1$ if obligor i is referenced in pool p and 0 otherwise.[2] $\mathbf{J}$ is therefore of dimension $n \times n_{\text{Pools}}$. Each pool can have any number of tranches defined on it, slicing up the loss distribution of that pool (e.g. 0–3%, 0–4%, 3–5.5% etc).

The connectivity of pools and tranches is denoted by the vector $\mathbf{T}$ (which is of dimension $n_{\text{Tranches}} \times 1$) where the element $T_\gamma \in [1, n_{\text{Pools}}]$ is a map denoting which pool tranche γ is defined on. Figure 13.3 gives an example structure of $\mathbf{J}$ and $\mathbf{T}$. Let the notional of obligor i in pool p be given by N_i^p. The total notional of pool p is given by

$$N^p = \sum_{i=1}^{n} J_{ip} N_i^p.$$

A particular tranche γ has attachment/detachment points given by $(K_{\text{L}}^\gamma, K_{\text{U}}^\gamma)$. The tranche notional is $N^\gamma = (K_{\text{U}}^\gamma - K_{\text{L}}^\gamma) N^p$.

[2] The motivation for this formulation derives from the statistical mechanical theory of disordered systems where the matrix $\mathbf{J}$ is referred to as the 'quenched disorder' (that is, disorder which is frozen into the system at inception and does not change) [Mezard *et al.* 1987].

As is the case for valuation of a single tranche, there is a correlation matrix **C** for the total universe of obligors (which is of dimension $n \times n$). This matrix is used to generate the correlated default times for each of the obligors. It is important to remember that correlation between obligors arises as a fundamental property of the interactions between the obligors, not the pools or the tranches. It is therefore correct conceptually firstly to generate the correlated default times of the obligors and then to aggregate the losses into the appropriate pools and then tranches. Let us assume that the correlation matrix is represented by a single homogeneous value ρ.

Although it is correct conceptually to think of the obligor default correlation as a property of the obligors and independent of pools or tranches, approaching the problem in this manner introduces a fundamental modelling problem. In practice in the market, as we have seen, there is correlation skew; different tranches defined on the same pool imply a different correlation. Consider we have two iTraxx tranches in the portfolio (referencing the current on-the-run index portfolio), a 0–3% and 3–6%. Each will have a different base correlation input which correctly reproduces the market observed price for the tranche. If we use this correlation in a Monte Carlo pricer we will (approximately) reproduce the observed price. But which correlation do we use to generate the correlated default times? This is an important issue and one that it is not clear how to resolve. In order to move forward we will assume that there is a single correlation input used to generate the default times. We will also use the same correlation subsequently to value the tranches.

We assume that for each simulation α we generate a set of correlated default times $\{\tau_i^\alpha : i = 1, \ldots, n\}$ for all of the obligors in the universe. If obligor i defaults on path α the loss to pool p is given by $J_{ip} N_i^p (1 - \delta_i)$ (assuming the obligor defaults before the maturity of the tranche). The total loss to pool p on path α at coupon date T_j is therefore

$$L_p^\alpha(T_j) = \sum_{i=1}^{n} J_{ip} N_i^p (1 - \delta_i) \mathbf{1}_{\tau_i^\alpha < T_j^\gamma}$$

and the cumulative loss to pool p on path α in the period $[0, T_j]$ is

$$\hat{L}_p^\alpha(T_j) = \sum_{k=0}^{j} L_p^\alpha(T_k).$$

Finally the loss to tranche γ (referencing pool$p$), on path α is

$$\tilde{L}_p^\alpha(T_j)^\gamma = \max\left\lfloor \min\left(\hat{L}_p^\alpha(T_j), K_{\mathrm{U}}^\gamma\right) - K_{\mathrm{L}}^\gamma, 0\right\rfloor$$

and the expected tranche loss is

$$\left\langle \tilde{L}_p^\alpha(T_j)^\gamma \right\rangle = \frac{1}{N_{\mathrm{Sims}}} \sum_{\alpha=1}^{N_{\mathrm{Sims}}} \tilde{L}_p^\alpha(T_j)^\gamma .$$

From this expression we can calculate the tranche fee and contingent legs and par spread and PVs. Let the value of tranche γ be given by

$$V^{\gamma} = -s^{\gamma} V^{\gamma}_{\text{Fee}} + V^{\gamma}_{\text{Cont}}$$

where s^{γ} is the contractual spread of the tranche. Each of the tranche positions can be long or short, denoting this by $\phi^{\gamma} = \pm 1$ (long/short). The total portfolio value is then given by

$$\Pi(\{s_i\}, \rho, \mathbf{J}, \mathbf{T}) = \sum_{\gamma=1}^{n_{\text{Tranches}}} \phi^{\gamma} V^{\gamma}(\{s_i\}, \rho, \mathbf{J}, \mathbf{T}).$$

In this expression we have explicitly indicated the dependence of the portfolio and tranche values on the various parameters. In the subsequent analysis to be presented all of the tranches are priced at par so that at zero perturbation to the market parameters the PV of the portfolio is zero. Note that we are making a simplifying assumption that all of the tranches have the same maturities (clearly this is an incorrect assumption in practice, but it will not impact the conclusions we draw from the analysis).

13.2.3 Algorithmic implementation

An algorithm for constructing the loss distributions for the different pools and then calculating the tranche losses and PVs is simply a generalisation of the Monte Carlo method introduced in Chapter 6. The general principle is to simulate the default times of the universe of obligors and then allocate losses to the different pools (taking into account the 'wiring' between the obligors, pool and tranches – similar to the CDO squared model presented earlier). For each simulation path the cumulative pool and tranche losses can be computed. The losses are then averaged over all the simulations and the tranche fee and contingent legs computed. Finally the overall portfolio value is computed as the straight sum of the individual tranche values. For vanilla synthetic CDOs the tranche loss (payoff) is a function only of the losses in the pool that that tranche is referenced to.

This valuation approach is referred to as a 'bottom-up look-through' approach. Bottom-up refers to the modeling of portfolio loss distributions at the individual obligor default level; look-through refers to the fact that the individual obligor losses are propagated through the individual tranches and up to the overall portfolio level.

The reader may justifiably be thinking that the formulation of the problem in such explicit detail is rather over the top for the problem. Surely all we need to do is value each tranche individually using one of the semi-analytic valuation models developed earlier. Under certain circumstances this is acceptable (making sure the correct obligors – recovery rates, spreads and notional amounts – are passed into the valuation algorithm). In particular if the payoff of each tranche is independent of other tranches (as is the case for vanilla CDOs) then it is feasible to apply the semi-analytic methods. This is because the conditional loss distribution of each of the tranches is independent of all the other tranches. This is a similar issue to that encountered for CDO squared trades where the payoff of the master tranche is a function of the losses to the slave tranches (hence overlap between the obligors in the slave tranche pools will make the calculation of the loss distributions problematic using semi-analytic methods). Please refer to Chapter 11 for more discussion of this point.

13.3 Portfolio risk measures

In general a portfolio of tranches will have the same sorts of sensitivities to fluctuations in market parameters as an individual tranche. On a day-to-day basis it is the responsibility of the portfolio/risk manager to measure, understand and control the risks and exposures. In particular a risk manager will want to characterise and monitor the following risks.

- The sensitivity of the portfolio to fluctuations in obligor par CDS spreads. This includes both systemic and idiosyncratic spread movements as well as small and large movements in spreads (the latter being referred to as gamma or gap risk).
- The default correlation sensitivity of the portfolio.
- The exogenous default sensitivity of the portfolio (VoD and running VoD).
- The time decay profile of the portfolio as the remaining maturity of instruments decreases.
- The impact over time of new trades being added to the portfolio and of existing trades rolling off.
- What-if analysis of the impact of new or proposed trades on the portfolio (before the trade is executed). For example does a new trade breach a portfolio concentration limit? Ideally this analysis should be available on an as-and-when-needed basis intra-day (although computational demands may make this impractical).
- The risk arising from multiple occurrences of the same obligor in different tranches (i.e. overlap/concentration risk)

The final point in this list is an additional important dimension of risk that must be quantified when considering multiple tranches: obligor concentration risk. This arises when the same obligor appears in multiple pools (and hence tranches). There is a limited universe of debt issuing obligors which a portfolio manager or CDO structurer/originator can choose from to include in a tranche. In addition to this if a desk has a view on the credit quality of a particular obligor or sector it would not be unusual for several different managers on the same desk to all want to invest (or not) in that obligor or sector. It is almost inevitable that the same obligor may appear in multiple tranches.

In general all of these risks should be monitored (and recorded) on a day-to-day basis. This allows a picture to be constructed of how the risk of the portfolio changes and evolves over time. To quantify the different portfolio risks the following quantitative measures may be used to understand, monitor and ultimately control the overall risk profile of a portfolio of structured credit instruments.

13.3.1 Spread sensitivities

Clearly the portfolio will be sensitive to fluctuations in the par CDS spreads of the underlying obligors. The following quantities (among others) can be defined:

- sensitivity of the portfolio to a bump in the spread of an individual obligor i, $\Delta_i^{\text{Port}} = \partial \Pi / \partial s_i$;
- sensitivity of the portfolio to a bump in the spread of a set of obligors $S \subseteq \Omega$ who constitute a particular sector, $\Delta_S^{\text{Port}} = \partial \Pi / \partial s_S$;
- sensitivity of the portfolio to a bump in the spread of all obligors in the total universe $S = \Omega$, $\Delta_\Omega^{\text{Port}} = \partial \Pi / \partial s_\Omega$;
- (correlation) skew aware versions of these sensitivities.

In addition to these measures we can define analogous second-order gamma risks. For example the second-order sensitivity of the portfolio to a bump in the spread of an individual obligor i can be calculated according to $\Gamma_i^{\text{Port}} = \partial \Delta_i^{\text{Port}} / \partial s_i$.

13.3.2 Correlation sensitivities

As was the case for a single tranche we can define a correlation vega according to

$$\nu_{ij} = \frac{\partial}{\partial \rho_{ij}} \Pi(\rho_{ij})$$

which is a matrix of sensitivities corresponding to independent bumps in each pairwise default correlation. In practice we will only consider the sensitivities to homogeneous (global) bumps in the default correlation.

13.3.3 Concentration/overlap risk

On a trading desk where the CDO portfolio is actively managed, managers will typically take a view on a particular obligor, and invest in it across all the structures (and tranches) they manage. This clearly increases concentration risk. Typically we would expect that the overlap between the totalities of tranches would be greatest between the same manager. For an investor to diversify their obligor exposure it becomes necessary to invest in tranches operated by different managers (an obvious point really). In addition to this there may be a trading limit to cap the overall degree of overlap of a manager with other managers from the same originator and there are typically individual obligor concentration limits that are set for a particular CDO. Concentrations are usually notional (exposure) weighted. Therefore in order to diversify the portfolio (keeping the notional weighted exposure below a specified limit) the manager must invest in many different names, further increasing the likelihood of obligor overlap amongst tranches.

In terms of concentration risk there are two competing effects that affect the overall portfolio exposure: idiosyncratic risk and systemic risk. Idiosyncratic risk refers to the impact upon the portfolio of an adverse movement in a single obligor. Systemic risk refers to the impact upon the portfolio of an adverse movement in a group (sector) of obligors. A portfolio of tranches will reduce idiosyncratic risk compared to a single-tranche position. This is because a portfolio will contain more separate obligors (chosen from the overall universe) than a single tranche and so the relative impact of a single obligor blow-up on the portfolio will be less than for a single tranche (assuming equal notional amounts etc.). On the other hand if the offending obligor is referenced in many different pools in the portfolio this will act to enhance the idiosyncratic risk. Similar considerations also apply to systemic risk; spreading exposures across a portfolio can reduce the impact, but multiple occurrences of the same obligors can enhance the impact.

There are two factors driving the concentration risk. On the one hand, the need to keep the individual obligor concentration to a specified level is a positive in terms of reducing single-name risk. On the other hand, if the manager can only invest a small amount in each name in each tranche, the same name may appear in many different tranches (because there are a limited number of obligors that the manager can choose from to place into a tranche). The first factor reduces the obligor concentration risk in a single tranche. The second factor increases the obligor concentration risk in the portfolio. It is not obvious a priori what the overall balance between these two risk contributors will be.

We can quantify obligor concentration risk by introducing an overlap matrix $\mathbf{O}$. The element O_{pq} of this matrix quantifies the degree of overlap between the constituents of pools p and q. We expect $0 \leq O_{pq} \leq 1$ representing the limits of no overlap and perfect overlap respectively and $O_{pq} = O_{qp}$. The dimensions of the overlap matrix are $n_{\text{Pools}} \times n_{\text{Pools}}$. A convenient way of visualising the concentration 'hot spots' is to colour code the elements O_{pq} of the matrix. Figure 13.4 shows a

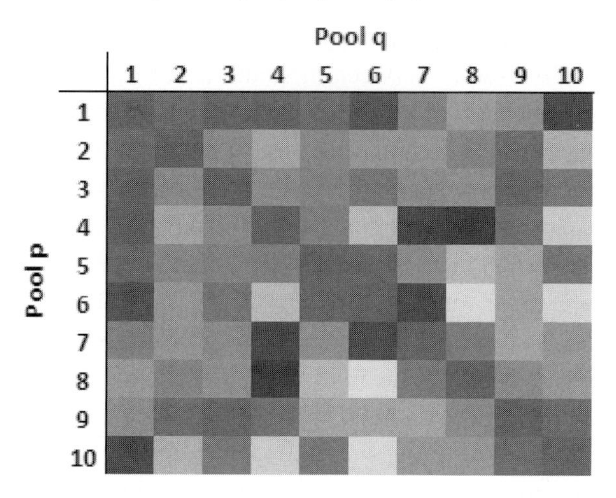

Figure 13.4 An example of the obligor/pool overlap matrix. The elements on the diagonal of this matrix are equal to 100% indicating that a particular pool has a perfect degree of overlap with itself. The figure shows this information represented in a colour coded format (the elements on the diagonal are coloured bright red indicating a high degree of overlap); white corresponds to 0% overlap, red to 100% overlap.

hypothetical example (in this example the overlaps are generated at random so there is no true information to be extracted from this plot). If we assume that $N_i^\gamma = N$, the elements of the overlap matrix can be written as

$$O_{\alpha\beta} = \frac{1}{\min(n^\alpha, n^\beta)} \sum_{i=1}^{n} \sum_{j=1}^{n} J_{j\beta} J_{i\alpha}.$$

Visual tools like this do not add any new quantitative information to the portfolio analysis. However, they can be very useful for quickly identifying potential areas of concern within a portfolio as they aggregate a lot of data into a readily interpretable format. In addition to looking at the pool concentrations, other overlap matrices can be constructed quantifying, for example, the degree of sector overlap between different pools etc.

13.3.4 Time decay sensitivity

Analogous to the case for a single tranche we can define the sensitivity to time decay of the portfolio as

$$\Theta_\Pi = -\frac{\partial}{\partial t} \Pi(t, \{T^\gamma\})$$

where we have explicitly included the dependence of the portfolio value on the maturity of the different tranches.

13.3.5 Obligor default sensitivity

Analogously to the single-tranche case, the sensitivity of the portfolio to defaults (idiosyncratic event risk) can be characterised via the following risk measures (see also De Servigny and Jobst [2007] and Douglas [2007] for other analyses similar to this):

- the marginal value-on-default (jump-to-default) value associated with the default of a single obligor i, $\text{VoD}_i^{\text{Port}}$ (for example which obligor contributes the largest marginal VoD);
- the running value-on-default (jump-to-default) value $\text{VoD}_X^{\text{Port}}$ when X of the obligors in the universe are simultaneously placed into default;
- default VaR.

These risk measures are computed analogously to the case for a single tranche (as was defined in Chapter 8). It is important to remember of course that an individual obligor may be referenced in multiple tranches and this must be accounted for.

How to choose which obligors to default for the running VoD measure is by no means clear. We could choose to default the riskiest obligors first (with the largest spreads). We could also choose to default them on the basis of their notional amounts (total notional across all the pools) – or the notional weighted spread – or by their degree of connectivity with the pools (i.e. first default the obligor which appears in the most pools, then the obligor with the next highest connectivity etc.). This is something that needs to be assessed carefully for each portfolio. Because the obligors have different connectivities etc., the order in which the defaults are introduced into the portfolio will impact the running VoD. This raises the question of what is the worst possible combination of defaults which gives rise to the maximum possible loss. From a computational point of view this is not an easy problem to solve as the possible combinations of default scenarios for a given number of defaults is large in a combinatorial sense.

13.4 Description of the sample portfolio

Having formulated the problem and identified the relevant risk measures we now come to the sample portfolio used to illustrate the concepts of tranche portfolio risk management. To characterise the sample portfolio to be analysed we need to specify the following parameters.

Total number of obligors in the universe	$i = 1, \ldots, n$
Number of separate pools referenced by the portfolio	$p = 1, \ldots, P$
Number of tranches defined on the reference pools	$\gamma = 1, \ldots, n_{\text{Tranches}}$
Maturity of each tranche γ	T^γ
Tranche attachment/detachment points	$\{(K_L^\gamma, K_U^\gamma) : \gamma = 1, \ldots, n_{\text{Tranches}}\}$
Tranche long/short flags	$\phi^\gamma = \pm 1$
Tranche portfolio weights	$0 \leq w^\gamma \leq 1$
Pool occupancy fraction	$0 \leq \psi_p \leq 1$
Tenor of coupons for each tranche	$\Delta^\gamma = T_j^\gamma - T_{j-1}^\gamma$
Notional of obligor i in pool p	N_i^p
Obligor/pool connectivity matrix	$\mathbf{J}$
Tranche/pool mapping vector	$\mathbf{T}$
Homogeneous obligor default correlation	ρ
Recovery rate of obligor i	δ_i
Par CDS spread of obligor i	s_i
Risk-free rate	r
Number of default simulations used to value the tranches	N_{Sims}

To compute the portfolio's spread sensitivity reported later the following steps are executed.

1. Specify the parameters $(n, P, n_{\text{Tranches}})$, N_i^p, (K_L^γ, K_U^γ), ϕ^γ, ψ_p, s_i.
2. Construct the obligor/pool connectivity matrix **J**.
3. Construct the tranche/pool mapping vector **T**.
4. Calculate the pool and tranche notional amounts.
5. Compute the par spreads for each of the tranches for the baseline par CDS spreads and default correlation.
6. Set the contractual spreads of the tranches to be equal to their par spreads (hence the PVs of the tranches and the overall portfolio will be identically equal to zero at zero perturbation to the model parameters).
7. Bump the par spreads of the obligors by a specified amount. Revalue the tranches and record the new portfolio value.
8. Repeat the above step for the appropriate spread range.

We set the size of the universe of obligors to be $n = 125$, from which we construct $P = 5$ pools. $n_{\text{Tranches}} = 10$ tranches are defined on the pools (so on average each pool has two tranches defined on it). The maturities of the (10) tranches are all assumed to be 5 years. In practice tranches will have different maturities but for the current purposes this simplifying assumption will not materially impact the analysis. The tranche coupon tenor is assumed to be quarterly. We assume that $N_i^p = \$10\,\text{m}$, i.e. homogeneous obligor notional amounts across all the pools and that $\delta_i = 40\%$. We assume that $\rho = 30\%$.

Table 13.1 *Summary of sample portfolio tranche properties*

Tranche number	Attachment point (% of pool notional)	Detachment point (% of pool notional)	Long (sell protection) or short
1	0	3	long
2	0	6	long
3	6	8	long
4	9	12	long
5	12	22	long
6	0	3	long
7	2	6	short
8	6	9	short
9	7	12	short
10	0	100	short

The obligor par CDS spreads are chosen at random according to $s_i = s_{\text{Min}} + (s_{\text{Max}} - s_{\text{Min}})u_i$ where $u_i \sim U(0, 1)$ are uniformly distributed random variables. We choose $s_{\text{Min}} = 20$ bps and $s_{\text{Max}} = 100$ bps (so that the average spread is approximately 60 bps). The pool occupancy fractions (the number of obligors in the pool divided by the number of obligors in the universe) are similarly chosen at random according to $\psi_p = \psi_{\text{Min}} + (\psi_{\text{Max}} - \psi_{\text{Min}})u_p$ where $u_p \sim U(0, 1)$ are uniformly distributed random variables. We choose $\psi_{\text{Min}} = 1/2$ and $\psi_{\text{Max}} = 1$. The elements of the connectivity matrix $\mathbf{J}$ are chosen at random according to $J_{ip} = \mathbf{1}_{u_i < \psi_p}$. The elements of the pool/tranche mapping vector $\mathbf{T}$ are chosen at random.

The tranche attachment/detachment points and long/short flags are shown in Table 13.1. Note that tranche number 10 is in fact an index-like position. To compute the spread and correlation sensitivities and also the marginal and running VoD measures we use 1000 default time simulations. For these calculations this small number of simulations is sufficient to obtain decent estimators for the tranche fee and contingent leg values. In the current context such a small number of simulations is adequate since we are only interested in obtaining a qualitative understanding of the behaviour of the portfolio. In a live environment it is important to quantify exposures precisely, and significantly larger numbers of simulations must be used.

In addition to the obligor characteristics listed above, each individual obligor will also have additional characteristics. These include

- agency credit rating (typically Moody's, S&P, Fitch or DBRS assigned ratings),
- geographical domicile (US, EMEA, Asia or emerging market),
- industrial sector classification.

For the purposes of the analysis to be presented, these obligor characteristics will not be of great relevance (since the portfolio, pool and tranche compositions are all artificially contrived). It is important to remember that in reality some of the most important risks arise from concentrations of obligors with similar characteristics.

Finally we remark that the sample portfolio constructed is not meant to be representative of a real portfolio held on the books of a financial institution. The purpose of simplifying the problem as we have done is to identify more easily what the drivers of the portfolio's behaviour are.

13.5 Basic analysis of the sample portfolio

We now analyse the sample portfolio detailed in the previous section. For simplicity all the tranches are priced at par so that in the baseline case where the spread and correlation inputs are not perturbed the total PV of the portfolio is zero. The risk measures reported here will all typically appear as part of a daily risk report that would be prepared overnight (reflecting the previous day's trading and market activity). The risk report is used to understand where the book might be over-exposed to certain types of risk and also as a tool for communicating to senior management the current exposure of the institution.

13.5.1 Descriptive statistics

Before analysing the portfolio in detail it is always instructive to consider some simple descriptive statistics of the portfolio. From a risk management perspective this information is sometimes the most valuable as it allows rapid identification of portfolio hot spots. For example, if we are concerned with spread widening in the financial sector then the first number to obtain is the exposure of the portfolio to this sector. Useful descriptive statistics to monitor as part of a risk report on a day-to-day basis include

- graphs of the main CDS indices (iTraxx and CDX) showing the history of the current on-the-run series (a graph is always a good tool for visually identifying developing trends),
- graphs of the history of CDS index tranche values (upfront payments, spreads, implied and base correlations),
- descriptive statistics of the main CDS indices (moments and rolling correlations) and the tranche exposures,
- a list of the top e.g. 10 widest and tightest credit spread obligors,
- the spread distribution of the obligors in the universe and the obligors in each of the tranches (how many obligors have spreads in the range 50–100 bps etc.),
- the credit rating distribution of the obligors in the universe and the obligors in each of the tranches,
- the geographical dispersion of the obligors in the universe and the obligors in each of the tranches,
- the industrial sector distribution of the obligors in the universe and the obligors in each of the tranches,

- the concentration of tranche and portfolio exposures to individual obligors (and checking that these concentrations do not breach specified limits),
- the degree of overlap between the composition of pools (and a list of the most frequently referenced obligors).

13.5.2 Marginal and systemic CS01s

The most obvious risk statistic to compute is the response of the portfolio to fluctuations in the spreads of the individual obligors. Table 13.2 shows a fragment of the marginal portfolio CS01s obtained by bumping individual obligor CDS spreads by 1 bp and revaluing the portfolio. In this table we report the marginal CS01s for each obligor for each tranche and for the overall portfolio. Also shown are the obligor spreads and the number of pools that each obligor appears in (for example obligor 3 appears in four of the five pools and has a spread of 54 bps). The first observation to make is that some of the tranches have zero sensitivity to bumps in some obligor spreads. For example the 0–6% tranche is insensitive to fluctuations in the spread of obligor 1. This is simply because this tranche is attached to a pool that does not include obligor 1.

The next important observation is that at an individual tranche level the widest spread obligors have the greatest sensitivities. For example, the first 0–3% tranche has sensitivities which are in direct proportion to the obligor spreads. At the portfolio level, however, the widest spread obligors do not necessarily translate into the largest Δ_Π. For instance obligor 117 has a spread of 65 bps, but an overall portfolio CS01 of only 280. Obligor 4 has a spread of 36 bps, but a CS01 of 2224. This is due to the increased connectivity of obligor 4 amongst the tranches. Also note that the marginal CS01s for each tranche can also be either positive or negative depending upon whether the position is long or short that tranche.

13.5.3 Portfolio spread sensitivity

Figure 13.5 shows the spread sensitivity of the individual tranches in the portfolio. In this figure all of the obligor spreads are simultaneously bumped and the tranches revalued. All of the tranches have contractual spreads set equal to the calculated par spread, so that at zero perturbation of the spreads from their initial values the tranche PVs are identically equal to zero. Figure 13.6 shows the overall portfolio value as a function of the systemic spread bump applied to all the obligors simultaneously. The portfolio value is plotted for long only positions, short only positions and the aggregate portfolio value. Also shown on this figure is the net portfolio exposure obtained using 100000 simulations. Clearly 1000 simulations is adequate for our current purpose of qualitatively understanding the portfolio sensitivity (although it is not recommended that so few simulations be used in practice). Although the

Table 13.2 *Marginal portfolio CS01s*

Obligor	Spread (bps)	Obligor/pool connectivity	Tranche CS01s										Portfolio CS01
			0–3%	0–6%	6–8%	9–12%	12–22%	0–3%	2–6%	6–9%	7–12%	0–100%	
1	48	3	−1910	—	533	—	−619	—	1735	—	—	—	−1326
2	98	3	—	—	−475	−371	—	—	1765	648	—	—	1566
3	54	4	−1967	−2979	—	−452	−589	−1917	1746	730	904	4274	−249
4	36	4	—	−2784	−546	−498	—	−1722	1696	767	974	4337	2224
5	48	3	−1912	—	—	−465	−617	—	1735	741	—	—	−518
6	96	5	−2265	−3224	−478	−375	−442	−2207	1766	652	776	4134	−1662
7	69	4	−2094	—	−508	−420	−525	—	1764	700	—	—	−1085
8	28	3	−1648	—	—	−523	−757	—	1655	783	—	—	−491
9	94	3	—	—	−480	−377	—	—	1766	655	—	—	1564
10	85	5	−2204	3179	−490	−391	−471	−2148	1768	670	805	4168	−1472
115	35	3	−1757	—	—	500	699	—	1693	768	—	—	494
116	81	5	−2176	−3156	−495	−399	−485	−2120	1768	679	818	4183	−1383
117	65	1	—	—	—	−429	—	—	—	708	—	—	280
118	68	4	−2087	—	−509	−422	−529	—	1763	702	—	—	−1083
119	30	4	−1683	—	−552	−516	−739	—	1668	779	—	—	−1042
120	41	4	−1828	−2847	−541	−484	−661	−1783	—	757	954	4319	−2115
121	69	4	—	−3091	−508	−420	—	−2041	1764	700	853	4222	1478
122	30	5	−1685	−2702	−552	−515	−737	−1645	1669	778	999	4356	35
123	43	4	−1856	−2875	—	−478	−646	1810	1722	752	944	4311	64
124	33	3	−1727	—	—	−506	−715	—	1683	773	—	—	−492
125	83	2	—	—	−493	−396	—	—	—	676	—	—	−214

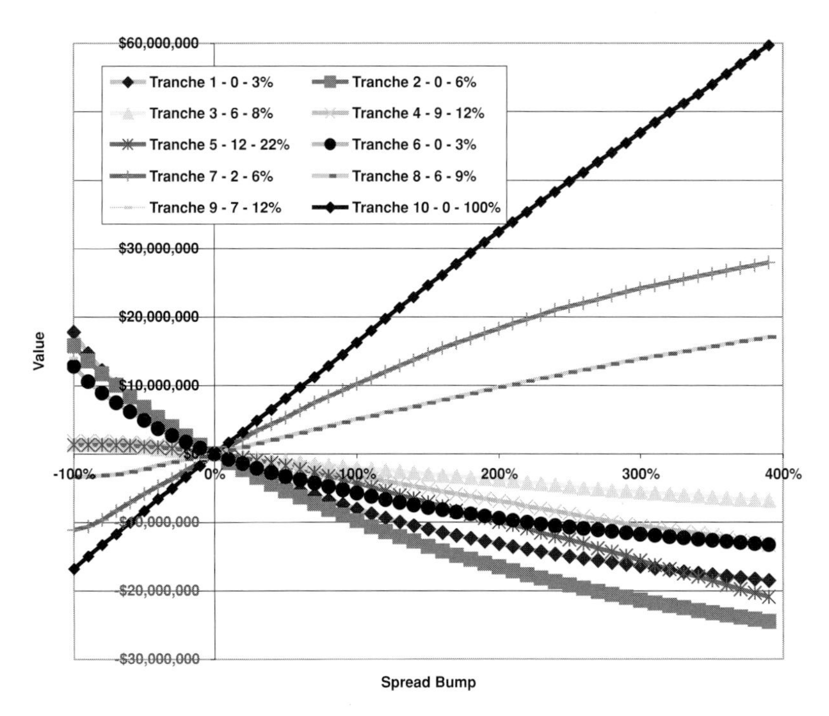

Figure 13.5 Spread sensitivity of the individual tranches in the portfolio.

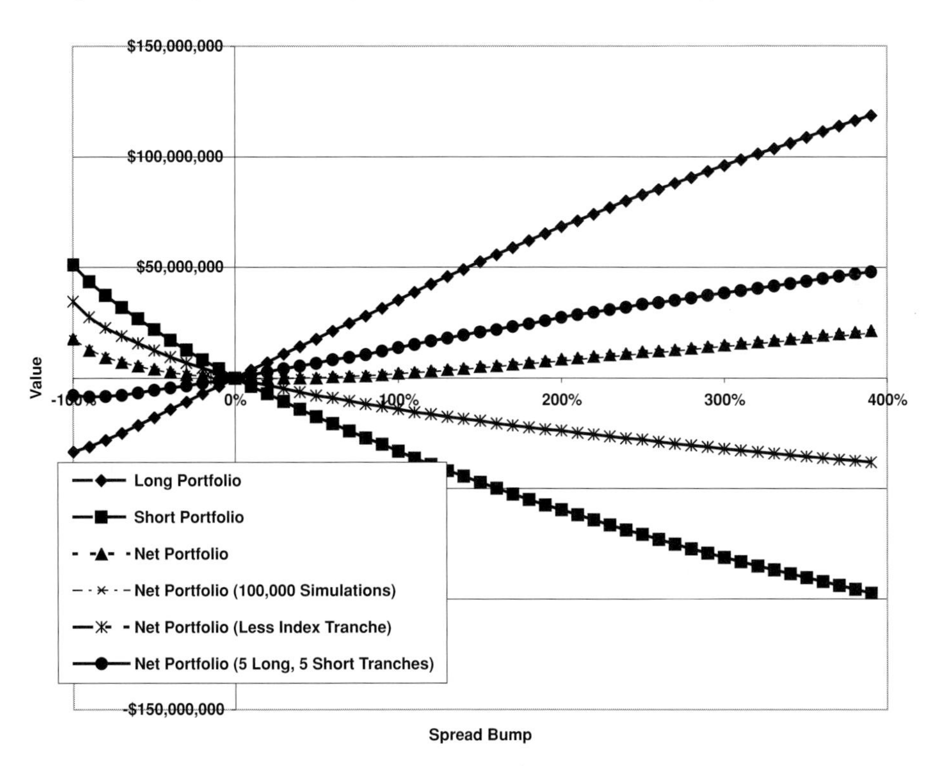

Figure 13.6 Spread sensitivity of the portfolio value. The long only, short only and net positions are shown, along with the net exposures when the index position is removed and also the net exposure when one of the long 0–3% tranches is turned short.

results are not shown on this graph, the results for tranche revaluations using semi-analytic models give quantitatively similar results.

Using the Monte Carlo model, for each simulation, we generate the set of correlated default times for all of the obligors. The defaults lead to losses to the pools and these losses are aggregated up into the tranches attached to these pools. It is very much a bottom-up approach. The semi-analytic model, however, treats each tranche in isolation and essentially calculates the loss distribution for each portfolio independently. That the two approaches give the same answer (within the tolerances of simulation noise) is a reflection of the fact that the payoff of each of the tranches is independent of the other tranches. In this case the loss distributions are themselves independent and the portfolio value can be computed as a straight sum of the individual tranche values. If one of the tranches had a payoff that was linked to another tranche (for example due to a complex structured product) the basic semi-analytic approach would break down and the only way to compute the correct joint loss distributions (and hence tranche losses) would be by Monte Carlo simulation.

From this graph we can conclude that the net portfolio position is well immunised against spread contractions (since the PV increases as the spreads decrease) and against large spread widening. For intermediate spread widening scenarios the portfolio P/L is slightly negative although not significantly so. We also plot on this graph the net portfolio exposure for the portfolio without the index position. It is apparent that the index position plays a significant role in ensuring the P/L is positive for large spread movements since the index exposure does not saturate as quickly as the other (smaller) tranches as the spreads widen. Finally also shown is the portfolio exposure when one of the long 0–3% tranche positions is turned short. As the spread widens the portfolio is now paying less on the contingent legs, hence the P/L increases. However the converse is true when spreads contract.

13.5.4 Correlation sensitivity

Figure 13.7 shows the correlation sensitivity of the portfolio as the homogeneous correlation is varied from its initial value. The results are plotted for the long, short and total positions. As the correlation increases the portfolio P/L is positive; as the correlation decreases the P/L is negative. To understand the drivers of the correlation exposure Figure 13.8 plots the correlation sensitivity of each of the tranches individually. The three equity tranches, which are all long (selling protection) are clearly the drivers of the positive P/L as correlation increases. As expected, the P/L of the equity tranches increases as correlation increases. Conversely, they are also responsible for the negative P/L as correlation decreases. The mezz and senior tranches do not display the same magnitude of correlation sensitivity as the equity

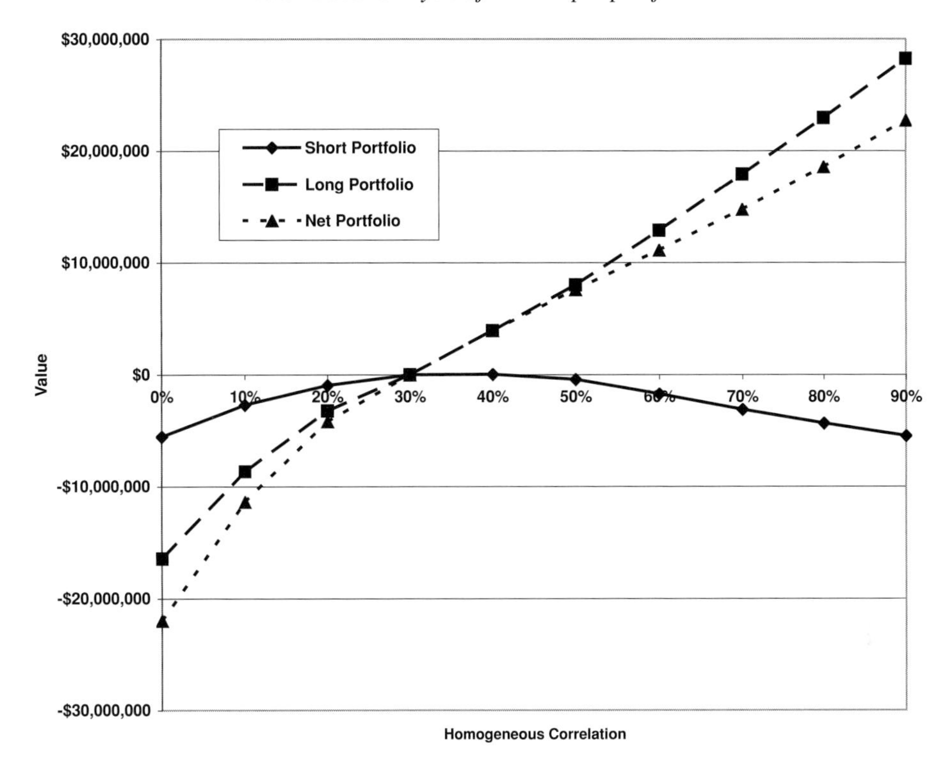

Figure 13.7 Correlation sensitivity of the portfolio value.

positions. Also the index position which saved the portfolio's spread sensitivity has no correlation sensitivity.

13.5.5 Marginal default sensitivity

Table 13.3 reports the marginal obligor VoD when each obligor is defaulted in turn. As was observed for the marginal CS01s, the magnitude of the marginal VoD for any particular obligor depends on its degree of connectivity amongst the pools. In this simplified example where all the obligors have identical notional amounts, the marginal VoD is directly proportional to the number of appearances of an obligor amongst the tranches. The same observations for marginal CS01s also apply here.

13.5.6 Running default sensitivity (running VoD)

We now turn our attention to the running VoD. In the previous running VoD examples studied in Chapter 8 we were only considering the response of tranches that were all attached to the same (identical) underlying pool. In the current context each

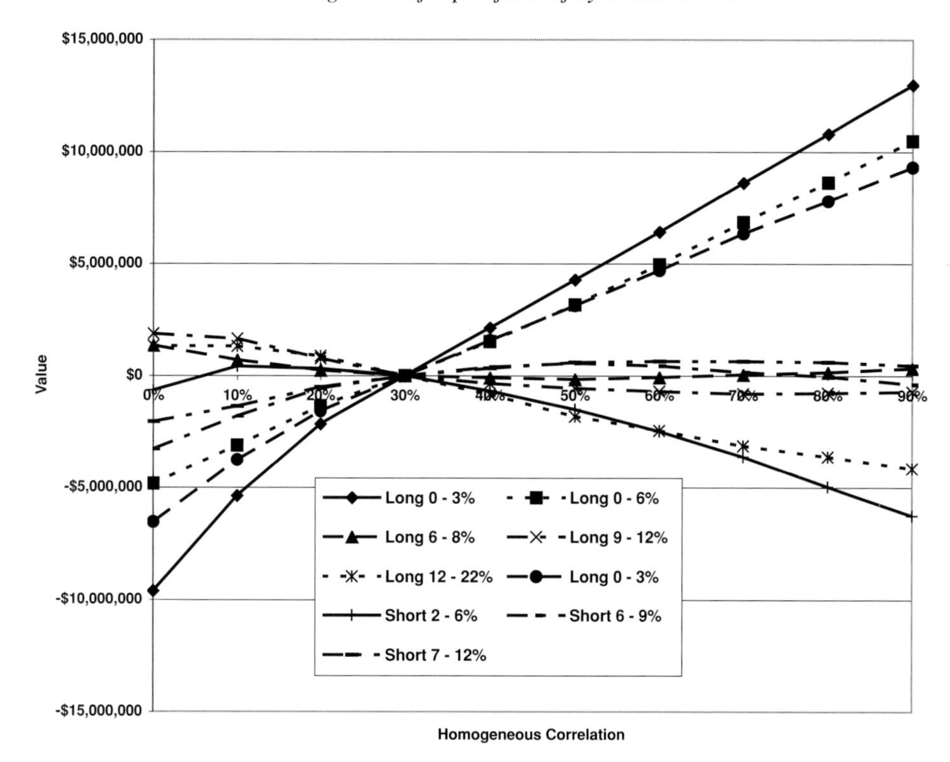

Figure 13.8 Correlation sensitivity of each of the individual tranches.

tranche is attached to a different underlying pool. Figure 13.9 shows the running VoD for each of the tranches in the portfolio. The obligors are defaulted in order of decreasing spreads (i.e. the first obligor defaulted is the one with the widest spread, the second the one with the next widest spread etc.). Consider the behaviour of the long 0–3% equity tranche. It was observed in Chapter 8 that the running VoD of an equity tranche becomes non-zero upon the introduction of the first default into the pool (because the equity tranche has no subordination). This, however, assumes that each obligor in the universe is present in the pool that the equity tranche references. In the current portfolio context this might not be the case. In the example studied it so happens that the first obligor defaulted (the one with the largest spread) does not appear in the pool that this tranche references. Therefore the default of this obligor has no impact upon this tranche. Eventually an obligor defaults which does impact this pool and a loss to the tranche is registered. As the defaults mount the tranche losses will eventually saturate as expected.

It is interesting to observe that the 6–9% tranche has a running VoD that is impacted by every default (bar one). This is because this tranche is connected to a pool that has a very high ($\sim$100%) occupancy fraction. This explains the smooth progression of the running VoD for this tranche (compared to the punctuated,

Table 13.3 *Marginal obligor VoD*

Obligor	Spread (bps)	Obligor/pool connectivity	Tranche VoDs										Portfolio VoD
			0–3%	0–6%	6–8%	9–12%	12–22%	0–3%	2–6%	6–9%	7–12%	0–100%	
1	48	3	−7 385 156	—	−303 103	—	−158 283	—	1 800 143	—	—	—	−6 046 399
2	98	3	—	—	−274 059	−148 613	—	—	1 704 691	340 896	—	—	1 622 915
3	54	4	−7 376 149	−7 189 407	—	−168 243	−154 421	−7 172 670	1 789 332	374 054	403 210	5 847 888	−13 646 407
4	36	4	—	−7 236 611	−310 704	−177 614	—	−7 196 542	1 822 823	388 833	422 712	5 926 875	−6 360 229
5	48	3	−7 384 805	—	—	−171 014	−158 127	—	1 799 717	378 503	—	—	−5 535 726
6	96	5	−7 303 540	−7 070 495	−275 327	−149 540	−130 990	−7 108 448	1 709 121	342 522	362 821	5 671 120	−13 952 756
7	69	4	−7 351 194	—	−290 566	−161 044	−145 101	—	1 760 533	362 216	—	—	−5 825 156
8	28	3	−7 414 018	—	—	−182 004	−173 630	—	1 836 864	395 476	—	—	−5 537 312
9	94	3	—	—	−276 266	−150 229	—	—	1 712 386	343 727	—	—	1 629 619
10	85	5	−7 322 713	−7 101 145	−281 195	−153 886	−136 219	−7 125 464	1 729 323	350 070	372 358	5 714 582	−13 954 288
115	35	3	−7 404 145	—	—	−177 931	−167 716	—	1 823 874	389 319	—	—	−5 536 598
116	81	5	−7 330 898	−7 114 373	−283 802	−155 849	−138 620	−7 132 716	1 738 136	353 437	376 637	5 733 722	−13 954 328
117	65	1	—	—	—	−162 995	—	—	—	365 463	—	—	202 468
118	68	4	−7 952 849	—	−291 138	−161 491	−145 668	—	1 762 397	362 963	—	—	−5 825 787
119	30	4	−7 411 163	—	−314 265	−180 777	−171 824	—	1 833 051	393 639	—	—	−5 851 338
120	41	4	−7 396 105	−7 223 853	−360 586	−174 914	−163 469	−7 190 208	—	384 654	417 153	5 904 749	−15 749 580
121	69	4	—	−7 147 325	−290 510	−161 001	—	−7 150 524	1 760 352	362 144	387 779	5 782 558	−6 456 528
122	30	5	−7 410 963	−7 250 312	−314 171	−180 692	−171 701	−7 203 228	1 832 786	393 512	428 984	5 951 488	−13 924 297
123	43	4	−7 392 595	−7 217 715	—	−173 664	−161 740	−7 187 127	1 809 274	382 697	414 562	5 894 334	−13 631 975
124	33	3	−7 407 136	—	—	−179 117	−169 414	—	1 827 754	391 129	—	—	−5 536 784
125	83	2	—	—	−282 864	−155 150	—	—	—	352 224	—	—	−85 780

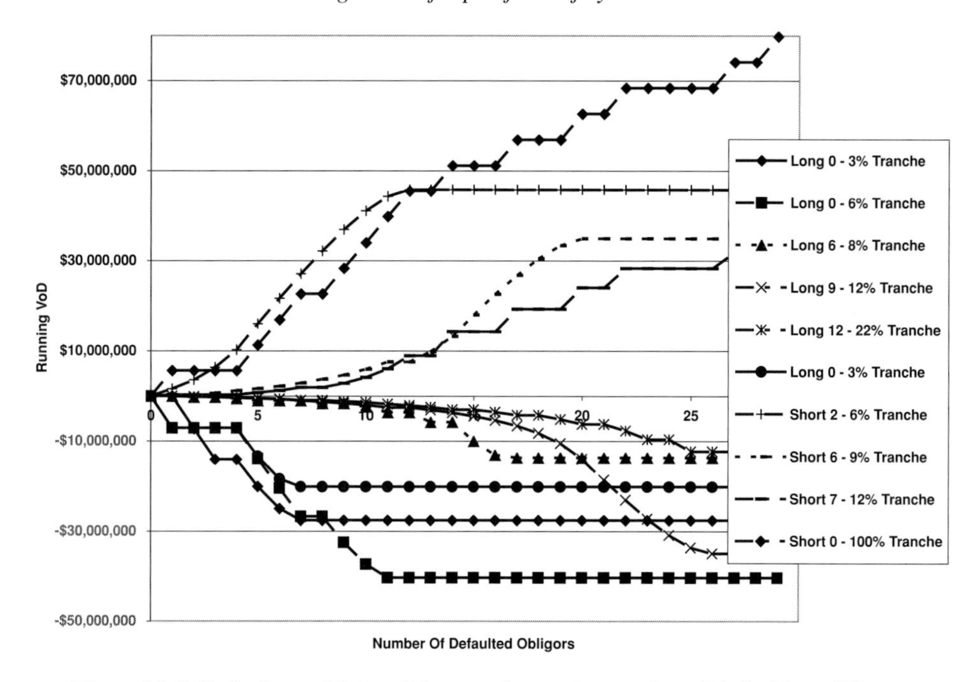

Figure 13.9 Default sensitivity of the tranches as the number of defaulting obligors is increased.

discontinuous jumps observed for some other tranches). The 2–6% and 9–12% tranches also display similar behaviour because they too are attached to high occupancy pools. For some tranches, such as the short 2–6% position, the running VoD saturates after approximately 10 defaults. Clearly this is because the tranche notional has been exhausted by the losses. In contrast the 0–100% (index) tranche has not saturated after 30 defaults.

Figure 13.10 plots the total long, short and net positions. As expected the long exposures have a positive VoD and the short exposures a negative VoD. The net exposure is initially negative but then becomes positive beyond approximately the first 10 defaulted obligors. The running VoD has been calculated by defaulting the obligors in order of decreasing spread. We can also default obligors according to notional amount (although in this case this will have no impact because we have assumed all the obligors have homogeneous notional amounts) and also according to the number of different pools an obligor appears in. Figure 13.11 shows the net portfolio running VoD where the defaults are ordered according to decreasing spreads, connectivity and at random. Note that the overall exposure can go up and down. Long or short exposures will always increase. As can be seen from this figure the order in which obligors are defaulted has an impact upon the cumulative losses (implying path dependence). This path dependence has an important consequence when attempting to compute the maximum portfolio loss.

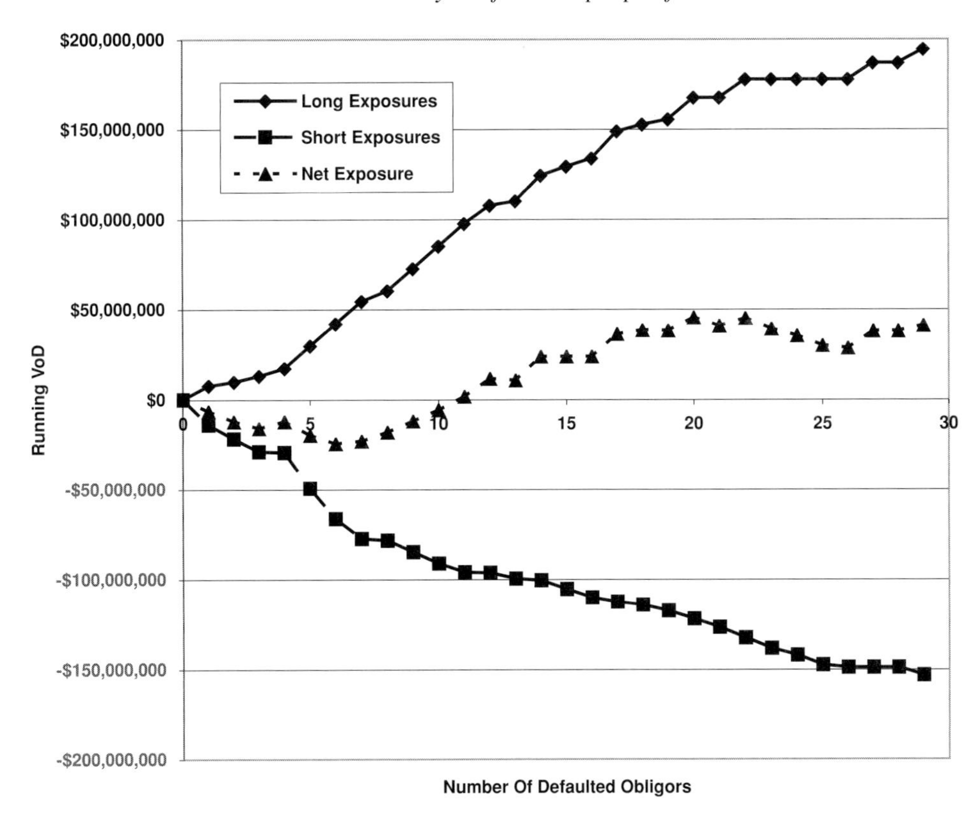

Figure 13.10 Default sensitivity for the long, short and net portfolio positions as the number of defaulting obligors is increased.

13.5.7 Computing the worst case portfolio loss scenario

The previous example demonstrated that the order of defaults has an impact on the running portfolio P/L. The obvious question to ask is the following. What is the particular combination of defaults which gives rise to the maximum portfolio loss? This is not as straightforward as simply choosing the default order based on spreads or connectivity. The problem, however, is the number of possible combinations of obligor defaults that are permissible. If there are n obligors in the overall universe there are $n!$ possible ways to order the obligor defaults; one of these combinations is going to give the maximum loss. Of course the possible number of combinations is too large to explore all of them exhaustively. Once again a stochastic search technique must be employed to search the possible combinations intelligently.

13.5.8 Credit spread and default VaR

Analogously for the case of a single obligor we can also define separate VaR measures to characterise the likely portfolio losses over a certain time horizon

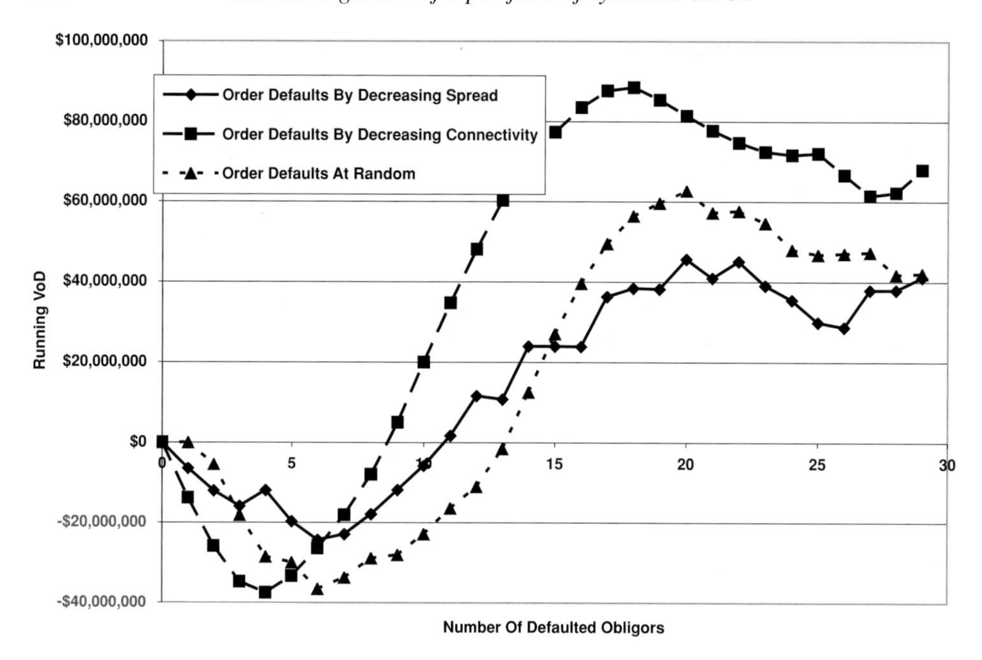

Figure 13.11 Portfolio running VoD where the defaults are ordered according to decreasing spreads, connectivity and at random.

due to spread movements and outright default. The analysis methodology is a straightforward extension of the models introduced previously and is left as an exercise for the reader to implement and analyse.

13.6 Adding new trades to the portfolio

The previous section analysed the existing portfolio. We now consider the impact of adding a new tranche to the portfolio. The motivation for a portfolio manager adding a new trade will typically be either

- hedging part of the portfolio sensitivity against adverse market movements (the manager may be taking a view that spreads will systematically widen and wish to immunise their portfolio's P/L against this scenario or position their portfolio to take advantage of it), or
- improving the spread return (carry) of the portfolio (without worrying too much about the risk exposure).

Let us assume for the sake of argument that the purpose of adding the tranche is to assist in hedging the portfolio's overall sensitivity to systemic fluctuations in the credit spreads of the obligors. The question is how do we choose the parameters (for example is it long or short, what are the attachment/detachment points etc.) and composition of the new tranche such that we achieve our objective of reducing spread sensitivity (or at the least, not increasing the spread sensitivity)?

To structure the optimal CDO we have to choose from the available universe of obligors the particular combination of obligors which best achieves our objective of minimising the spread sensitivity. We are in essence creating a new pool which has the new tranche attached to it. The problem is not as straightforward as simply choosing the obligors with the smallest spreads since we have to take into account the 'wiring' of the possible universe of obligors into the pools. Note that the tranches are not hedged per se at an individual level (as described in Chapter 12). We are attempting to structure the new tranche in such a way that it provides a natural hedge for the portfolio as a whole while satisfying the imposed constraints (perhaps this is an impossible task!).

For an overall universe of n obligors there are an astronomical number of possible choices of pool (even if we constrain the pool to have, for example, no more than $3n/4$ and no less than $n/2$ obligors). Even for a good portfolio manager who will have an expert opinion on the potential obligors to include (and will therefore be able to discount some obligors immediately), there are far too many possible combinations to analyse all of them systematically. It is necessary to utilise an appropriate optimisation methodology. Possible choices of optimisation approach could include genetic algorithms or simulated annealing (Appendix B provides a simple example of this technique). Both these methods are stochastic optimisation techniques that are well suited to problems with this type of combinatorial complexity. A downside to stochastic search algorithms, however, is the computational burden. Typically many iterations of the algorithm are required in order to explore adequately the problem phase space (i.e. the parameter space) but they usually do a good job of finding the true global extremis of the function being analysed. For the current application of adding an exposure to an existing portfolio, we are not necessarily facing severe time constraints (structuring a trade can take weeks or months to execute) so these methods are well suited to the problem (particularly as a simulated annealing algorithm may also take weeks or months to execute!). We now outline how to apply simulated annealing to the problem of finding the optimal tranche to add to the portfolio to minimise the overall spread sensitivity.

The original portfolio structure is characterised by the obligor/pool connectivity matrix $\mathbf{J}$ (of dimension $n \times n_{\text{Pools}}$) and the tranche/pool connectivity vector $\mathbf{T}$ (of dimension $n_{\text{Tranches}} \times 1$). The new portfolio will therefore have $n_{\text{Pools}} + 1$ pools and $n_{\text{Tranches}} + 1$ tranches. The new portfolio structure is characterised by a connectivity matrix $\hat{\mathbf{J}}$ and vector $\hat{\mathbf{T}}$. The additional column of $\hat{\mathbf{J}}$ will specify the composition of the new pool and the additional element of $\hat{\mathbf{T}}$ will specify the pool the new tranche is attached to (which is the newly created pool).

In order to apply simulated annealing to this problem we need to specify an objective function that is minimised. Our overall objective is to minimise the portfolio spread sensitivity over a particular range of systemic spread bump $s_i \to s_i + \beta s_i$ where $\beta_{\text{Min}} \leq \beta \leq \beta_{\text{Max}}$ (e.g. $\beta_{\text{Min}} = -1$ corresponding to a spread contraction of

100% and $\beta_{\text{Max}} = 4$ for a spread widening of 400%). In this case a suitable objective function might be

$$E(\hat{\mathbf{J}}) = \sum_{\forall\beta} \Pi(\{s_i\}, \beta, \hat{\mathbf{J}}) + \Delta E,$$

i.e. the sum of the portfolio P/L over the range of spread fluctuations. The objective of the optimisation is to find the structure of the new tranche that gives the minimal average spread profile. If we want to add constraints to the optimisation these are added into the objective function as an additional term ΔE. For example, if we require the number of obligors in the new pool $\hat{n}$ to be in the range $n_{\text{Min}} \leq \hat{n} \leq n_{\text{Max}}$ we need an extra term in the objective function of the form

$$\Delta E = \lambda \mathbf{1}_{n_{\text{Min}} \leq \sum_{i=1}^{n} \hat{J}_{in_{\text{Pool}}+1} \leq n_{\text{Max}}}$$

and we have to choose $\lambda > 0$ (and 'large') so as to penalise any potential system configuration that violates the constraint. The simulated annealing algorithm is as follows. In this algorithm we denote by $\Gamma = \{s_i, \delta_i, N_i, l, u\}$ a configuration of the system:

> generate initial configuration Γ at random
> calculate objective function $E(\Gamma)$
> set $E_{\text{Min}} = E(\Gamma)$ and $\Gamma_{\text{Min}} = \Gamma$
> for $T = T_{\text{Start}}$ to T_{Stop} step ΔT
>   for $\alpha = 1$ to N_{Sims}
>   generate new configuration Γ'
>   calculate $\Pi(\Gamma', \beta)$ for all β
>   calculate $E(\hat{\mathbf{J}}) = \sum_{\forall\beta} \Pi(\{s_i\}, \beta, \hat{\mathbf{J}})$
>   generate $p \sim U(0, 1)$
>   if $E(\Gamma') < E(\Gamma)$
> $E_{\text{Min}} = E(\Gamma')$, $\Gamma_{\text{Min}} = \Gamma'$
>   else if $p < \mathrm{e}^{-(E(\Gamma')-E(\Gamma))/T}$
> $E_{\text{Min}} = E(\Gamma')$, $\Gamma_{\text{Min}} = \Gamma'$
>   end if
>   end loop over α
> end loop over T.

For such a sophisticated technique (the fundamental science behind simulated annealing is amazingly complex, being a direct descendant of the statistical mechanical theory of disordered systems [Goldbart *et al.* 2004]) the algorithmic implementation is remarkably simple! The really subtle point about simulated annealing is the test $p < \mathrm{e}^{-\Delta E/T}$ (and the temperature dependence). This is what allows the algorithm to jump probabilistically out of local minima in the hope of finding global minima. As the temperature decreases the probability of jumping out of local minima decreases, meaning that the system essentially freezes into a particular configuration. At high temperatures the algorithm is able more easily to explore the system phase space; as the temperature cools the system freezes.

The trick with simulated annealing is to know how to choose new system configurations Γ' and how slowly the system needs to be cooled. If there are a large number of nearly degenerate (equal) minima then the cooling schedule must be sufficiently slow to allow the algorithm to explore its phase space adequately. The new configurations must be chosen so as to evolve the system configuration gradually rather than jumping discontinuously between very different configurations. For example, each evolution of the system configuration could be generated by flipping a certain percentage of the 1s to 0s (and vice versa) in the extra column of the $\hat{\mathbf{J}}$ matrix. The bottom line is that some experimentation must be undertaken to determine the optimal algorithm parameters.

The simulated annealing algorithm outlined may prove to be computationally very demanding (requiring a slow cooling schedule and a large number of iterations per temperature). One of the advantages of stochastic search techniques (i.e. techniques where system configurations are chosen at random and each choice of configuration should be independent of all other choices), like other Monte Carlo methods, is that they are very amenable to implementation on parallel processing architectures. Figure 13.12 shows a schematic view of how the tranche optimisation problem could be implemented on multiple processors. In this formulation we have a master processor which coordinates the activities of all the other processors. The master processor controls the temperature loop and at each temperature distributes the configuration iterations amongst the slave processors (passing to the slaves the current temperature and the currently obtained optimal system configuration). Each slave performs their iterations and when these are finished returns to the master processor the optimal solution obtained. The master processor collates all this information and keeps the best solution. The process then begins again at the next temperature. It is important to ensure that each of the slaves generates configurations that are independent of the other slaves. Some care must therefore be taken with the choice of random number seed for each of the slave processors [Jackel 2002].

13.7 Origination of synthetic CDOs

The previous sections have analysed the properties of a pre-existing portfolio of synthetic CDO tranches. In this section we will consider a slightly different problem: how to construct (originate) a synthetic CDO from scratch to satisfy a client's requirements.

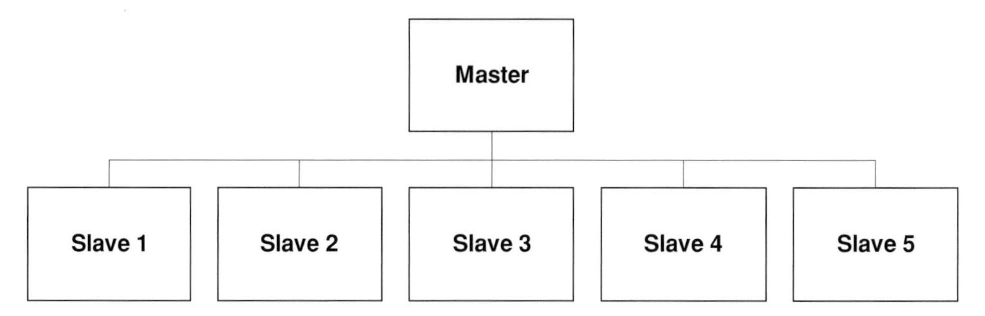

Figure 13.12 Schematic diagram of a possible parallel processing architecture for simulated annealing.

A CDO originator's task is to construct a synthetic CDO essentially from scratch subject to certain constraints. The constraints might be to limit the concentration of the pool to, for example, no more than 50% located in any one geographical region. Part of the role of the originator (and part of the justification for their associated fees!) is to leverage their knowledge and expertise of the markets and the obligors in the market in order to fulfil all the constraints and requirements of the CDO investor.

Consider the following problem. We have a universe of $n = 125$ obligors from which we can choose a fraction ψ (subject to certain maximum and minimum limits) to place into a pool. Each obligor has a spread s_i, notional N_i (assumed homogeneous) and recovery rate δ_i (also assumed homogeneous). Let us assume that the maturity of the tranche is specified to be T and that the (homogeneous) obligor default correlation is ρ. Finally we assume that we wish to invest in a 0–3% equity tranche.

To construct the tranche we need to choose the obligors from the overall universe. As for the portfolio case let us define a connectivity matrix (or in this case vector of dimension $n \times 1$) $\mathbf{J}$ where $J_i = 1$ if obligor i is included in the pool and zero otherwise. Given a particular choice of pool composition we can characterise the 'risk' of this pool by computing the weighted average spread (in this particular example all the notional amounts are identical hence a notional weighted average will have no impact on the calculation). This is calculated according to

$$\bar{s} = \sum_{i=1}^{n} J_i s_i \bigg/ \sum_{i=1}^{n} J_i.$$

In this example the obligor par CDS spreads are chosen at random according to $s_i = s_{\text{Min}} + (s_{\text{Max}} - s_{\text{Min}})u_i$ where $u_i \sim U(0, 1)$ are uniformly distributed random variables. We choose $s_{\text{Min}} = 20$ bps and $s_{\text{Max}} = 200$ bps. A random sample from this population (irrespective of the size of the sample as long as it is not too small) will

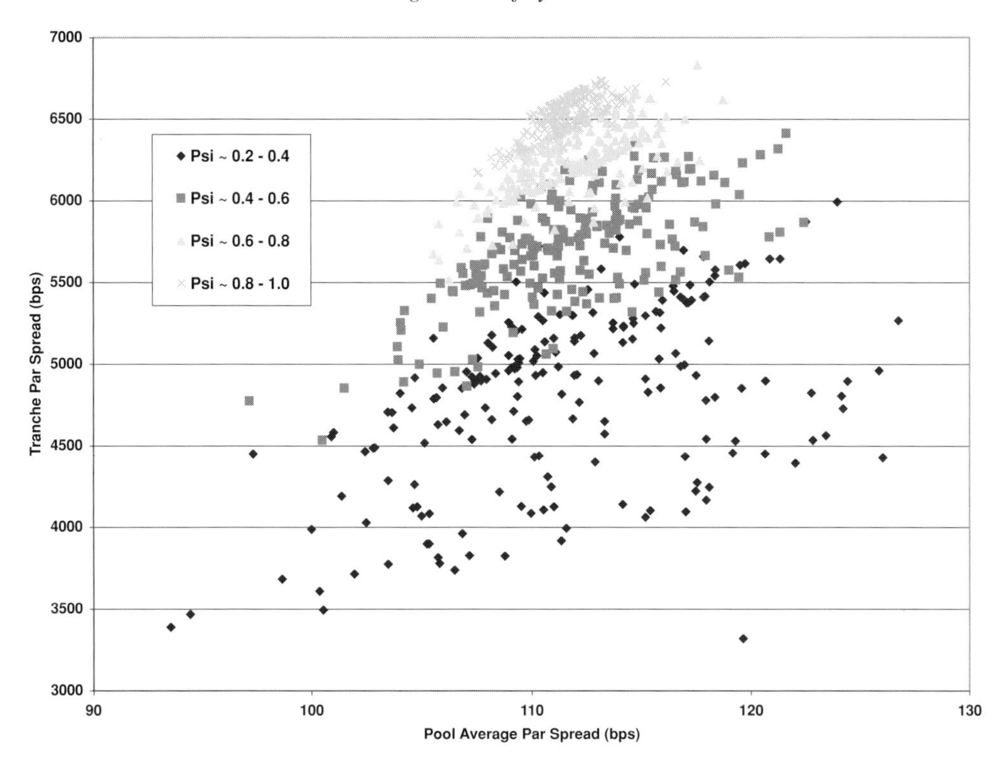

Figure 13.13 Tranche risk/return profile for randomly chosen pool compositions.

tend to an average spread of 110 bps. For a given pool composition characterised by $\mathbf{J}$ the 'return' is the tranche par spread $s(\mathbf{J})$ which is the carry earned by the seller of protection.

From the perspective of a seller of protection the requirement is to determine the optimal choice of pool composition such that the par spread is maximised for a specified level of risk (average pool spread). Figure 13.13 shows a crude optimisation approach (based on the technique of iterative convergence – otherwise known as trial-and-error). This figure is generated as follows. Firstly a range for the pool occupancy fraction ψ is specified. Then a sample pool is constructed at random with this occupancy fraction (which we allow to take values of 20–40%, 40–60%, 60–80% and 80–100%). For this sample pool the tranche par spread and pool average spreads are computed. This procedure is repeated multiple times for different random choices of pool.

When the pool occupancy fraction is low (e.g. 40 obligors chosen from a universe of 125) there are many different possible combinations of obligors that can be chosen to include in the pool. This implies that there is a wide dispersion of possible spreads in the pool (although the average will always tend to 110 bps). Conversely, if $\psi \equiv 1$

there is only one choice of pool composition (all the obligors included) and hence only one possible value for the average spread and also the tranche par spread.

Figure 13.13 is consistent with this expectation. At low pool occupancy fractions there is a wide dispersion of pool average spreads. At higher occupancy fractions the dispersion of pool average spreads decreases. By a similar argument the tranche par spreads will be widely dispersed for low occupancy fractions and more tightly clustered at higher occupancy fractions. There is a general upward sloping trend to the clusters of points, indicating that as the average portfolio spread increases (i.e. as the pool becomes riskier), the tranche par spread increases. Having characterised the risk/return profile, an originator could now decide on the level of risk that they wanted (by choosing an average spread) in the pool and choose the pool composition which gave the maximum return for this risk.

In this simple example a lot of the tranche parameters have been kept fixed or specified exogenously (such as the tranche attachment/detachment points). The only parameter that was allowed to vary was the precise composition of the pool. In the more general case these additional parameters will also be allowed to vary. This increases the volume of parameter space that needs to be searched in order to determine the solutions that achieve the optimal risk/return ratio. A simple Monte Carlo approach whereby points in the parameter space are sampled at random will cease to be effective at finding the optimal solution. In these circumstances more sophisticated approaches such as simulated annealing or genetic algorithms are once again recommended.

13.8 Chapter review

Managing any large multi-instrument portfolio is always going to be a difficult task. Different instruments have different characteristics and are sensitive to different risk factors. This means that as market observables fluctuate, some instruments will move in one direction and others will move in another direction. In this chapter we have set out some of the issues related to risk managing a portfolio of synthetic CDO tranches. The key risk sensitivities of a portfolio of tranches are similar to the sensitivities of a single tranche: spreads, correlations and defaults. However, for a portfolio there is also the issue of overlap of obligors amongst different tranches and pools. This introduces an additional interaction between tranches which is difficult to characterise at the portfolio level.

14

Hedging simulation of structured credit products

14.1 Introduction

In previous chapters we have introduced and analysed the standard market models for the valuation and risk management of synthetic CDOs. In this final chapter we consider the application of a relatively advanced simulation methodology for assessing the practical usefulness of a model: hedging simulation. For an investment bank trading and structuring complex financial derivatives on behalf of clients and for their own proprietary trading purposes, the quantification and management of risk is of paramount importance. In an ideal world the bank would be able to take positions in complex derivatives on behalf of clients and simultaneously enter into offsetting trades which would eliminate their risk. The spread that the bank earns as a commission for being the intermediary and structuring the deal then represents a risk-free profit. Life, unfortunately, is never that simple.

A significant risk that the investment banks run is the need to come up with prices for the structured instrument they construct for their clients. If the instrument is vanilla and liquidly traded there will probably be a market price that can be observed. It is more likely, however, that the instrument is bespoke and highly exotic to satisfy a particular client's risk/return requirement. To determine a price typically requires a sophisticated model. The uncertainty now is that the model fails to capture adequately all of the relevant risks that the product is sensitive to, leading to a price that could be arbitraged or a model that will not prove to be very effective at risk managing the exposure. To mitigate the risks an investment bank will construct an appropriate strategy for hedging the exposure (which requires calculating the sensitivities of the model to variations in parameter inputs). But if the hedging strategy is poor (either because it is an inherently poor hedge or because the model is a poor representation of the true behaviour of the instrument) then the investment bank will 'leak' P/L as time passes and the market evolves. Hedging simulation is a technique which can be used to assess the effectiveness of a model and its associated hedging strategy.

This chapter is structured as follows. In Section 14.2 we briefly describe the concept of hedging simulation and what its potential uses are. Section 14.3 then discusses the application of this technology to structured credit products and outlines what the additional difficulties are. As a precursor to a more advanced model, Section 14.4 applies hedging simulation to a simplified problem of hedging a first-to-default basket in the limit of zero default correlation amongst the obligors in the basket. In this case the basket can be perfectly hedged with offsetting CDSs. Following this Section 14.5 applies the same techniques to synthetic CDOs, which as we will see offer a wide range of interesting behaviours. Also described in this section are two very recent examples of models which attempt to remedy some of the problems with the standard market model when it comes to hedging.

The related topic of simulation of potential future exposure is discussed in Section 14.6 (this is a topic that deserves a chapter all of its own, so the level of detail in the current chapter is kept quite simple). Finally in Section 14.7 we summarise the key points that have been introduced in this chapter. It is important to add the following caveat to this chapter. Hedging of synthetic CDOs is an issue which currently does not have a widely accepted solution. The material presented in this chapter is exploratory to some degree and the reader should adopt a critical viewpoint when reading this chapter. Indeed, the reader is encouraged to implement their own simulation models in order to test out the validity of the ideas discussed, and to also test out their own hypotheses.

14.2 What is hedging simulation?

Consider a complex structured derivative product that an investment bank has sold to an investor, for example, an interest rate derivative structured so as to allow a directional bet to be taken on underlying rates. In return for constructing this bespoke product and selling it to the investor, the investor pays the investment bank a premium. The investment bank is therefore short the position in the product and as a consequence exposed to market and credit risk because they have an obligation to deliver the payoff of the product to the investor. In order to mitigate these risks the investment bank can construct a portfolio of instruments which replicate the payoff profile of the structured product. The combination of the structured product and the replicating portfolio effectively (temporarily) immunises the investment bank against the risk of being short the structured product. Note that the simplest mitigation against the risks of being long an instrument is simply to go short the same instrument; if the short position requires funding at a slightly lower level than the long position receives, a risk-free profit could be had. However, market efficiency implies that such simple arbitrages will not persist in the market for very long (if existing at all). In addition to this, if we are trying to hedge an exotic (and therefore probably illiquid) instrument, an offsetting position in all likelihood does not exist.

By the fundamental theorem of asset pricing (see for example the classic texts [Baxter and Rennie 1996, Musiela and Rutkowski 1998, Etheridge 2002, Joshi 2003] for extensive details on no-arbitrage pricing theory) the cost of the derivative security is equal to the cost of constructing the replicating portfolio. Determining the value of the product (and of the simpler, more liquid hedging instruments) requires the specification, parameterisation, calibration and usage of a model. If the investment bank can charge the investor this amount plus an additional spread (it is reasonable for the bank to expect to earn a spread since they are the one applying resources and using their expertise to solve the client's problem), then the additional spread represents the profit for the bank.[1] The task of a model is therefore to prescribe an appropriate hedging strategy for a particular payoff, and to price the replicating portfolio. Generally the price can be determined by evaluation of the expected value of the payoff under the risk-neutral measure. Note that this prescription only applies for a complete market. In a complete market all risks can be perfectly hedged, i.e. all variations in price of a derivative security due to fluctuations in the underlying market and credit risk factors can be eradicated by an appropriate choice of a replicating strategy. Model risk exists if the model does not fully capture all the factors that the product is sensitive to. Using a specific model to value a complex product introduces a certain amount of model risk; if the model is a poor reflection of market reality then it will give values that are not replicable with a set of (real) hedging instruments and may lead to arbitrage opportunities if the model continues to be used. On the other hand, a 'good' model for the complex product and the hedging instruments will produce an overall portfolio P/L that fluctuates around zero over time since the model accurately captures the risks associated with the complex product. The P/L should be approximately zero since the overall combination of product plus replicating portfolio will, if the model is a good reflection of reality, be immunised against all fluctuations in market variables. It is important to note that we are modelling the combination of product plus replicating portfolio. It is necessary to capture any interactions between the different instruments. For example, if the value of the product and replicating instruments all depend on the value of a particular interest rate, the models should reflect this.

[1] The investor could also purchase the replicating portfolio, but investment banks are typically better placed to exploit economies of scale and thus undertake this sort of task more efficiently.

Derivative trading therefore requires an understanding of how to hedge a position dynamically (by varying the amounts of the hedging instruments held) such that at the maturity of the contract, the issuer of the derivative holds exactly the right amount of 'stuff' needed to replicate the payoff. To achieve this it is necessary to model accurately the dynamics of the risk factors. This usually includes modelling sudden, discontinuous jumps in financial variables which diffusion based models find difficult to capture. Hedging simulation is a procedure whereby the through life behaviour over time (from product inception to maturity) of a product and its associated portfolio of hedging instruments is modelled. To capture the through life behaviour it is necessary to model the following.

- The evolution of the market risk factors that the product and hedging instruments are sensitive to (either as stochastic processes or using historical data).
- An appropriate valuation methodology for the product and the hedging instruments (as a function of the underlying economic environment).
- The sensitivities of the product and hedging instruments to fluctuations in the market risk factors (which are typically computed from the model since these are unlikely to be observed in the market).
- The time-decay of the value of the product and the hedging instruments.
- The occurrence of periodic coupon payments (which contributes to the time decay in value of the product) and the procedure for dealing with these (typically placed into a cash-account which accrues at the risk-free rate).

For each time step the values of the product and hedging instruments are calculated and the value of the portfolio determined. The cumulative P/L can then be monitored. A 'good' model will have a cumulative P/L that does not deviate significantly from zero. Simulating the hedging process can check the model's effectiveness as a risk managing tool (since if the model sensitivities are poor the P/L leakage will be large). In addition to this, different models will yield different sensitivities to fluctuations in the market risk factors. A comparative hedging simulation using different models with the same input (market) data can provide some insight into what is the most appropriate model for risk managing a position (assisting in the justification for a particular model choice).

Where the instrument pricing model is built on an underlying stochastic process, the parameters of this process cannot in general be observed directly from the market but are inferred from the prices of traded instruments. For example, caplets and swaptions in the interest rate market are used to calibrate the parameters of the stochastic differential equation governing the interest rate dynamics [Brigo and Mercurio 2001, Rebonato 2002]. In an ideal case, once the parameters have been determined, they should not change over the life of the trade. The model should be capable of explaining price movements by relating them to the stochastic factors only and not to the parameters. However, in practice the parameters have to be re-calibrated from time to time in order to retrieve the market prices. During the course of a hedging simulation, should the parameters be changed? On the one hand, changing the parameters ensures that the model reflects the market price of the derivative. On the other hand, changing the parameters demonstrates that the model contains some risk that is not hedged, and one should therefore expect to leak (or gain) a certain amount of money due to that risk (at a re-calibration point the P/L will jump discontinuously). A successful hedging simulation can only conclude that *given* stable parameters, the risks returned by the model are adequate.

Other complications with hedging simulation relate to practical issues. The no-arbitrage principle says that the *expected* value of the hedging portfolio and the derivative should be zero. When using historical data, only one path exists, which is inadequate to evaluate an expected value. However, using different periods and/or different hedging frequencies can mitigate this problem. On the other hand, simulated data could be used. However, this would raise the question how to create the simulated data. The best advice, as always, is to experiment with different choices of model for generating the simulated data to understand the product sensitivities.

14.3 Hedging of structured credit products

Most of the synthetic CDO modelling literature is concerned with the problem of pricing (as evidenced by the discussions in the previous chapters). There is little consideration given to either the theoretical or practical issue of hedging and replication of positions. In other markets and asset classes (for example, interest rates) the concepts of pricing and hedging are difficult to separate. Part of the reason for this situation is the rapid adoption by the market of the normal copula model as the market standard for pricing. Hedging strategies that are used by practitioners are therefore based on empirical rules which are observed to work under certain market conditions.

Typical hedging strategies in other asset classes dynamically and instantaneously remove the market risk due to day-to-day fluctuations in market risk factors such as credit spreads and interest rates (the latter being of less importance for credit derivative positions). Credit derivatives are certainly sensitive to these market risks. However, in addition to these risks, they are also exposed to the risk of instantaneous or sudden default of an obligor. Fundamentally, hedging of credit derivative positions requires immunising the position against market and credit risk events, in particular the following.

- Day-to-day fluctuation in the credit spreads of the obligors in the underlying pool (which lead to an MtM shift of CDS and tranche positions).
- Day-to-day fluctuations in the implied or base correlations of the tranche positions (leading to an MtM shift of tranche positions).
- Idiosyncratic risk associated with large 'blow-outs' (possibly including default) in the credit spreads of individual obligors, for example, due to a leveraged buy-out (LBO) rumour. It is important to note that these blow-outs may in principle be quite rapid in nature (for example if the market timing of the LBO rumour is unexpected). In practice the research function of a credit desk should monitor the credit quality and economic performance of the obligors in the portfolio constantly and be able to identify potential problems (e.g. deteriorating performance following a series of profit warnings) well in advance, allowing positions to be modified accordingly.
- Systemic risk associated with significant spread widening of all the obligors in a particular sector or sub-sector (e.g. the home building sector).

Focusing on either of the market or credit risk factors in isolation, it is possible to define a delta which measures the sensitivity of the position to fluctuations in the underlying variable. For example, it is a simple matter to delta hedge a particular index tranche (at the current spread levels) using an index position. But as seen in Chapter 12 this hedge requires dynamically rebalancing as soon as an individual obligor's credit spread moves to maintain the efficacy of the hedge. On a day-to-day basis this is both impractical (potentially due to liquidity issues) as well as prohibitive in terms of costs. Focusing on default risk, protection against the default of a single obligor from a small subset of obligors (e.g. five) in the overall pool could be obtained by an offsetting position in an FtD basket (or alternatively separate positions in individual CDSs). But what about a second default, third etc. These defaults could in principle be hedged by a second- and third-to-default basket. But in practice the costs in terms of carry would again be uneconomical (to say nothing of the fact that we would need to find a counterparty to sell the appropriate protection on the appropriate basket of obligors).

Hedging each type of risk is therefore difficult in isolation. The problem becomes even harder when one tries to hedge both types of risk simultaneously. In practice, traders view the behaviour of CDO tranches as similar to that of ordinary call spread options and immunise positions against fluctuations in credit spreads, but not against idiosyncratic default risk. Default risk is sometimes partially mitigated by purchasing FtD basket protection referencing a pool of obligors from a sector that is of particular concern. Default risk can also be mitigated by utilising managed tranches. The portfolio manager's task then is to monitor the credit quality of the obligors in the tranche and remove those which they believe to exhibit unacceptable default risk. Typically this type of strategy would be a buy-and-hold strategy where the amount of active trading day-to-day to exploit short-term fluctuations in credit spreads would be minimal. However, employing a portfolio manager has an associated management cost (typically a flat fee based on the notional amount of assets under management and a percentage of profits). This reduces the overall carry of the trade. It is clear that trading in synthetic CDO tranches is not for the faint hearted. Volatility of credit spreads can lead to large P/L swings, there is the ever present danger of unexpected default (and worse – contagion of defaults), and hedging strategies are based on empirical rather than model based rules.

In terms of the market standard Gaussian copula model where do we stand? Even though from a pure pricing perspective this model is incapable of matching, out-of-the-box, market observed prices, it has been suitably extended to enable calibration to the market. These 'successes' notwithstanding, copula models have a number of significant and fundamental weaknesses (see for example Rebonato and Kainth, Hull *et al.* 2005, Frey and Backhaus 2007 for further discussion of these points).

- There is no clear economic rationale for choosing one copula (which, recall, characterises the dependence structure between the obligors in the portfolio) over another, save for the empirical observation that one particular copula may fit the market better than another at a particular point in time. Indeed one of the attractions of copula models is that they allow the separation of the marginal behaviour of the obligors from the specification of their joint behaviour.
- Copulas represent a static snapshot of the joint default time distribution of the obligors in the portfolio (meaning that the default environment is static). This distribution is also static through time, i.e. $\mathrm{d}p(l)/\mathrm{d}t = 0$. This is related to the fact that the inputs to the model such as the correlation and spreads are also assumed to be static.
- In the most popular one-factor implementations of the Gaussian copula model, the full dependence structure, represented by a full correlation matrix $\mathbf{C}$, is approximated with a single value. This is a significant simplification. Indeed, some research indicates that introducing more texture into the correlation matrix allows the model to generate correlation smiles and skews.
- Related to the above point, in the standard model credit spreads do not evolve dynamically (neither diffusing nor undergoing large instantaneous jumps – without defaulting). Only jumps-to-default are captured in the model. Volatility is not an input into the standard model. Most pricing models in other asset classes consider the dynamics of financial variables to be continuous diffusions (however most models do not include jump components to the diffusion process which would render the market incomplete).
- Copula models have no theoretical linkage with formal no-arbitrage pricing theory (implying that any hedging strategy used will be based upon empirical quantification of the product sensitivities).

In addition to these modelling difficulties, synthetic CDOs have a number of intrinsic features which make them difficult to replicate with vanilla products.

- The payoff of the underlying CDSs is binary (default/no-default) – therefore a suitable model must incorporate jumps-to-default.
- The payoff of the tranche depends on the joint, correlated behaviour of multiple obligors.
- The payoff is tranche dependent which is non-linear in the underlying pool losses.

These separate facts make it unlikely that a synthetic CDO tranche can be perfectly replicated with vanilla CDS instruments. In general, a full dynamic hedging model of synthetic CDOs requires a coupling between the CDS par spread dynamics and incidences of obligor default. It is this coupling that is currently lacking in modelling approaches.

On a practical level there is also the complication that there are so many different obligors in a particular CDO pool. For an index tranche with 125 obligors, each obligor will have its own separate term structure of par CDS spreads used to compute the marginal survival curve. Each spread term structure may be composed of 10 separate protection maturities meaning that in terms of spreads alone there are 125×10 separate risk factors to hedge against since the movements of each of these different spreads will impact the valuation of the synthetic CDO tranche. On the other hand it can be argued that a 1 bp move of the 5 year point of an individual obligor's term structure will not have a material impact on the overall tranche valuation. We could therefore approximate the sensitivity to, say, only looking at parallel shifts of the credit spreads for each obligor (reducing the number of sensitivities from 1250 to 125). In addition to this there is correlation sensitivity and default sensitivity to hedge against.

14.4 Hedging simulation of default baskets

As a simple first example we will consider hedging a first-to-default basket (of $i = 1, \ldots, n$ obligors) in the limit where the default correlation is zero and where the protection and fee payments are made continuously. We assume that a long position in a default basket is hedged by short positions in individual CDSs (held in the time-varying amounts $\Delta_i^{\text{Hedge}}(t)$). In the limit of $\rho_{\text{Def}} = 0\%$ the CDSs and FtD can be valued analytically (as shown in Chapter 5). We also assume that the cash account is ignored. In order to hedge the FtD we construct a portfolio of CDSs each referencing one of the obligors contained in the basket. For simplicity we assume that the maturity of the FtD and hedging CDSs are all identical and given by T.

The portfolio value $\Pi(t)$ at some time t is given by

$$\Pi(t) = V^{\text{FtD}}(t, T) + \sum_{i=1}^{n} \Delta_i^{\text{Hedge}}(t) V_i^{\text{CDS}}(t, T)$$

where the hedging CDSs fee and contingent legs and PVs are given by

$$V_i^{\text{Fee}}(t, T) = \frac{1}{r + h_i^{\alpha}(t)} \left[1 - \mathrm{e}^{-(r+h_i^{\alpha}(t))T}\right],$$
$$V_i^{\text{Cont}}(t, T) = (1 - \delta_i) h_i^{\alpha}(t) V_i^{\text{Fee}}(t, T),$$
$$V_i^{\text{CDS}}(t, T) = \phi_i N_i^{\text{CDS}} \left\lfloor -s_i^{\text{CDS}} V_i^{\text{Fee}}(t, T) + V_i^{\text{Cont}}(t, T) \right\rfloor,$$

respectively (the symbols in these expressions have their usual meanings). The sensitivity of the CDSs is given by $\Delta_i^{\text{CDS}}(t, T) = -\phi_i N_i^{\text{CDS}} V_i^{\text{Fee}}(t, T)$ (per basis point). The value of the FtD is given by

$$V^{\text{FtD}}(t, T) = \phi_{\text{FtD}} N^{\text{FtD}} \left[-s^{\text{FtD}} + \sum_{i=1}^{s} s_i(t) \right] \frac{1 - \mathrm{e}^{-(r+\sum_{i=1}^{n} h_i(t))T}}{r + \sum_{i=1}^{n} h_i(t)}$$

and the sensitivity of the FtD to fluctuations in the spreads is calculated by bumping the spreads of each obligor. The portfolio hedge ratios for each obligor are computed from

$$\Delta_i^{\text{Hedge}}(t) = -\frac{\Delta_i^{\text{FtD}}(t)}{\Delta_i^{\text{CDS}}(t)}.$$

The stochastic variables in this simulation are the par CDS spreads for each obligor.[2] For each obligor $i = 1, \ldots, n$ a variable is modelled corresponding to the market observed par spread rate for a CDS referencing this obligor. We have assumed that each obligor's term structure of par CDS credit spreads is characterised only by a single par CDS spread rate. For each obligor we assume a simple geometric Brownian motion (GBM) to describe the stochastic evolution of its market observed par CDS spread. Each GBM is of the form

$$s_i^{\alpha}(t_j) = s_i^{\alpha}(t_{j-1}) \exp\left[\left(\mu_i - \frac{1}{2}\sigma_i^2 \right)(t_j - t_{j-1}) + \sigma_i \sqrt{t_j - t_{j-1}} \phi_i^{\alpha} \right]$$

for simulation path α where the random deviates ϕ_i^{α} are correlated according to a user-defined spread correlation matrix (via Cholesky decomposition of the

[2] These variables correspond to the actual spreads quoted in the market for liquidly traded protection and are therefore the natural variables to consider since it is fluctuations in these values which drive the daily changes in MtM of credit products.

correlation matrix) and μ_i and σ_i are the drift and volatility of the par CDS spreads of obligor i. Once the simulated spread has been determined the implied par CDS hazard rate can be calculated from $s_i^\alpha(t) = h_i^\alpha(t)(1 - \delta_i)$. This model is similar to that presented by Rebonato and Kainth (except there the basket has a non-zero default correlation and uses Monte Carlo simulation to perform the basket revaluations).

The scenario to be analysed has the following parameters.

Obligor parameters	
Par CDS spreads	60, 120, 180, 240, 300 bps
Contractual CDS spread	par spread
Recovery rate	40%
Spread drift	0%
Spread volatility	10%
Buy/sell protection	sell
Environment parameters	
Days per year	250
Timenode grid spacing	1 day
Risk-free rate	0%
Spread bump size	1 bps
Spread correlation	0%
Basket parameters	
Basket order	first-to-default
Number of obligors in basket	5
Contractual spread	900 bps
Basket maturity	5 years
Coupon tenor	quarterly
Notional	\$10 m
Buy/sell protection	buy
Hedging CDS parameters	
CDS maturities	5 years
Coupon tenor	quarterly
Notional	\$10 m
Rehedging frequency	daily

Because we are considering the zero default correlation case, the contractual spread of the FtD is computed as the sum of the spreads of the underlying CDSs. The contractual spread of each of the CDSs is set to be equal to their initial par spread. This means that at $t = 0$ all of the CDSs will have a PV of zero. The FtD PV is also zero at this time.

As the simulation progresses for $t > 0$ the obligors' par CDS spreads will evolve according to the GBM away from their initial values. The MtMs of the FtD and

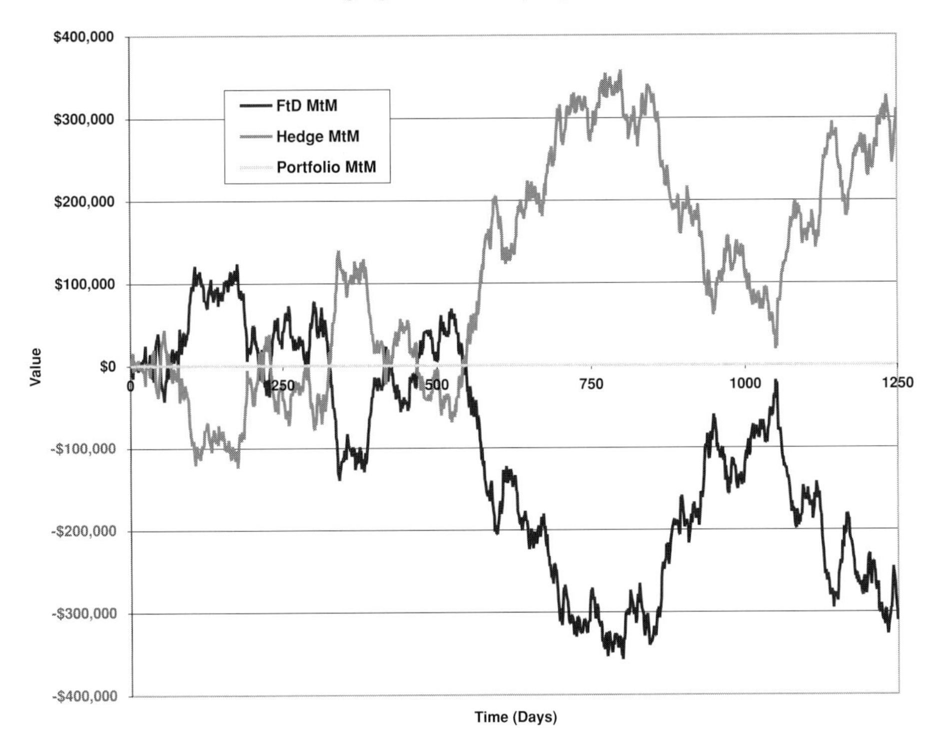

Figure 14.1 Simulated values of the FtD, hedge and portfolio MtMs for a single simulation of the CDS par spreads.

hedging CDSs will also fluctuate. The task of the hedging simulation is to compute the overall portfolio P/L over time.

Figure 14.1 shows the MtM values of the FtD, hedge (defined as $\sum \Delta_i^{\text{Hedge}}(t) V_i^{\text{CDS}}(t, T)$) and portfolio for a single simulation of the par spreads. Clearly the FtD and hedge values are a mirror image of one another, meaning that the overall portfolio value is zero. Similar results are obtained for different parameters, for example higher spread volatilities or correlations, basket sizes, maturities etc. The reason for this perfect hedging performance is because we are re-computing the hedge ratios every day. This means that for every move in the underlying spreads we calculate a new amount of obligor CDS protection to buy/sell to offset exactly the change in MtM of the FtD. Reducing the frequency of re-hedging leads to the results shown in Figure 14.2. This shows the portfolio value where the hedge ratio $\Delta_i^{\text{Hedge}}(t)$ is recalculated every day, 5 days, 25 days and 250 days. As seen in the figure, daily re-hedging enables the FtD position to be perfectly hedged. However, as the frequency of hedging decreases the constant day-to-day movements of the simulated CDS par spreads means that the amounts of protection sold on the CDSs are no longer adequate to offset the FtD MtM fluctuations. As expected, as the

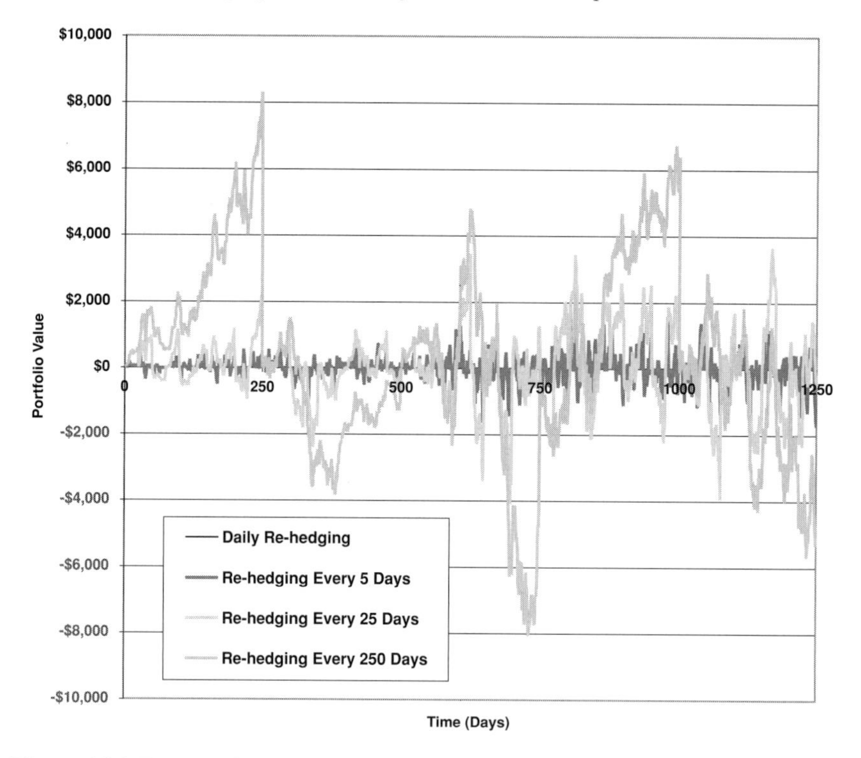

Figure 14.2 Impact of the frequency of re-hedging upon the portfolio value.

frequency of re-hedging decreases the fluctuations of the portfolio P/L away from zero increase.

14.5 Hedging simulation of synthetic CDO tranches

14.5.1 Introduction

Having introduced the concept of hedging simulation and applied it to a simple example of a default basket (where we know we should be able to hedge the exposure with CDSs), we now come to consider hedging simulation of synthetic CDO tranches. In principle hedging simulation of tranches is no different to that for simpler instruments: revalue the tranche for the current market parameters. In practice the value of a tranche is sensitive to fluctuations in a large number of parameters (all the par spreads of the underlying obligors, interest rates, recovery rates, implied correlation). One of the biggest challenges even before any modelling decisions are made is to manage all the market data. The objective of the current analysis, however, is to gain insight and understanding into the behaviour of the tranches. To facilitate this we will make some simplifying assumptions to reduce the complexity of the problem.

14.5.2 Simulation algorithm

The algorithm that we apply for the hedging simulation is as follows.

1. Specify all the initial conditions, compute the tranche par spread and set the contractual spread equal to this (so that the tranche PV at the beginning of the simulation is zero). The contractual spread is kept constant for the remainder of the simulation.
2. Update the time from time 0 to Δt (in our example Δt is 1 day) and evolve all the obligor par CDS spreads according to the chosen stochastic differential equation (described later).
3. Revalue the tranches (with the new spreads and reduced maturity).
4. Repeat steps 2 and 3 for each time t.
5. Repeat steps 1–4 for a large number of spread scenarios.

The tranche time decay is computed using the method outlined in Section 8.7. There is therefore an implicit assumption that none of the obligors has defaulted during the course of the simulation. Because we are performing multiple revaluations of the tranches (360 days per year for 5 years revaluations per tranche per simulation in the current example) speed of revaluation becomes important. In the analysis of this chapter we are using the normal proxy model of Section 6.5 (with a Gauss–Hermite [Press *et al.* 2002] integration scheme for computing the unconditional loss distribution). This model was chosen because of its very good combination of speed of valuation and accuracy. The only other possible model which is quicker in execution is the LHP model, but this does not have the same accuracy as the normal proxy model.

14.5.3 Simulating credit spreads

The par CDS spreads of each obligor are assumed to evolve via simple geometric Brownian motion

$$\frac{\mathrm{d}s_i^\alpha(t_j)}{s_i^\alpha(t_j)} = \mu_i \mathrm{d}t + \sigma_i \mathrm{d}W_i^\alpha(t_j).$$

The Brownian motions are correlated according to $\mathbf{E}[\mathrm{d}W_i^\alpha(t_j)\mathrm{d}W_k^\beta(t_l)] = \rho_{ik}^S \delta_{\alpha\beta}\delta(t_j - t_l)\mathrm{d}t$ where ρ_{ik}^S is the correlation between the spread movements of obligors i and k (this is not to be confused with the default correlation represented simply by ρ). By application of Ito's lemma this SDE is easily solved to yield

$$s_i^\alpha(t_j) = s_i^\alpha(t_{j-1})\exp\left[\left(\mu_i - \frac{1}{2}\sigma_i^2\right)(t_j - t_{j-1}) + \sigma_i\sqrt{t_j - t_{j-1}}\phi_i^\alpha\right]$$

(where the symbols have the same meanings as described previously). The simulation algorithm described here is similar to that in Rebonato and Kainth.

14.5.4 Default correlation

The implied default correlation is the market's view of the likelihood of multiple defaults amongst the obligors in the underlying pool. As we have seen many times throughout the preceding chapters, default correlation has a very significant impact upon the tranche valuation. Implied correlation can jump suddenly, as demonstrated during the correlation crisis. It can also switch between quasi-static 'regimes' (for example, when there is a systematic re-pricing of risk in the market as observed during the credit crunch).

How should this be modelled? The index tranche market is still relatively young and so it is not altogether clear from the market data what the dynamics of the correlation market are. Is it GBM, mean-reverting Ornstein–Uhlenbeck etc? The only things we can say with certainty is that it has to be bounded in the range [0, 1] and that there are probably ill-defined minimum and maximum values (that will change over time) that the default correlation can take. What we can do, of course, is understand how systematic increases or decreases in the correlation impact the tranche P/L. For simplicity we will consider default correlation to be constant.

14.5.4 Baseline scenario

In addition to the modelling assumptions outlined previously, we make our usual assumption that the term structure of par CDS spreads is flat and characterised by a single spread value. As has been noted on a number of occasions this is not an assumption that should be made in practice since the full term structure does have a significant impact on tranche pricing (particularly the equity tranche). But for the purposes of understanding the phenomenological behaviour of tranches it is a justifiable assumption to make. We also assume that recovery rates are constant through time and that interest rates are also constant.

The parameters used for this analysis are as follows.

Number of obligors in pool	125
Capital structure	iTraxx
Long/short tranche	long (sell protection) on all tranches
Tranche maturities	5 years
Coupon tenor	quarterly
Spread simulation timenode spacing	daily
Homogeneous obligor notional	\$10 m
Homogeneous obligor recovery rate	40%
Homogeneous (flat) obligor par CDS spread	60 bps
Homogeneous (default) correlation	0%
Risk-free rate	5%
Homogeneous drift of obligor par CDS spread	varies
Homogeneous volatility of obligor par CDS spread	varies
Homogeneous correlation of obligor par CDS spread	varies
Number of spread simulations	1 or 1000

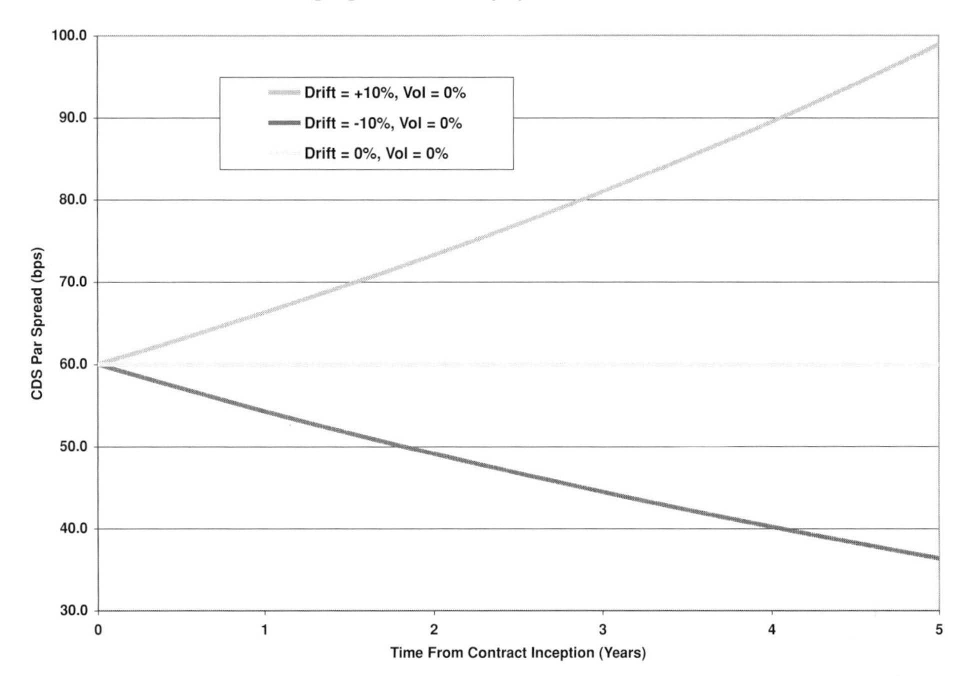

Figure 14.3 Static spread and spread widening and contracting scenarios.

For all the tranches, we are selling protection on the tranche (meaning we receive the fee payments and make any contingent payments).

14.5.5 Results

14.5.5.1 Spread scenarios

In order to build up some intuition as to the likely behaviour of the tranches we will consider a number of simple limiting cases where we systematically turn on and off different aspects of the spread simulation process (when the spreads are static the tranche PV behaviour as time progresses was discussed in Section 8.7). The figures are generated using a single spread simulation (one scenario).

The simplest limiting cases to consider are when $\mu_i = 0, \pm\mu$ and $\sigma_i = 0$ (in this case the correlation between the spreads is irrelevant since there is no volatility). The spreads are either static or widen or contract systemically and deterministically by $s_i^\alpha(t_j) = s_i^\alpha(t_{j-1})e^{\mu(t_j - t_{j-1})}$ for each obligor. This is shown in Figure 14.3. The next simplest case to consider is when $\mu_i = 0\%, \sigma_i > 0\%$. Figure 14.4 plots a sample evolution of the $i = 1, \ldots, n$ separate spread processes for each of the obligors. In this figure we have set $\rho_{ik}^S = \rho^S = 0\%$, i.e. the correlation amongst the spread processes is assumed to be homogeneous and zero. In contrast Figures 14.5 and 14.6 plot the same spread simulation as Figure 14.4 (using the same random

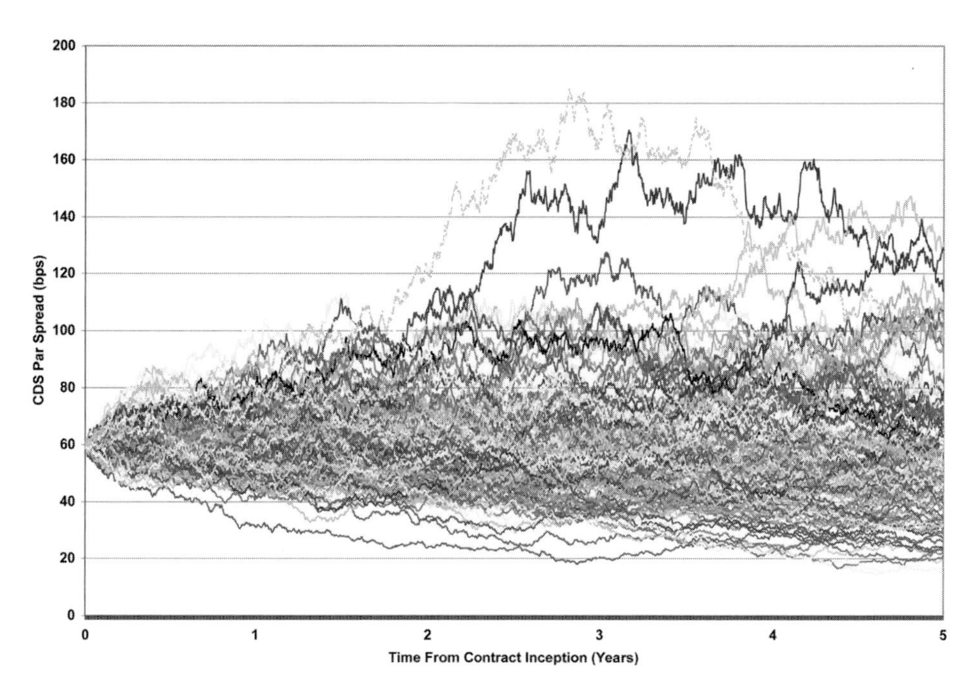

Figure 14.4 A stochastic spread scenario where $\mu_i = 0\%, \sigma_i > 0\%$ and $\rho^S_{ik} = \rho^S = 0\%$. The daily evolution of the spreads for each of the $i = 1, \ldots, n$ obligors is plotted.

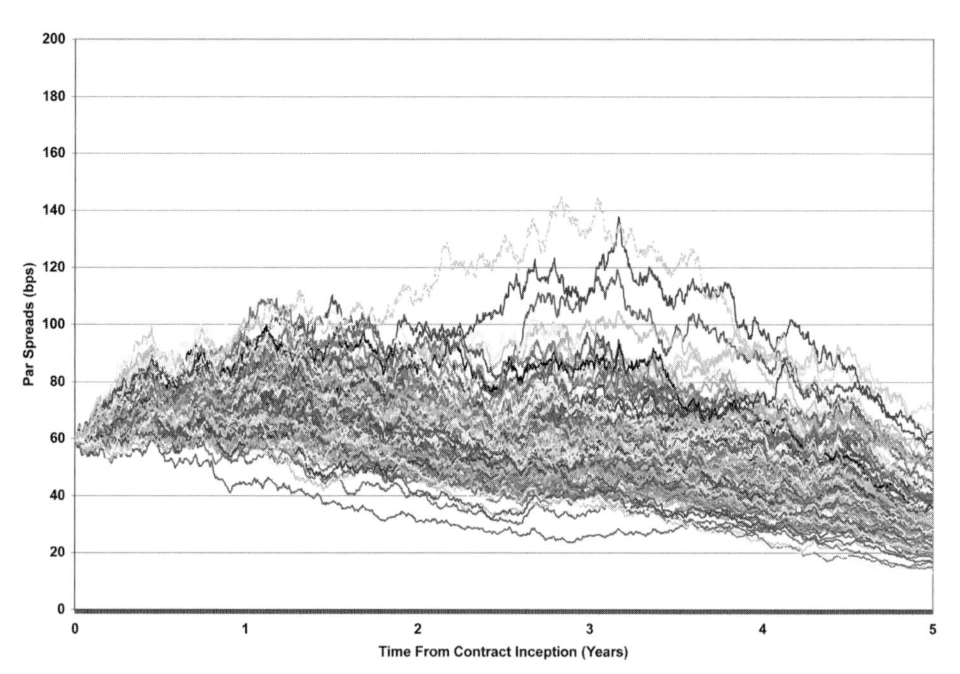

Figure 14.5 A stochastic spread scenario where $\mu_i = 0\%, \sigma_i > 0\%$ and $\rho^S = 40\%$. The daily evolution of the spreads for each of the $i = 1, \ldots, n$ obligors is plotted.

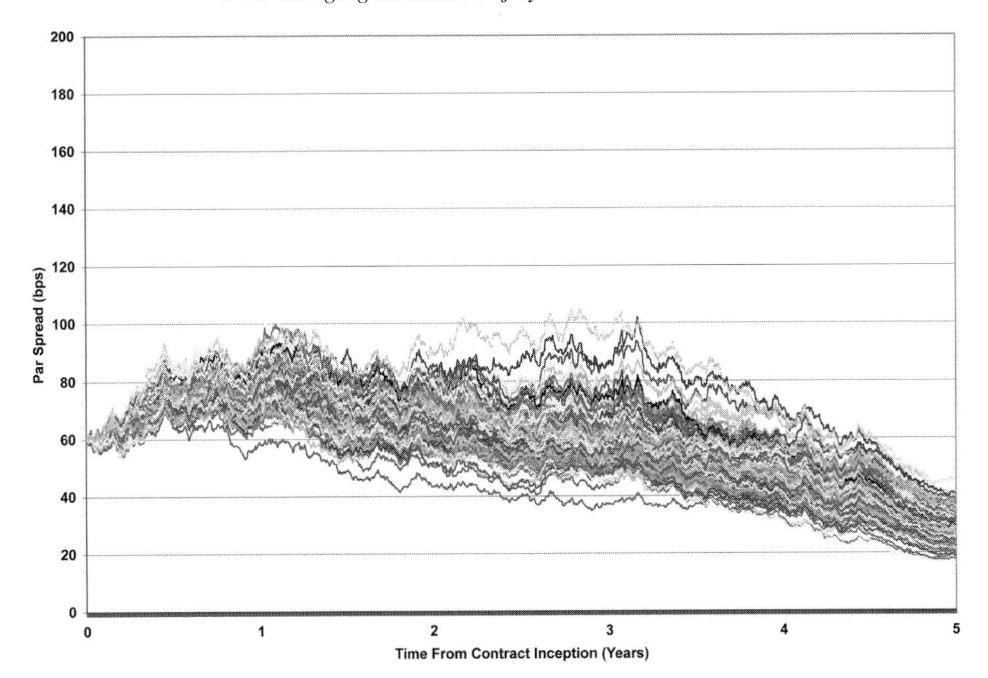

Figure 14.6 A stochastic spread scenario where $\mu_i = 0\%, \sigma_i > 0\%$ and $\rho^S = 80\%$. The daily evolution of the spreads for each of the $i = 1, \ldots, n$ obligors is plotted.

number seed to generate the scenarios), but this time with $\rho^S = 40\%$ and $\rho^S = 80\%$ respectively (the vertical axes are all plotted on the same scale to highlight the impact of the higher degree of correlation on the spreads). As expected, increasing the correlation between the spread processes results in less dispersion amongst the simulated spreads. Finally Figure 14.7 plots the simulated obligor spreads for the parameters $\mu_i = 0\%, \sigma_i = 60\%$ and $\rho^S = 0\%$. As expected, the increase in volatility results in a far greater dispersion of the simulated scenarios (evident from the scale of the vertical axis).

14.5.5.2 Tranche exposures

Now we consider the behaviour of the tranches under these different obligor spread scenarios. Figure 14.8 plots the PV of the 0–3% tranche as we move forward in time from $t = 0$ to $t = T$ for the cases where $\mu_i = 0\%, \sigma_i = 0\%, \mu_i = \pm 10\%, \sigma_i = 0\%$ and $\mu_i = 0\%, \sigma_i = 20\%, 60\%$ for all the obligors. The scenario where $\mu_i = 0\%, \sigma_i = 0\%$ is the baseline case where the spreads of the obligors do not evolve from their initial values (of 60 bps). For all of these cases we have set $\rho^S = 0\%$ (we will assess the impact of varying this parameter separately).

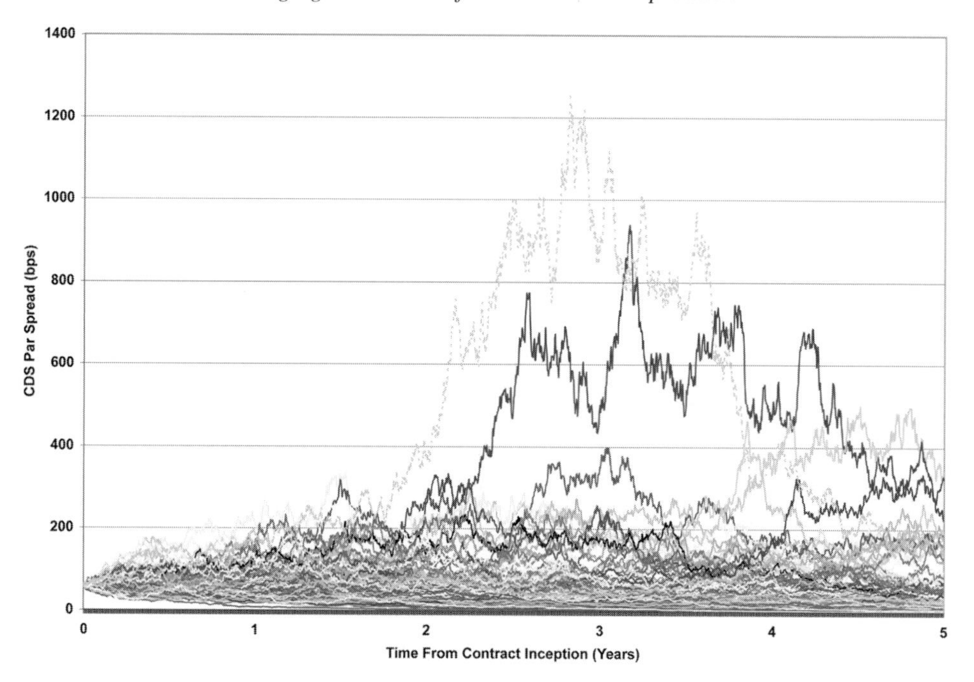

Figure 14.7 A stochastic spread scenario where $\mu_i = 0\%$, $\sigma_i = 0\%$ and $\rho^S = 0\%$. The daily evolution of the spreads for each of the $i = 1, \ldots, n$ obligors is plotted. Note the vertical scale compared to the other simulated scenarios.

Considering the baseline case, the first observation to make is that the tranche PV is zero at $t = 0$ and $t = T$. The tranche PV is zero at contract inception by construction (since the contractual spread is set equal to the par spread which is calculated by the algorithm at $t = 0$). At the maturity of the tranche, there are no further cashflows so the fee and contingent legs are zero. The behaviour in-between these end-points was discussed in Section 8.7 and will not be repeated here. What should be noted is that the tranche exposure has a 'stepped' nature. This is due to the discrete nature of the coupon payments (and the approximation of the continuous contingent integral payment with a discrete sum). This was not apparent in Section 8.7 because we were taking large time-steps (of 1 year) there. In this case the tranche is being revalued daily and so the impact of cashflows dropping out of the summation is immediately apparent. The sharp-eyed will observe that the tranche PVs are not quite identical to those in Section 8.7, despite the obligor pool and tranches being identical. This is because we are using a different model for the tranche revaluations in this chapter (the Monte Carlo model was used in Section 8.7). The fee and contingent legs are very close in value when computed using the two different models, but of course yield a slightly different par spread. This translates into slightly different PVs.

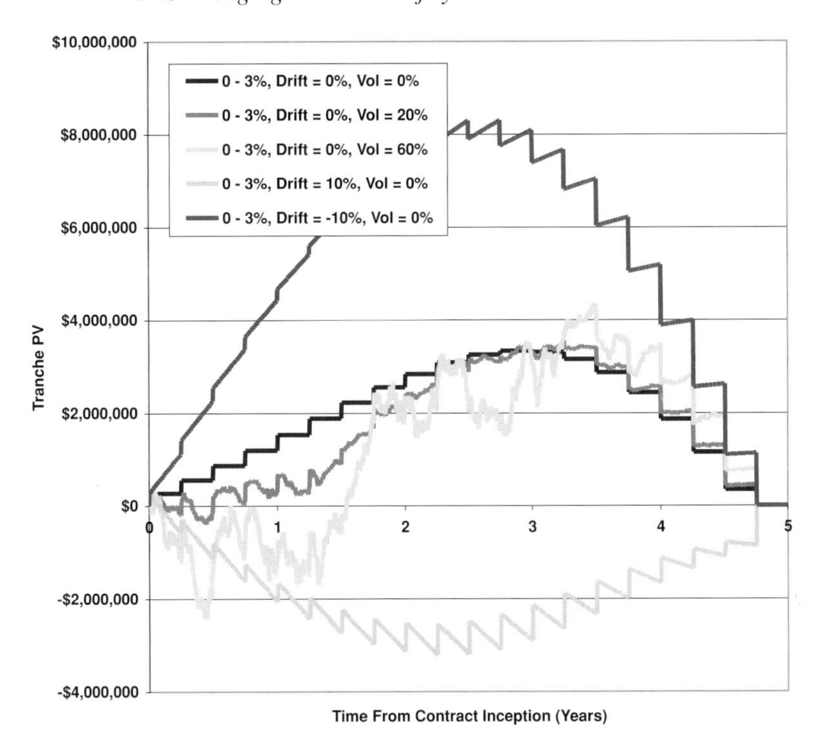

Figure 14.8 PV of the 0–3% tranche for various combinations of obligor spreads and volatilities. For all of these cases we have set $\rho^S = 0\%$.

Next we consider the impact of turning on the drift of the stochastic processes for the spreads in Figure 14.8. We consider two cases where the drift is positive (spread widening from the initial value) and where the drift is negative (spread contraction). In both cases the spread volatility is zero. Figure 14.8 shows that when the drift is positive, the PV of the tranche flips from being positive to negative. When the drift is negative, the PV remains positive but is increased in magnitude compared to the zero drift case. To explain this, Figure 14.9 plots the total fee (fee leg value multiplied by the par spread to give the actual $ amount of the protection payment) and contingent legs for the cases $\mu_i = 0\%, \pm 10\%$ and $\sigma_i = 0\%$. For the baseline case the fee leg value is greater than the contingent leg. Since we are providing protection, this results in a positive MtM for the tranche. Adding a positive drift to the credit spreads means all the obligors become simultaneously more risky. All things being equal, this should increase the expected losses in the 0–3% tranche resulting in a reduced fee leg compared to the zero drift case (since the coupon at each payment date is proportional to $N^\gamma - \bar{L}^\gamma(T_j)$ and the expected losses are a function of the obligor spreads). This is what is observed. In this case the fee leg is less than the contingent leg, resulting in a negative PV. By the same argument

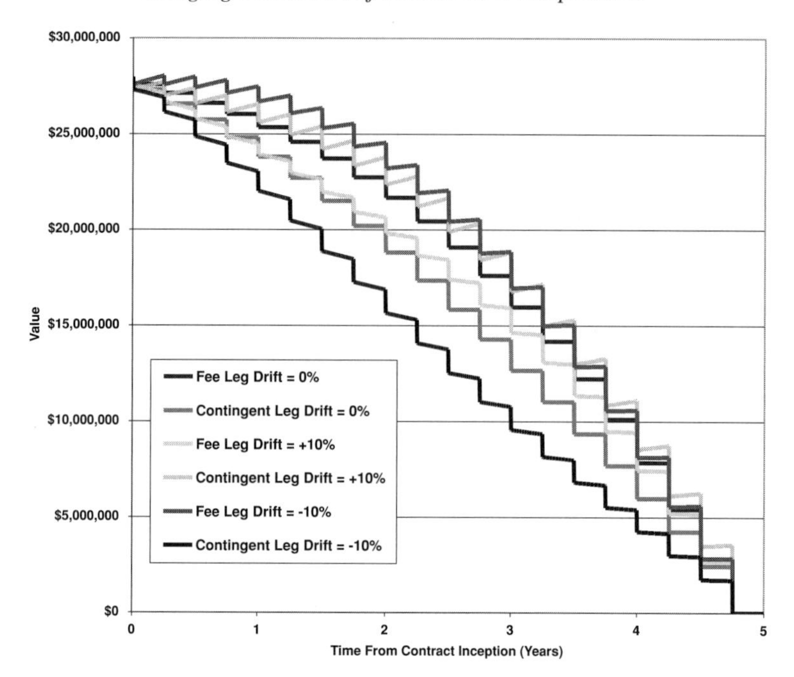

Figure 14.9 Fee and contingent legs for the 0–3% tranche for the cases $\mu_i = 0\%$, $\pm 10\%$ and $\sigma_i = 0\%$.

if the expected losses increase the contingent leg should also increase. Conversely, when the spreads are contracting (negative drift) the fee leg should increase and the contingent leg decrease relative to the baseline case of static spreads. This is what Figure 14.9 shows. In this case the fee leg is greater than the contingent leg, resulting in a positive PV.

Returning to Figure 14.8, we also plot the tranche PV for the cases $\mu_i = 0\%$ and $\sigma_i = 20\%, 60\%$. In these cases the obligor spreads are no longer (deterministically) drifting but have their own idiosyncratic behaviour. In the $\sigma_i = 20\%$ case, there is a clear impact upon the tranche but it is not that severe. In the $\sigma_i = 60\%$ case however the impact on the tranche is quite severe, resulting in a PV which can go negative by a large amount. What these figures show is that depending upon the particular realisation of the obligor spreads, the tranche PV can be either positive or negative. The P/L of a naked tranche position is therefore sensitive to the spreads; this makes it important to hedge a tranche exposure against spread movements.

This behaviour is to be contrasted with the 3–6% tranche PVs. These are shown in Figure 14.10. As for the 0–3% tranche, turning the deterministic drift on and off has an intuitively expected impact on the tranche. The effect of introducing greater spread volatility is also as expected. It is observed, however, that the impact of

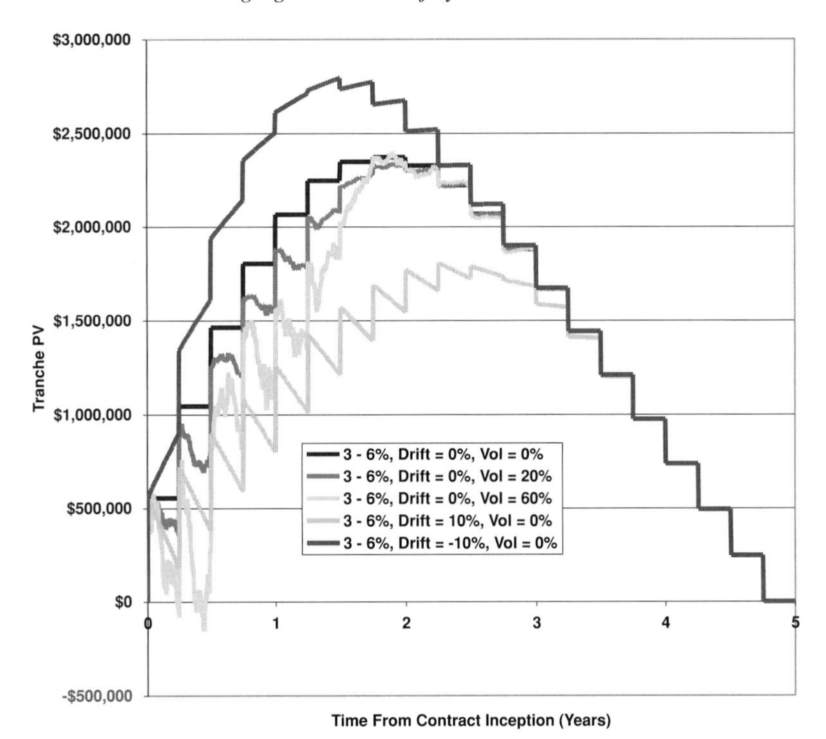

Figure 14.10 PV of the 3–6% tranche for various combinations of obligor spreads and volatilities.

the spread volatility is not so apparent for the 3–6% tranche as it is for the 0–3% tranche. This is due to the increased subordination that this tranche has.

In all of these cases the overall exposure is easy to understand. There are two competing forces driving the exposure profile of the tranche. On the one hand we have the diffusion of the spreads away from their initial values (in this case 60 bps). The average dispersion of a Brownian motion as time progresses scales as $\sigma \sim \sqrt{t}$ with time t (for a normal Brownian motion). Therefore as we move away from time zero the uncertainty in exposure increases. The other competing force is the total amount of cashflows left before the end of the contract. As cashflows drop out of the trade the total notional of the trade gradually decreases. The first effect increases the exposure, the second decreases it.

Finally, Figure 14.11 shows the impact on the 0–3% tranche of increasing the obligor spread correlation. The obligor spread correlation controls the degree to which obligor spread movements are dispersed. Zero correlation means the spread movements are dominated by idiosyncratic behaviour. Perfect correlation implies systemic effects are dominant. In all of these cases we keep the drift and volatility of the spread processes constant. When $\rho^S = 0\%$ the spread processes for all the

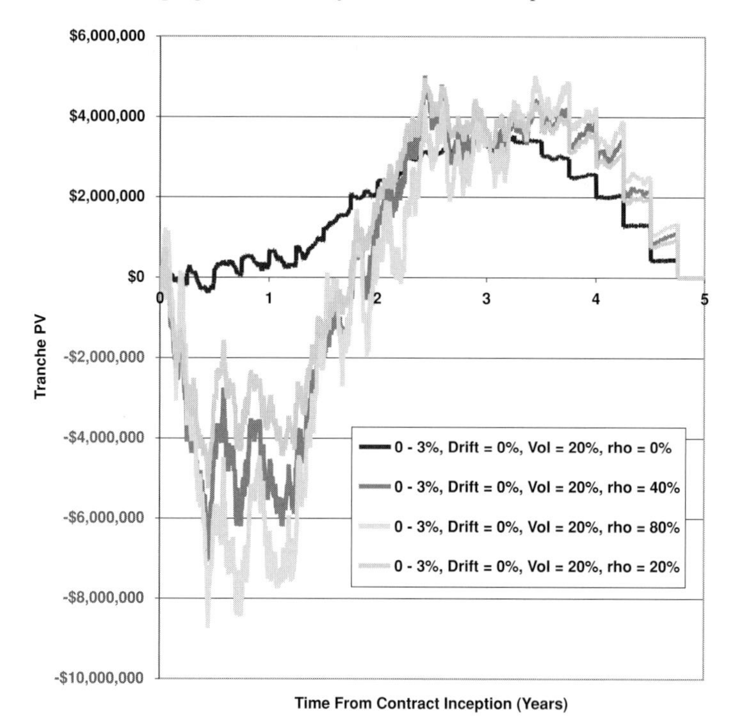

Figure 14.11 PV of the 0–3% tranche for the case where $\mu_i = 0\%$, $\sigma_i = 20\%$ and $\rho^S \geq 0\%$.

obligors are independent. This case therefore corresponds to an idiosyncratic scenario. When $\rho^S \geq 0\%$ the obligor spreads behave more synchronously, corresponding to a systemic credit event. To explain these results we need to look in more detail at the particular spread scenarios shown in Figures 14.3–14.7 (remember that this is only a single realisation of the spread scenarios – in practice we should generate many different possible scenarios in order to build up some statistics about tranche P/L). Figure 14.12 plots $\bar{s}(\{\mu_i, \sigma_i\}, \rho^S, t) = \sum s_i(\mu_i, \sigma_i, \rho^S, t)/n$, i.e. the average obligor spread at each t. For $t < T/2$ (approximately), the spreads are on average wider than the baseline case (where there is no correlation). As we saw in Figure 14.8, the impact of wider spreads on the 0–3% tranche is to flip its PV from positive to negative. This is also observed in Figure 14.12. Conversely, when $t > T/2$, the average spreads are tighter than the baseline case, meaning that the PV of the 0–3% tranche should revert back to being positive (and perhaps also larger than the baseline case depending on the values of the spreads).

14.5.6 Discussion

In this section we have considered the through life behaviour of tranches under different spread scenarios. To do this we have made some simplifying assumptions

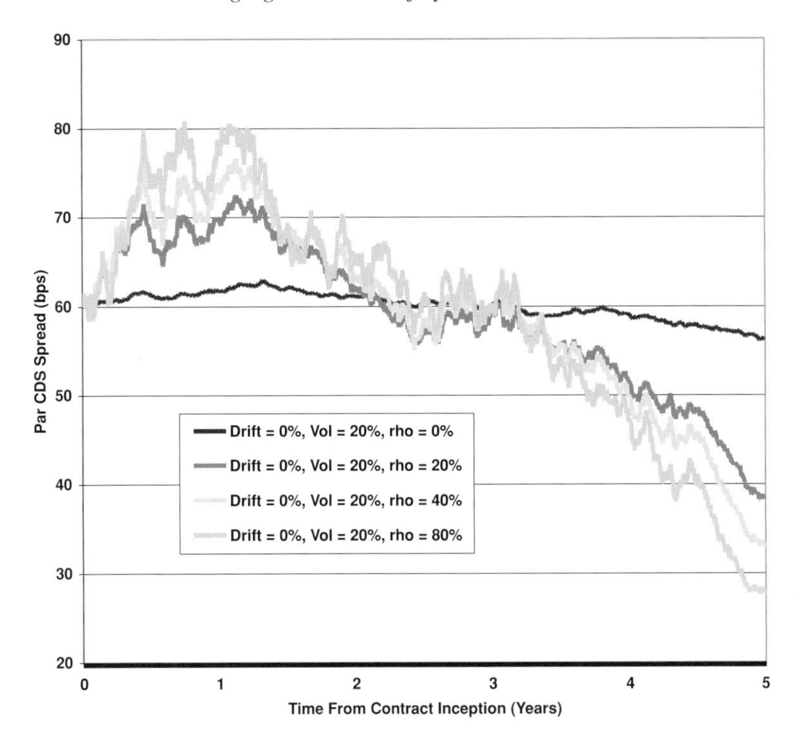

Figure 14.12 Average spreads (at each time t) for differing values of the spread correlation. The baseline case where there is no correlation does not really deviate much from the initial value of 60 bps. However, when $\rho^S > 0\%$ we observe (for this particular scenario) that for $t < T/2$ (approximately), the average spread is wider than the baseline case, and for $t > T/2$ the average spread is tighter.

(such as a flat spread term structure) as well as some modelling assumptions (for example, modelling the spread evolutions as a set of correlated geometric Brownian motions and assuming that the tranche correlations are constant). It is reasonable to question all of these assumptions. Indeed the reader is encouraged to build their own models, analyse the data and come to their own conclusions as to the most appropriate modelling methodology.

14.5.7 Optimal static hedging of defaults

The previous approach looked at simulating the through life behaviour of a synthetic CDO and understanding its behaviour as a function of credit spread fluctuations. Defaults amongst the underling pool were explicitly excluded from the analysis. We briefly describe an alternative methodology developed by Petrelli *et al.* [2006] to assess static hedging of a synthetic CDO against defaults.

In this approach the hedge portfolio is assumed to be reference bonds and is established at $t = 0$ and kept constant until $t = T$. The total change in wealth of the tranche $\Delta W_{\text{Tranche}}(T)$ and hedging portfolio $\Delta W_{\text{Hedge}}(T)$ over the period $[0, T]$ combine to give the total change in wealth $\Delta W(T) = \Delta_{\text{Tranche}}(T) + \Delta W_{\text{Hedge}}(T)$. If the hedging strategy is good we would expect that $\Delta W(T) \sim 0$, i.e. the P/L over the period from the tranche is balanced by that from the hedging portfolio.

We assume there are $i = 1, \ldots, n$ obligors in the pool. The notional of obligor i is given by N_i. The total notional of the pool is given by $N_{\text{Total}} = \sum N_i$. If obligor i defaults, the recovered amount is δ_i of the notional of the obligor. Therefore the pool loss at time t is

$$L(t) = \sum_{i=1}^{n} (1 - \delta_i) N_i \delta \mathbf{1}_{t<\tau_i}$$

and the cumulative pool loss at this time is

$$L_{\text{Cum}}(t) = \sum_{i=1}^{n} (1 - \delta_i) N_i \Theta(t - \tau_i)$$

(where $\Theta(x) = 1$ if $x > 0$ is the Heaviside step function). The cumulative recovered amount is

$$\delta_{\text{Cum}}(t) = \sum_{i=1}^{n} \delta_i n_i \Theta(t - \tau_i).$$

Consider a tranche which has an upfront payment u and a running spread of s. The attachment/detachment points are given by $(K_{\text{L}}, K_{\text{U}})$ and these are fixed at time zero. At time t the tranche notional is $N_{\text{Tranche}}(t) = k_{\text{U}}(t) - k_{\text{L}}(t)$ where

$$k_{\text{L}}(t) = \min\left[\max\left[k_{\text{L}}, L_{\text{Cum}}(t)\right], k_{\text{U}}\right],$$
$$k_{\text{U}}(t) = \max\left[\min\left[K_{\text{U}}, N_{\text{Total}} - \delta_{\text{Cum}}(t)\right], K_{\text{L}}\right],$$

are the dynamically moving attachment/detachment points due to the occurrence of obligor defaults in the underlying pool. The expression for $k_{\text{L}}(t)$ amortises the tranche from the bottom due to defaults (with associated contingent payments from the protection seller to the protection purchaser), and that for $k_{\text{U}}(t)$ amortises the tranche from the top due to recoveries from defaults (ensures the outstanding tranche notional matches the undefaulted reference pool notional).

The change in wealth of the tranche position (which is assumed to be unfunded) over $[0, T]$ is therefore given by

$$\Delta W_{\text{Tranche}}(T) = u(K_{\text{U}} - K_{\text{L}}) + s \int_0^T N_{\text{Tranche}}(t') \mathrm{e}^{-rt'} \mathrm{d}t' + \int_0^T \mathrm{e}^{-rt'} \frac{\mathrm{d}}{\mathrm{d}t'} k_{\text{L}}(t') \mathrm{d}t',$$

i.e. the net of the upfront payment, tranche coupon payments and tranche contingent payments. The change in value of the hedging portfolio over the period $[0, T]$ is given by

$$\Delta W_{\text{Hedge}}(T) = \sum_{i=1}^{n} \Delta W_i(T)$$

where

$$\Delta W_i(T) = h_i \left[-f_i + \mathrm{e}^{-rT} + c_i \int_0^T \mathrm{e}^{-rt'} \mathrm{d}t' \right] \mathbf{1}_{\tau_i > T},$$

$$\Delta W_i(T) = h_i \left[-f_i + \delta_i \mathrm{e}^{-r\tau_i} + c_i \int_0^{\tau_i} \mathrm{e}^{-rt'} \mathrm{d}t' \right] \mathbf{1}_{\tau_i < T}.$$

In these expressions h_i is the notional amount held in the bond of obligor i, $h_i f_i$ is its market price and c_i is its periodic coupon.

The hedging problem is to determine the set of bond notional amounts h_i and the tranche parameters $\{(K_L, K_U), u, s\}$ which minimise the overall wealth balance of the portfolio $\Delta W(T) \sim 0$. For a perfectly hedged position we would expect $\langle \Delta W(T) \rangle_\alpha \equiv 0$ where we are averaging over all possible realisations (simulations) α of the default environment (see Bouchaud and Potters [2000] and Voit [2005] for more details of discrete non-optimal hedging). In reality the position may not be perfectly hedged. Therefore we want to choose the optimal parameters/hedging the position (bond notional amounts) such that the hedging error/residual Θ is minimised. The hedge error can be chosen in any particular way. For example

$$\Theta^2 = (\Delta W - \langle \Delta W \rangle)^2$$

(measuring the dispersion of the hedging error in the body of the distribution) is a suitable choice. Alternatively an expected shortfall type measure of the form

$$\Theta = \mathbf{E}\lfloor \Delta W | \Delta W < -\mathrm{VaR}_q \rfloor,$$

where $P(\Delta W < -\mathrm{VaR}_q) = 1 - q$ defines the VaR (left-tail) at the qth quantile level, can also be used to characterise the hedging error. The expected shortfall (which is the expected wealth change, conditional on the wealth change being below the VaR threshold) characterises the tails of the wealth change distribution.

If hedging error can be eliminated altogether and perfect replication is theoretically possible then the model pricing results are not sensitive to the choice of hedge error minimisation objective function, which is the central attraction of the risk-neutral pricing theory's postulate of a derivative contract value being equal to the cost of perfect replication. However, even when replication is not theoretically possible, fitting prices to purported risk-neutral pricing models takes place in the day-to-day practice of marking-to-market/model complex derivative contracts.

The optimisation process is performed by simulating the default times of the obligors (using either a normal copula or variance gamma process). This is performed multiple times and the optimal tranche and hedge portfolio parameters determined. The reader is referred to the original research paper for a discussion of the results obtained by the authors using this model.

14.5.8 Markovian models for dynamic hedging of synthetic CDO tranches

The standard hedging technique applied to synthetic CDOs is to perturb the spreads of the obligors manually, revalue the tranches and compute the sensitivities. There is a crucial element missing from this approach. Assume that a particular obligor i defaults. In a global, interconnected economy, one obligor defaulting will have an impact on other obligors. For example, the credit spreads of obligors in the same sector of the economy may widen to reflect the increased risk to their own survival. It is possible therefore for a default to 'ripple' through a portfolio of obligors, the effects of the default spreading like a contagious disease. Contagion models were first introduced into the credit arena by Davis and Lo [2001], Jarrow and Yu [2001] and Yu [2007]. In the model of Frey and Backhaus [2007] the effects of default contagion are introduced into a novel CDO pricing model (which also incorporates stochastic spread dynamics). We briefly describe the formulation of the model as it brings together a number of different threads that have been introduced throughout the text. For further details of the model the reader is referred to the original research paper (and also to the model of Laurent *et al.* [2007]).

Assume that we have a filtered probability space $(\Omega, \Im, Q)$ (where Ω is the set of possible outcomes, $\Im$ the filtration and Q the probability measure). Let there be $i = 1, \ldots, n$ obligors in the portfolio. The default/stopping times for each of the obligors are denoted by $\tau_i \in (0, \infty)$. We define a default vector $\mathbf{Y}(t) = (Y_1(t), \ldots, Y_i(t), \ldots, Y_n(t))^T$ where $Y_i(t) = \mathbf{1}_{\tau_i < t}$ is the default indicator for obligor i. If we assume that the measure Q represents a risk-neutral measure for pricing, then the value of contingent claims is given by

$$\frac{H_t}{B(t,T)} = \mathbf{E}_Q\left[\frac{H_T}{B(T,T)}|\Im_t\right]$$

where $B(t,T) = \exp\{-\int_t^T r(s)\mathrm{d}s\}$. We also assume that the portfolio credit model is constructed under the same probability measure.

We now introduce a state variable process $\Psi = \{\Psi_t\}_{t\geq 0}$ representing the state of the macro-economic environment. This state variable is modelled as a finite state Markov chain with state space $S^\Psi = \{\psi_1, \ldots, \psi_k\}$. The overall state of the system is therefore characterised by $\Gamma = (\Psi_t, \mathbf{Y}(t))$. This is a finite state Markov chain with state space $S^\Gamma = \{0, 1\}^n x S^\Psi$. An element of this state space is represented by $\gamma = (y, \psi)$.

The default intensity of obligor i is given by $\lambda_i(\Psi_t, \mathbf{Y}_t, t) \geq 0$. The default intensity is therefore a function of the overall economic environment, Ψ_t, and also the default state of the portfolio $\mathbf{Y}(t)$. Changes in the default states of other obligors therefore result in changes in the spreads of obligor i (default contagion); fluctuations in the economic environment also impact the spreads (modelling systemic factors in spread fluctuations).

The dynamics of the system are characterised by the transition matrix. The entries of this matrix are given by $q^\Gamma((y, \psi), (\tilde{y}, \tilde{\psi}), t) = q^\Psi(\psi, \tilde{\psi})$ if $\tilde{y} = y$ (the default state does not change) and $\psi \neq \tilde{\psi}$ (therefore the only transitions are those of the economy), $q^\Gamma((y, \psi), (\tilde{y}, \tilde{\psi}), t) = ((1 - y_i)\lambda_i y, \psi, t)$ if $\tilde{y} = y_i$ (for $i \in [1, n]$) and $\psi = \tilde{\psi}$ and zero otherwise.

When the obligors are homogeneous $\lambda_i(\Psi_t, \mathbf{Y}_t, t) = h(M_t, \mathbf{Y}_t, t)$ where $M_t = \sum_{i=1}^n Y_i(t)$. The authors model the obligor default intensities as

$$h(l, \psi, t) = \lambda_0 \psi + \frac{\lambda_1}{\lambda_2}\left[\mathrm{e}^{\lambda_2 (l - \mu(t))^+ / n} - 1\right]$$

where $\lambda_0, \lambda_1, \lambda_2 > 0$. $\mu(t)$ is the expected number of defaults in the period $[0, t]$ and is calibrated to the CDS index level. l is the portfolio loss. λ_1 determines the strength of the default contagion. If $\lambda_2 \to \infty$ then $h = \lambda_0 \psi$, i.e. no default contagion. To specify the model further it is necessary to give the state space $S^\Psi = \{\psi_1, \ldots, \psi_k\}$ and the transition matrix $\mathbf{q}^\Psi$. The model is then calibrated to the index data to determine the parameters $\lambda_0, \lambda_1, \lambda_2 > 0$.

The market value of a CDS is written as

$$V_i^{\mathrm{CDS}}(t) = \mathbf{E}_Q\left[-s_i \sum_{j=1}^N Z(t, T_j)\Delta_j \mathbf{1}_{\tau_i > T_j}\mathbf{1}_{t < T_j} + (1 - \delta_i) Z(t, \tau_i)\mathbf{1}_{\tau_i \leq T} | \Im_t\right]\mathbf{1}_{\tau_i > t}.$$

If $t > \tau_i$ then $V_i^{\mathrm{CDS}}(t) \equiv 0$. The market value of a CDO is given by

$$V^{\mathrm{CDO}}(t) = \mathbf{E}_Q\left[s_{\mathrm{CDO}} \sum_{j=1}^N Z(t, T_j)\Delta_j N^{\mathrm{CDO}}(T_j)\mathbf{1}_{t < T_j} - \int_0^t Z(t, t')\mathrm{d}L_{t'}^{\mathrm{CDO}} | \Im_t\right].$$

For both CDSs and CDO there is a discontinuous jump in their value as coupon payment dates are crossed. The gains processes for the CDSs and CDO represent the change in value as time progresses. This comprises cashflows due to coupon payments, recovery payments and also the change in MtM value. For the CDSs the gains process is written as

$$\mathrm{d}G_i^{\mathrm{CDS}}(t) = -s_i[1 - Y_i(t)]\mathrm{d}t + (1 - \delta_i)\mathrm{d}Y_i(t) + \mathrm{d}V_i^{\mathrm{CDS}}(t).$$

This states that if $\tau_i \in [t, t + \mathrm{d}t]$ then $Y_i(t) \to 1$, causing subsequent coupon payments to cease and triggering the contingent payment. The gains process for the CDO is given by

$$\mathrm{d}G^{\mathrm{CDO}}(t) = +s_{\mathrm{CDO}}N^{\mathrm{CDO}}(t)\mathrm{d}t - \mathrm{d}L^{\mathrm{CDO}}(t) + \mathrm{d}V^{\mathrm{CDO}}(t).$$

With these components in place we can now consider how this model (which incorporates both spread fluctuations and default contagion) compares to the standard market model. Specifically the authors compute the sensitivity based hedge ratios of this model with those of the Gaussian copula. The hedge ratios needed to immunise the position against default risk are computed from

$$\Delta G^{\mathrm{CDO}}(t)|_{\tau_i=t} + \sum_{k=1}^{n} \Delta_k^{\mathrm{Def}}(t)\Delta G_k^{\mathrm{CDS}}(t)|_{\tau_i=t} = 0$$

for $i = 1, \ldots, n$. If an obligor has already defaulted their hedge ratio is set to zero. The authors report that the hedge ratios from the Markov chain model are less than those from the copula model for the equity tranche, but greater for mezz and senior tranches. The reason for this behaviour is postulated to be the impact of default contagion. In the standard market model when obligor i defaults, the spreads of all the other obligors remain unaffected. However, in the Markov chain model the default of obligor i results in an impact on the valuations of the other obligors j. The results for spread hedge ratios (determined as outlined in Chapter 12) for the two models are more closely aligned.

In the presence of default (jumps) and spread (diffusion) risk, it is to be expected that the market is incomplete [Musiela and Rutkowski 1998, Joshi 2003]. This means that it will not be possible to replicate perfectly the tranche payoff by dynamic trading in the hedging instruments (CDSs). To quantify the residual hedging error the authors use risk minimisation (incomplete market theory) techniques to determine the optimal hedging strategy. Let the hedging strategy be represented by the vector $\vec{\theta}(t) = (\theta_1(t), \ldots, \theta_n(t))^T$ representing the time-dependent holdings in each of the hedging instruments. In addition to this there is a cash amount held. Over the time interval $[0, t]$ the change in portfolio value is given by

$$G^{\mathrm{CDO}}(t) - G^{\mathrm{CDO}}(0) + \sum_{i=1}^{n} \int_0^t \theta_i(t')\mathrm{d}G_i^{\mathrm{CDS}}(t') + G^{\mathrm{Res}}(t) = 0$$

for each $0 \le t \le T$. The quantity $G^{\mathrm{Res}}(t)$ is the residual hedging error that results from lack of perfect replication. If the hedging strategy were perfect we would expect $G^{\mathrm{Res}}(t) = 0$. A strategy is risk minimising if the hedging error given by $\mathbf{E}_Q\lfloor(G^{\mathrm{Res}}(\vec{\theta}, T) - G^{\mathrm{Res}}(\vec{\theta}, 0))^2|\Im_0\rfloor$ is minimised over all suitable strategies simultaneously for all $0 \le t \le T$. The authors report that risk minimising hedging strategies interpolate between the hedging of spread and default risk in an endogenous fashion (the reader is referred to the original research paper for more discussion of the results obtained).

14.6 Portfolio exposure measurement

14.6.1 Introduction

The previous section looked at the through life simulation of a single synthetic CDO tranche. This information, when combined with the through life behaviour of a proposed hedging strategy, is useful in order to assess a hedging strategy's effectiveness under different market scenarios. If different valuation models are also available this methodology can be used to assess the hedging effectiveness of each of them.

This type of simulation methodology is also used for the purposes of counterparty credit exposure measurement. Credit exposure to a particular counterparty (obligor) is the dollar amount that an institution stands to lose if a counterparty they do business with defaults (meaning the counterparty will not be able to honour their outstanding obligations). One function of a credit risk management group in an investment bank is to assess the exposure of the bank to an obligor and to compare this exposure against set credit limits. If the trading desk wants to do more business with this obligor, credit risk management must assess the potential impact of the proposed new business on the bank's overall exposure and take action accordingly. Credit risk needs to estimate the potential future exposure (PFE) of the proposed trade over the lifetime of the contract. If the new business completely blows pre-set trading limits, credit risk management may not sign-off on the trade, or may require the desk to make modifications to the trade or their exposures elsewhere.

14.6.2 Counterparty exposure simulation model

Counterparty exposure arises principally through the trades that the institution enters into with that counterparty. To calculate counterparty exposures it is necessary to model the through life behaviour of the trades and also to model the through life behaviour of the actual counterparties themselves. Specifically, a counterparty can default at some point in the future. If that happens, the institution will not receive the full positive benefit of trades with that counterparty going forward. An exposure measurement model must therefore also capture obligor default. The components of a simplified portfolio exposure measurement simulation are now outlined. Note that this is a very simplified model indeed. There are many additional bells and whistles that can be added to this basic framework.

Table 14.1 *Illustrative rating transition matrix, showing probabilities (in per cent) of transitions between different rating states*

Initial rating state	Final rating state								
	AAA	AA	A	BBB	BB	B	CCC	Default	Sum
AAA	90.00	5.00	3.00	1.00	0.50	0.50	0.00	0.00	100
AA	1.00	90.00	7.00	1.00	0.50	0.25	0.25	0.00	100
A	1.00	3.00	87.00	5.00	2.00	1.00	0.75	0.25	100
BBB	0.50	1.00	6.00	83.00	7.00	1.00	0.50	0.50	100
BB	0.00	1.00	2.00	7.00	80.00	8.00	1.50	0.50	100
B	0.00	0.50	0.50	1.00	3.00	85.00	5.00	5.00	100
CCC	0.00	0.00	0.50	1.50	3.00	5.00	75.00	15.00	100
Default	0.00	0.00	0.00	0.00	0.00	0.00	0.00	100.00	100

We assume that there are $k = 1, \ldots, N_{\text{RF}}$ market risk-factors (credit spreads, interest rates, FX rates, equity levels etc.). There are $\gamma = 1, \ldots, N_{\text{Trades}}$ trades that the institution is sensitive to. For example the valuation of a synthetic CDO will depend on the credit spreads of all the obligors in the underlying pool. Let there be $i = 1, \ldots, N_{\text{CP}}$ separate counterparties that the institution has exposures to. To identify which trades are associated with each counterparty we introduce a connectivity matrix $\mathbf{J}$ (similar to that introduced in Chapter 13). $J_{i\gamma} = +1$ if trade γ is associated with counterparty i and zero otherwise. Note that $\sum_{\gamma=1}^{N_{\text{Trades}}} J_{i\gamma} = 1$ for all counterparties i since a particular trade is only associated with a single counterparty.

The objective of the credit exposure simulation is to calculate the exposure of the institution (via their trading activity) to each of the different counterparties. It is also desirable to compute the overall credit exposure profile of the institution. We assume that the simulation is conducted at a set of timenodes $\{T_j : j = 1, \ldots, N\}$ and that there are $\alpha = 1, \ldots, N_{\text{Sims}}$ simulations. The simulation timenodes need not necessarily be evenly spaced. This is an important point when considering the issue of so-called 'credit spikes' to be discussed later.

We assume that counterparties are characterised by their rating state as described in Chapter 2. The probabilities of transitions between different rating states are represented by a rating transition matrix $\mathbf{T}$. For example, Table 14.1 shows an illustrative transition matrix. The rows give the initial rating state at the beginning of the period, and the columns give the rating at the end of the period. This matrix is typically estimated from historical data and gives the probability of rating transitions over a particular time horizon, usually 1 year. For the purposes of simulation it is necessary to rescale this matrix to provide the rating transition probabilities over an arbitrary time horizon Δt (because the simulation timenodes might be set to be quarterly or monthly etc.). This is done by firstly calculating the generator (of the Markov chain [Truck and Ozturkmen 2003, Frydman and Schuermann 2007]) matrix $\mathbf{G} = \ln \mathbf{T}_{1\text{yr}}$. The transition matrix for a time horizon Δt can then be computed from $\mathbf{T}_{\Delta t} = \mathrm{e}^{\Delta t \mathbf{G}}$. The thresholds for determining counterparty rating transitions can be determined from the rescaled transition matrix (see below).

A simple schematic diagram of the credit exposure engine model is shown in Figure 14.13. This figure outlines the calculation steps for a single simulation. The input data are the simulation data (number of simulations, timenodes etc.), market data (parameters for the stochastic processes for the risk factors, correlation matrix, rating transition matrix etc.) and the portfolio positions (contract parameters, notional amounts, long/short etc.). The input data are fed into the scenario generation engine. This is a set of stochastic differential equations for the correlated evolution of the risk factors. The scenario generation also includes the rating migration generator. Both of these components rely on the provision of good quality random numbers from an appropriate generator. As noted in Chapter 5, it is advisable to have a whole suite of random number generators to hand that can be quickly interchanged. It is also highly recommended that great care be given to the implementation and testing of random number generators. If a poor generator is used, then the sophistication of the ensuing analytics is irrelevant.

Once the scenarios have been generated, the next step is to revalue all the instruments in the portfolio. To do this we require appropriate valuation models. It is proposed that there are two types of analytics available: fast analytic models that can be used in 'simulation mode' (for $t > 0$) and slower, accurate front office P/L standard models that can be used in 'MtM mode' (at $t = 0$). Once the instruments have been revalued, the credit and market exposures can be determined. From the credit exposures, and the rating generation engine, portfolio losses can be computed. This in turn leads to the calculation of credit risk statistics such as economic capital. From the market risk exposures we can compute statistics such as VaR and expected shortfall. All of these different calculation steps will now be discussed in more detail.

14.6.2.1 Evolve the market risk factors from time T_{j-1} to T_j for simulation α

Each risk factor is assumed to have stochastic dynamics of the form (more discussion is given of this later)

$$\mathrm{d}s_k^\alpha(t) = \cdots \mathrm{d}t + \cdots \mathrm{d}W_k^\alpha(t)$$

where the risk factors are correlated according to

$$\mathbf{E}\big[\mathrm{d}W_k^\alpha(t)\mathrm{d}W_l^\beta(t')\big] = \rho_{kl}\delta(t - t')\delta_{\alpha\beta}\mathrm{d}t.$$

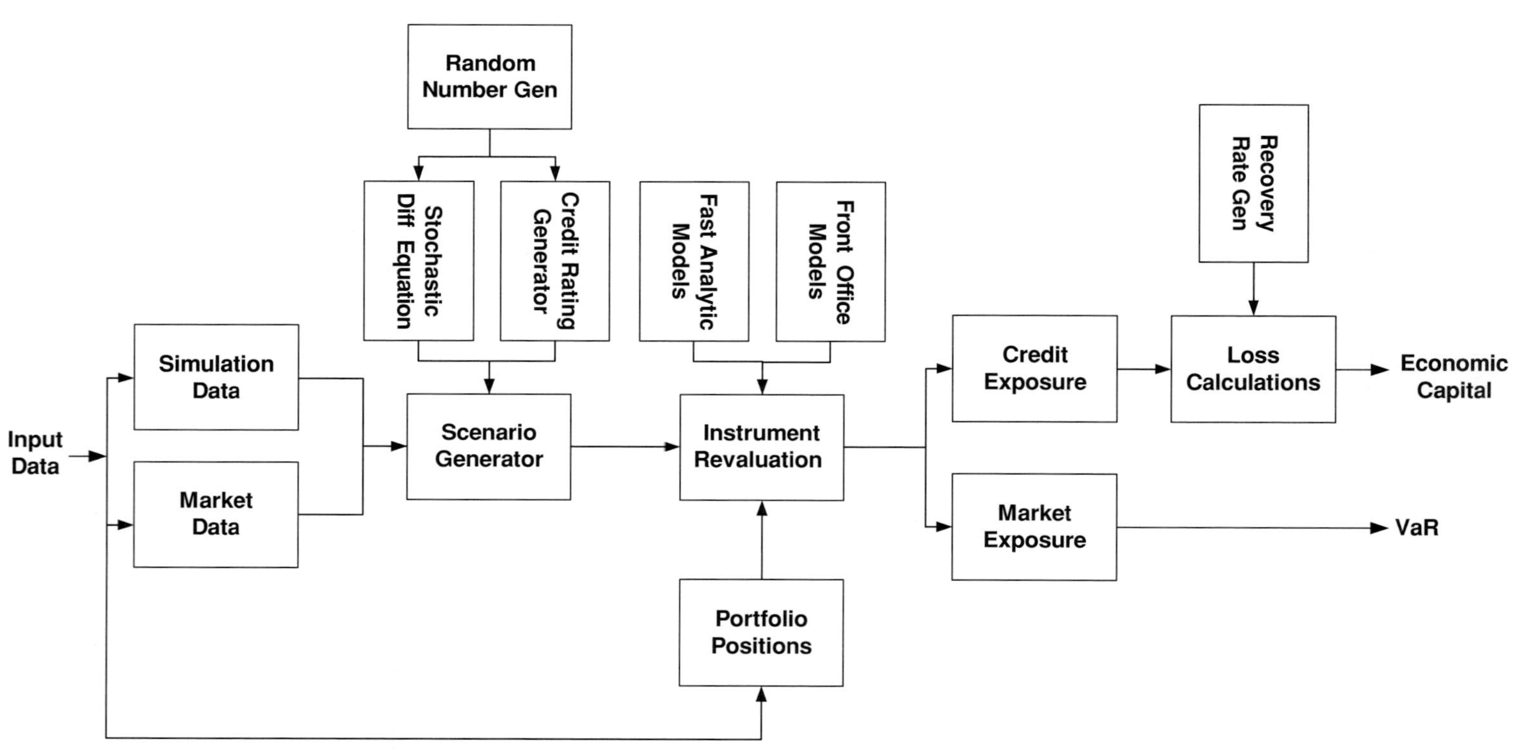

Figure 14.13 Schematic diagram of the credit exposure engine architecture. See the text for a description of the different components.

Correlated shocks are calculated, as usual from $\phi^\alpha = \mathbf{A}\varepsilon^\alpha$ where $\mathbf{C} = \mathbf{A}\mathbf{A}^{\mathrm{T}}$. The correlation matrix $\mathbf{C}$ is assumed to be an input.

14.6.2.2 Evolve the counterparty rating states at time T_j for simulation α

To evolve the counterparty credit ratings a factor model representation can be used [CreditMetrics 1997]. Specifically we model the asset value of the counterparty to be a superposition of systemic and idiosyncratic factors. We assume there are m systemic factors. The asset value is given by

$$X_i^\alpha(t) = \sum_{j=1}^{m} w_{ij} s_j^\alpha(t) + \sigma_i \varepsilon_i^\alpha(t).$$

In this expression $\{w_{ij} : j = 1, \ldots, m\}$ are a set of weights coupling the counterparty asset value to different systemic factors. $\varepsilon_i^\alpha(t) \sim N(0, 1)$ is an idiosyncratic factor and σ_i is the coupling strength of the asset value to the idiosyncratic factor.

The (suitably rescaled) transition matrix $\mathbf{T}_{\Delta T}$ (where for this timenode $\Delta T = T_j - T_{j-1}$) is used to determine the evolution of the obligor's rating state. For a particular row i, the elements T_{ij} represent the probabilities of rating transitions from state i to j. Because $\sum_{j=1}^{N_{\text{Ratings}}} T_{ij} \equiv 1$, the elements of T_{ij} can be used to set thresholds for rating transitions from the current state. The rating migration thresholds for an obligor in an initial state i to move to state j are computed from the appropriate column of the rating transition matrix (see CreditMetrics [1997] for a more detailed example of this calculation) by mapping them onto the quantiles of a standardised normal distribution by the transformation $Z_j^i = N^{-1}(\sum_{k=\text{Default}}^{j} T_{ik})$. Depending on the value of the asset the new simulated obligor rating state can be determined. If the counterparty rating is 'Default' then this is recorded in the simulation and appropriate losses computed. The default state is an absorbing state in the Markov chain. Once an obligor defaults it stays defaulted.

14.6.2.3 Revalue trades at time T_j for simulation α

Each trade has a maturity of T_γ. If $T_\gamma < T_j$ then the new value of the trade, given the current values of the market risk factors, $V_\gamma^\alpha(\{s_k^\alpha T_j)\}, T_j - T_\gamma)$, is computed. If $T_\gamma > T_j$ the trade has expired and it does not contribute any further to the counterparty credit exposure. The payoff of the trade may be paid into a cash account (which for simplicity we have ignored).

14.6.2.4 Calculate the counterparty credit exposures at time T_j for simulation α

The counterparty exposures $C_i^\alpha(T_j)$ are computed according to

$$\text{Gross exposure } C_i^\alpha(T_j) = \sum_{\gamma=1}^{N_{\text{Trades}}} \max\left(V_\gamma^\alpha(T_j), 0\right) J_{i\gamma}$$

$$\text{Net exposure } C_i^\alpha(T_j) = \max\left(\sum_{\gamma=1}^{N_{\text{Trades}}} V_\gamma^\alpha(T_j) J_{i\gamma}, 0\right)$$

Credit exposures can be sliced-and-diced according to the institution's reporting requirements (e.g. aggregating at counterparty, desk, business level).

14.6.2.5 Calculate the counterparty credit losses at time T_j for simulation α

For each obligor that has defaulted (determined in a previous step) we can compute a loss. A loss is registered if the counterparty credit exposure is positive. The magnitude of the loss is taken as the exposure at the time of default, less the fraction of the exposure that is expected to be recovered (the recovery rate). Recovery rates are assumed to be stochastic, and can be sampled from a beta distribution (which is only defined on the range [0, 1] making it suitable as a proxy for a recovery fraction). The beta distribution is defined according to Loffler and Posch [2007]

$$B(z, w) = \int_0^1 t^{z-1}(1-t)^{w-1} \mathrm{d}t = \frac{\Gamma(z)\Gamma(w)}{\Gamma(z+w)}$$

where

$$\Gamma(z) = \int_0^\infty x^{z-1} \mathrm{e}^{-x} \mathrm{d}x$$

is the gamma function. The pdf of the beta distribution is given by

$$\beta(x, a, b) = \frac{1}{B(a, b)} x^{a-1}(1-x)^{b-1} \mathbf{1}_{0<x<1}.$$

For a random variable sampled from a beta distribution $X \sim \beta(x, a, b)$, the first two moments of the distribution are given by

$$\mathbf{E}[X] = \frac{a}{a+b},$$

$$\mathrm{var}(X) = \frac{ab}{(a+b+1)(a+b)^2}.$$

Therefore if we know the moments of the (beta) distribution we wish to sample from, we can calculate the distribution parameters from

$$a = \frac{\mathbf{E}[X]}{\mathrm{var}[X]} \left[\mathbf{E}[X](1-\mathbf{E}[X]) - \mathrm{var}[X\}\right],$$

$$b = \frac{1-\mathbf{E}[X]}{\mathrm{var}[X]} \left[\mathbf{E}[X](1-\mathbf{E}[X]) - \mathrm{var}[X\}\right].$$

The mean and variance of recovery amounts are typically estimated from historical data for each particular rating state. Assume that the sampled recovery rate for counterparty i on simulation α is δ_i^α.

For a simulation α, the losses at timenode T_j for counterparty i are represented as $L_i^\alpha(T_j)$. The losses over a time horizon T_{Loss}, $L^\alpha(0, T_{\text{Loss}})$ are computed by summing up the total losses for each counterparty over the period $[0, T_{\text{Loss}}]$. That is

$$L^\alpha(0, T_{\text{Loss}}) = \sum_{i=1}^{N_{\text{CP}}} \sum_{j=1}^{N} \sum_{\gamma=1}^{N_{\text{Trades}}} Z(0, T_j)\left(1 - \delta_i^\alpha\right) V_\gamma^\alpha\left(\{s_k^\alpha(T_j)\}\right) J_{i\gamma} \mathbf{1}_{T_{j-1} \leq \tau_i^\alpha < T_j} \mathbf{1}_{T_j < T_{\text{Loss}}}$$

where τ_i^α is the default time for counterparty i on simulation α (a default is recorded when an obligors rating state is evolved and the new state is found to be the default state). The overall simulated expected losses can be computed from

$$\bar{L}(0, T_{\text{Loss}}) = \frac{1}{N_{\text{Sims}}} \sum_{\alpha=1}^{N_{\text{Sims}}} L^\alpha(0, T_{\text{Loss}}).$$

We can also of course determine the full loss distribution $P[L(0, T_{\text{Loss}})]$ from the ensemble of simulations. Typically, loss reserves are taken as the mean of the loss distribution (i.e. expected losses) and economic capital is computed as the difference between an upper quantile and the mean. Because of the discounting performed throughout the simulation, the inferred capital represents the cash that must be set aside *today* as a cushion against future losses.

We have only focused on the calculation of losses in the example model described. Of course, once we have the simulated values of the assets we can compute the total portfolio value, for example, for each simulation at each timenode. The distribution of simulated portfolio values can be used to compute simple VaR or expected shortfall risk measures. In the next few sections we provide some more discussion of issues that the example model presented raises.

14.6.3 Risk factor evolution

In a realistic simulation model it will be necessary to model many different risk factors (a risk factor is taken to mean a fluctuating financial observable, such as an equity price etc., which the trade valuations are a function of). These will include

- equity values,
- interest rates (typically will require a complete term structure of rates – spot, 3 month, 6 month etc. – for each currency for which there are trades),
- FX rates (although these can be computed from the interest rates as well),
- credit spreads.

In the previous section on hedging simulation, CDS par spreads were evolved using a GBM

$$\frac{ds_k^\alpha(t)}{s_k^\alpha(t)} = \mu_k dt + \sigma_k dW_k^\alpha(t).$$

This modelling choice was made only for simplicity and for the purpose of understanding the phenomenology of tranche behaviour. A nice feature of this model of course is that the SDE can be solved exactly by applying Ito's lemma [Baxter and Rennie 1996, Oksendal 1998]. This means that it is possible to evolve the SDE between arbitrarily spaced timenodes and still sample from the correct distribution [Glasserman 2003].

It is observed empirically that credit spreads can undergo sudden 'blow-outs'. For example, if a corporate is the recipient of an unanticipated LBO rumour, then their credit spreads can jump by quite huge amounts in a short space of time, reflecting the market's revised view of the corporate's default risk. To incorporate this type of phenomenon we can add to the plain GBM (which is a nicely behaved diffusion process) a jump component $J_i(t_j)$

$$\frac{ds_k^\alpha(t)}{s_k^\alpha(t)} = \mu_k dt + \sigma_k dW_k^\alpha(t) + dJ_k^\alpha(t)$$

(and we assume that the diffusion and jump processes are uncorrelated). See Glasserman [2003] for more details on how to simulate this process. All of the different risk factors are correlated according to $\mathbf{E}[dW_k^\alpha(t)dW_l^\beta(t)] = \rho_{kl}\delta_{\alpha\beta}dt$, i.e. for different times and different paths the shocks are uncorrelated. It is a very important part of the simulation that the shocks to the different risk factors are correlated. This enables phenomena such as 'wrong-way' exposure (where there is an adverse relationship between exposure size, driven by the market risk factors and counterparty default) to be captured as potential market scenarios.

It is a matter of debate what the correct characterisation of CDS spread dynamics is. One condition that must be obeyed is $s_i(t) > 0$ for all obligors i and all times t, i.e. the spread must always be positive. This rules out models such as the Hull–White model for evolving the spreads since this is not guaranteed always to generate a positive value, although models such as Cox–Ingersoll–Ross and Black–Karasinski would be suitable [Rebonato 1998, 2002, Brigo and Mercurio 2001]. There is also evidence [Longstaff and Schwartz 1995, Prigent *et al.* 2001] to suggest that spreads might be mean-reverting over a long enough observation period (also ruling out GBM since $\sigma \sim \sqrt{t}$ for a GBM meaning that the dispersion of the simulated values becomes increasingly large as time progresses). These authors also observe that the speed of mean-reversion is credit rating dependent. Also, spread volatilities are observed to be increasing in level. A possible model formulation to capture

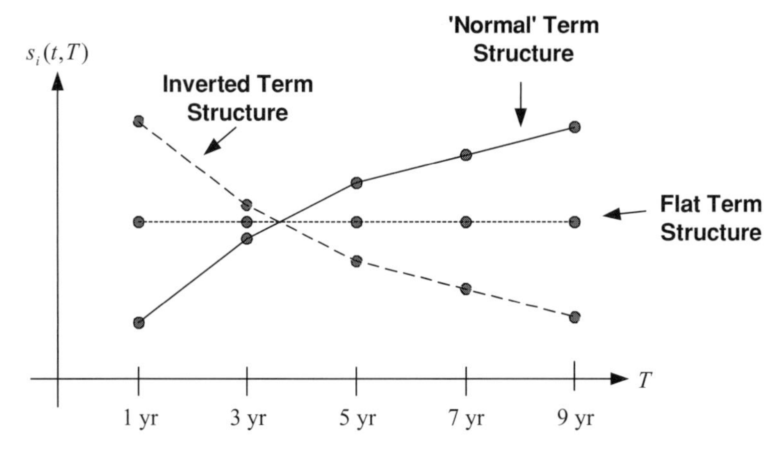

Figure 14.14 Schematic diagram showing the different types of spread curve term structure ('normal', inverted and flat) that a suitable stochastic model for credit spreads must capture.

this is the Cox–Ingersoll–Ross model [Cox *et al.* 1985]. Finally, another important characteristic of credit spreads that is observed is so-called spread inversion. This is illustrated schematically in Figure 14.14. Under 'normal' market conditions the term structure of spreads is upward sloping as the protection term increases. However, under some circumstances the term structure can invert, meaning that short-term default risk exceeds long-term risk. This was indeed observed during summer 2007. A suitable stochastic model for spreads must be able to generate this type of scenario. To capture all these features another possible model for credit spreads would be a mean-reverting (Ornstein–Uhlenbeck) process [Risken 1996] such as [Longstaff and Schwartz 1995]

$$\mathrm{d}\ln s_k^\alpha(t) = a_k \left\lfloor \mu_k - \ln s_k^\alpha(t) \right\rfloor \mathrm{d}t + \sigma_k \mathrm{d}W_k^\alpha(t)$$

where a_k is the speed of mean-reversion, μ_k is the long-term level about which the spread reverts and σ_k is the volatility of the spread (to which Prigent *et al.* [2001] also add a jump diffusion component). This is a mean-reverting SDE and has an explicit solution given by

$$s_k^\alpha(t_{j+1}) = s_k^\alpha(t_j)\exp\left\{ \left[\mu_k - \ln s_k^\alpha(t_j)\right]\left[1 - \mathrm{e}^{-a_k(t_j - t_{j-1})}\right] + \sigma_k \sqrt{\frac{1 - \mathrm{e}^{-2a_k(t_j - t_{j-1})}}{2a_k}}\varepsilon_k^\alpha \right\}$$

(meaning that spreads can be evolved across arbitrarily large time steps while still sampling from the correct distribution). For this stochastic process the long-term

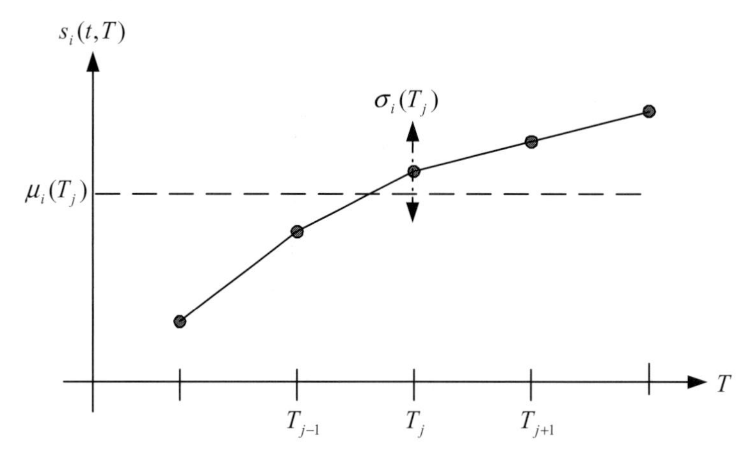

Figure 14.15 Schematic diagram showing the model for evolving each point on the credit spread term structure as a separate stochastic equation.

volatility is constant and tends to a value of $\sqrt{\sigma_k/2a_k}$ meaning the exposure is limited.

This formulation only models the spot credit spread (taken to mean the shortest maturity spread observable in the marketplace). We can generate from the spot spreads the entire term structure of spreads from an expression of the form $s_k^\alpha(t,T) = h_T + k_T s_k^\alpha(t)$. Alternatively we can simulate each separate point on the credit spread term structure according to, for example,

$$\mathrm{d}\ln s_k^\alpha(t,T_j) = a_k(T_j)\left\lfloor \mu_k(T_j) - \ln s_k^\alpha(t,T_j)\right\rfloor \mathrm{d}t + \sigma_k(T_j)\mathrm{d}W_k^\alpha(T_j)$$

where the points on each term structure are each correlated, i.e. $\mathbf{E}[\mathrm{d}W_k^\alpha(t,T_j)\times \mathrm{d}W_l^\alpha(t,T_k)] = \tilde{\rho}_{kl}\mathrm{d}t$. The volatilities σ_k and other parameters can be estimated from the time series of, for example, the 5 year spread point. This is shown schematically in Figure 14.15.

Finally, single-factor models use only a single stochastic factor to shock the spreads. This means that all points on the term structure move in parallel. To be able to 'twist' and 'rotate' the curve to generate more complex term structure shapes, it is necessary to introduce more stochastic factors into the model formulation.

Thus far we have only discussed possible formulations for modelling the evolution of credit spreads. Similar models must also be constructed for the other risk factors in the simulation. Parameters for all the stochastic processes can be estimated by a variety of methods. The simplest is maximum likelihood estimation [Hamilton 1994]. Alternatively, a more sophisticated technique such as simulated annealing could be used. Scenario generation and parameter estimation is done under the objective (historical) measure. Valuation is done under the risk-neutral measure.

14.6.4 Instrument revaluation

A typical credit exposure simulation might have a time horizon of decades (swap contracts in particular can extend for this type of maturity). If we assume that the simulation timenodes are distributed at monthly intervals this means there are $12 \times$ time horizon timenodes where the portfolio is revalued per simulation. An institution will typically have many thousands of instruments on their books. To compute reliable Monte Carlo estimators of the portfolio value may require many thousands of different scenarios to be generated. Whichever way you look at it, it is necessary to revalue the different instruments in the portfolio many times.

For front office MtM purposes accuracy in valuation is paramount. Full Monte Carlo simulation can be used to value, for example, a path dependent interest rate derivative (such as a range-accrual note – although these notes can in fact be valued analytically) to arrive at a single PV and sensitivities. It is not too important if it takes 10 seconds to revalue a tranche and calculate its delta values using a semi-analytic approach such as the recursive or normal proxy models. What is more important is that the valuation and sensitivities from the model are accurate so that the trade can be executed at the right price and the correct hedge put in place. For a credit exposure measurement engine, however, 10 seconds is an eternity. Typically sub-millisecond revaluation times are required in order to ensure that a Monte Carlo simulation is completed in a timely manner (usually overnight). Speed of instrument revaluation is therefore of crucial importance.

In the previous section on hedging simulation we used the normal proxy model for tranche revaluations. This is a very good model as it is accurate and also fast. For the purposes of assessing the behaviour of a single tranche it was appropriate. But as a full credit exposure engine, it is not fast enough. An alternative model to use would be the LHP model outlined in Chapter 6 which computes the portfolio loss distribution via a few calls to statistical distribution functions. Even this may not be fast enough for certain portfolios. In general, if the instrument valuation model is too slow, then it is necessary to find fast, analytic approximations to revalue the instrument. This is a highly non-trivial undertaking, often requiring a great deal of research on an instrument-by-instrument basis.

An obvious, although brute force, solution to this problem is to use more processors to do the simulations. As noted in Chapter 13, Monte Carlo simulation is extremely amenable to parallel processing architectures since each separate simulation is independent of all the others. With the advent of grid computing and the availability of cheap processing power on everyone's desktops (which is often not utilised overnight) this is a very simple and foolproof solution to the problem. The technological problems to be overcome in organising all the data, farming out the simulations, collating and processing the results are non-trivial. However, these problems can be overcome. For a serious credit exposure engine a parallel processing solution is highly recommended.

Finally another important requirement for the instrument revaluation engine is that it should reproduce at time $t = 0$ the MtM values produced by the front office system. This ensures consistency of the credit exposure engine PVs with the reported P/L for the institution. Although this sounds simple, from an implementation point of view this requirement may add a lot of extra complexity to the overall system architecture. For example, the front office may value their tranches using a semi-analytic model. But for speed purposes the credit exposure engine may use some quicker analytic approximation. The credit exposure engine would need to incorporate both models into its architecture and be able to switch between them.

14.6.5 'Credit spikes'

In Section 14.5 we performed the simulation where the spacing between the simulation timenodes was equal to $\Delta t = 1$ day. This is the most granular level of simulation timenode density possible (there is no point whatsoever in simulating intra-day). Of course this means the computational time required for a simulation is impractical. Using weekly, monthly, quarterly or even annual timenodes solves this problem. However, this is at the expense of introducing another problem: spikes in credit exposure that are not captured by the simulation.

Credit spikes can under- or over-estimate potential future exposure. The danger is that credit spikes might breach predefined limits and would thus influence business decisions regarding potential trading activity. In general credit spikes can arise for the following reasons:

- cashflows occurring between discrete timenodes (if the cashflow is large it may have a material impact on the PFE),
- trades maturing between discrete timenodes (the potential exposure will therefore drop abruptly,
- options being exercised between discrete timenodes,
- cashflow mismatches between offsetting trades (e.g. offsetting swaps),
- irregular payments.

Credit spikes are a difficult problem to solve in all generality since each separate portfolio will have a different distribution of significant cashflow dates and significant cashflows. A simple solution is to run a presimulation sweep to determine where all the cashflow dates are. For example, if a single instrument has cashflows at the set of dates $\{T_j : j = 1, \ldots, N\}$ (for example a swap of some sort), then to capture the full credit exposure (including the dips in the exposure when the cashflows drop off) we would need timenodes at $\{T_j - 1, T_j + 1, (T_{j-1} + T_j)/2 : j = 1, \ldots, N\}$. For a single swap of maturity 10 years with quarterly coupon payments, this represents a large number of timenodes. For a whole portfolio of instruments (meaning many thousands of trades) with differing maturities and tenors, the number of timenodes becomes impractical. It is recommended that experimentation on each portfolio is required so that the optimal set of simulation timenodes can be determined (remember that if the SDEs all have analytic solutions, then the timenodes can be placed anywhere).

14.7 Chapter review

In this final chapter we have introduced the technique of hedging simulation and discussed its application to structured credit products. The purpose of the chapter has been to introduce the concept as a potentially useful tool for addressing questions such as whether a model is fit-for-purpose (does it provide good risk management of exposures to fluctuations in market risk factors), and for comparing the performance of different models in various market environments (which model provides the best explanation of P/L given the prevailing market conditions).

The premise of hedging simulation is simple: track the through life behaviour of an instrument and its associated hedging strategy. The changes in P/L as the market variables fluctuate are recorded. In order to make progress in understanding hedging simulation of synthetic CDOs we have made a number of simplifying assumptions. For example, we assumed that the term structure of credit spreads was flat.

We also made a number of modelling assumptions regarding the stochastic processes driving the evolution of the obligor par spreads as well as assuming that default correlation remained constant throughout time. All of the modelling simplifications and assumptions can be relaxed at the expense of greater data management and computational complexity. The simplified model that we have presented, however, appears to produce results that are consistent with our knowledge of how tranches behave. Specifically factors such as the tranche subordination, obligor spread volatilities and the degree to which spreads move together (systemic versus idiosyncratic behaviour) have a strong bearing on the tranche P/L. The analysis that has been presented in this chapter is by no means comprehensive or complete and the reader is strongly encouraged to implement their own hedging simulator to test their own ideas and assumptions. The only way really to understand a model and its behaviour is to build one. We also discussed briefly some very recent models that have been introduced which attempt to rectify some of the modelling limitations of the standard market model.

Finally we introduced some basic modelling issues regarding counterparty credit exposure measurement. This was introduced at this point as the methodology is a natural fit with that described for the hedging simulation. To illustrate some of the modelling issues a simple counterparty credit engine was described. Counterparty credit management is an extensive topic and the simulation models typically used in a live banking environment are far more sophisticated than the simple model described here. Once again the reader is encouraged to construct their own models, using what is described here as a framework to test out their own ideas.

During the course of this text we have gradually added to the complexity of problem analysed. From simple CDSs through default baskets to synthetic CDOs defined on the standardised credit indices, the focus has been on postulating a scenario (typically simplified to assist in understanding the key issues), specifying a model, generating the model outputs and understanding them. Later chapters then built on this foundation and looked at topics such as correlation trading, the characterisation and behaviour of portfolios of synthetic CDOs and finally hedging simulation to quantify the potential P/L implications of a model. We also introduced a number of models which are the current state-of-the-art in synthetic CDO modelling. In what direction will the next stage of credit derivative evolution take us? Only time will tell. However, it is the author's hope that the material presented in this text will provide future researchers with a solid understanding of the current state-of-the-art in modelling credit derivatives, and that this in turn will help the development of more advanced and better models.

Appendix A
Explanation of common notation

Every effort has been made throughout the text to keep notation consistent, concise and logical. However, in such a technically detailed work there are inevitably occasions when it is not possible to adhere to these principles. It is hoped that on the occasions when this happens, the deviations have been adequately explained. It is also acknowledged that the notation is at times quite dense (although nowhere near the level of concision to be found in a typical textbook on financial mathematics). In this brief appendix we will give some examples of the notation and explain the rationale for it.

For the most part we are dealing with situations where there are multiple obligors. An individual obligor is typically tagged by an index i (and always represented as a subscript). Usually the total number of obligors is denoted by n. Time is represented by t. Fixed-time points such as a schedule of coupon payments are denoted by the uppercase version of this. For example, the maturity of an instrument is represented as T. A fixed coupon payment date is represented by T_j where j represents the index of the coupon payment (e.g. the first coupon $j = 1$ occurs at time T_1 and so on and $T_j \leq T$ for all j). A time in-between two fixed coupon dates would be represented as $t \in [T_{j-1}, T_j]$. The time of obligor defaults is always represented by τ (so obligor i defaults at time τ_i).

An obligor's recovery rate is always represented by δ_i (for obligor i). Their par CDS spread is represented as s_i and hazard rate as h_i. The marginal survival curve for this obligor is always represented as $S_i(t, T)$ representing the survival probability between time t and T. Obligor correlations are represented by ρ and a full correlation matrix by $\mathbf{C}$ (and its Cholesky decomposition by $\mathbf{A}$).

The number of Monte Carlo simulations is written as N_{Sims} and an individual simulation identified by α (with the simulation identifier normally written as a superscript, so for example the time of default of obligor i on simulation α is denoted by τ_i^α). Uncorrelated normal deviates are represented as ε and correlated normal deviates as ϕ.

A tranche is typically represented by the greek letter γ. This is taken to represent both the attachment and detachment points that are written as K_{L}, K_{U} respectively. Notional amounts are typically written as N_i for the notional amount of obligor i or N^γ for the notional amount of tranche γ. The tranche loss is usually written as L (or sometimes l).

Unfortunately N is used for the number of coupon payments, the number of simulations and a notional amount! It should be clear from the context what the meaning is.

Values of 'things' are usually represented by V (V is also used as the value of the systemic factor in the factor model). So the value of a tranche γ is given by V^γ. The values of the fee and contingent legs for this tranche would be written as V_{Fee}^γ and V_{Cont}^γ respectively.

Appendix B

Simulated annealing

In the text several references are made to the technique of simulated annealing for the purpose of finding optimal solutions to a particular multivariate problem. Here we provide an introduction to this technique and describe why it is such a powerful and intuitively appealing tool for a wide range of problems.

Consider that we have an instrument whose value is a function of a set of variables $\{s_i\}$. Denote this by $V(\{s_i\})$. In the context of synthetic CDOs the variables $\{s_i\}$ could correspond to the spreads of the underlying obligors, for example. Our objective could be to determine the optimal composition of the underlying pool (choosing, for example, 100 obligors from the overall universe of obligors) such that we obtain the maximal par spread of the tranche while stipulating that the weighted average spread of the underlying pool is a minimum, i.e. maximising the carry (return) for minimal risk.

To visualise the problem it is helpful to think of the independent variables $\{s_i\}$ as defining a multi-dimensional parameter space. Each choice of a set of values for the variables defines a point in the space (a particular point in parameter space will be referred to as a 'configuration' of the system, adopting the language of statistical mechanics). The objective of the optimisation is to find the point(s) in the parameter space that meet the specified objectives, while at the same time satisfying any constraints that are imposed on the solution (e.g. the pool weighted average spread cannot exceed 100 bps).

The problem could be solved by systematically searching every possible combination of obligors. However, this is an astronomically large number of possibilities which renders this approach unfeasible from a computational perspective. An alternative strategy would be to pick an initial configuration at random and then gradually evolve this configuration (e.g. by choosing an obligor at random to remove from the portfolio and replacing it with another randomly chosen obligor). If the new configuration provides a better solution than the previous configuration, the new configuration is accepted. This process is repeated a large number of times.

The drawback of this approach is that a poor initial choice of system configuration will lead to a poor result. For example, if the parameter space is characterised with many local minima/maxima interspersed amongst the true global minima/maxima then there is a high degree of likelihood that the algorithm would only find a local extremum. This could be partially rectified by repeating the process several times, each time starting from a different initial random choice for the system configuration.

This process may find the global maxima but it is a scattergun approach. Simulated annealing is a refinement to this process that allows some control to be placed over the randomised choices of the system configuration. The simulated annealing algorithm gradually 'cools' the system from an initial (high) temperature to a final (low) temperature. The key idea is that at high temperatures the system is essentially free to explore all of its parameter space at random, thereby allowing the algorithm to jump out of local minima that it may find (it ignores the local microscopic features of the energy landscape). As the system is cooled it becomes gradually harder for the algorithm to escape from local minima and the solution becomes 'frozen' into the current minimum focusing on smaller and smaller regions of the parameter space. The initial temperature should be such that $e^{-\Delta V/T_{\text{Start}}} \sim 1$ and all new configurations are accepted to ensure that the system is not stuck initially in a local minimum. The schedule should end at $T_{\text{Stop}} = 0$ or close to this. The schedule should be sufficiently slow that the algorithm is able to explore systematically a large enough proportion of the parameter space. It is advisable to perturb the schedule periodically by, for example, introducing random jumps (upwards) or by introducing a sawtooth form for the cooling schedule.

For each temperature and iteration the quality of the solution obtained is characterised by the objective function. The objective function computes the value of the system for the current configuration. To incorporate constraints into the problem an additional energy penalty term can be added to the objective function. This should be such that a solution which violates the constraint introduces a large positive contribution to the objective function. For example, if we require the weighted average spread $\bar{s}$ to be less than some critical value s_{Max} this can be incorporated by adding a term such as $\lambda \mathbf{1}_{\bar{s} > s_{\text{Max}}}$ (where $\lambda > 0$ and is 'large') to the objective function. A solution which violates the constraint therefore becomes energetically unfavourable.

At each temperature T_k a Monte Carlo simulation (with N_{Sims} simulations) is performed to search stochastically the local neighbourhood of the current best minimum. At each iteration of the Monte Carlo simulation the new configuration (which determines the objective function in terms of the system parameters) is chosen to be quite close energetically to the current configuration. This is to ensure that the parameter space is searched in a relatively smooth fashion (without large fluctuations in the objective function).

The change in energy between the two configurations is represented by ΔV. If $\Delta V < 0$ then the new configuration of the system is accepted with probability $p = 1$ since the new configuration does represent an improvement in the search for the global minima. If $\Delta V > 0$ then the new configuration of the system is accepted with a probability $p = e^{-\Delta V/T}$. Therefore even if the new configuration is unfavourable in terms of the cost function, there is a finite probability that the algorithm will accept it anyway. This prevents the system becoming stuck in a local minimum of the objective function.

When $T_k \to \infty$ then the probability of accepting the new configuration approaches unity, meaning that any new configuration is accepted. Therefore at high temperatures the system will wander aimlessly around the phase space of the problem, only coming across minima by chance. As $T_k \to 0$ then $p(\Delta V) \to 0$ irrespective of ΔV. Therefore at low temperatures the new configuration is never accepted if $\Delta V > 0$, even for infinitesimally small increases in the energy. The system is frozen into a minimum energy configuration. As $T_k \to 0$ the system increasingly only favours rearrangements that go 'downhill'. When $T_k = 0$ the algorithm only ever accepts downhill moves (it reduces to the gradient descent algorithm which greedily follows its way down an incline irrespective of whether it is in a local or global minimum). As $\Delta V \to \infty$ then $p(\Delta V) \to 0$ irrespective of T_k. Therefore for any two configurations with $\Delta V > 0$ the algorithm will tend to favour the one with the lower change in the cost function.

The simulated annealing algorithm used to find the minima of a function (referred to as the objective or cost function) is as follows.

1. Choose an initial configuration of the system at random $\{s_i\}$.
2. Calculate the value of the objective function for this choice of configuration $V(\{s_i\})$. Note that any constraints to be imposed on the solution should be included in the cost function as an energy penalty.
3. Set the value of the temperature for iteration k to be $T_k = r^k T_{\text{Start}}$ where $r < 1$ controls the rate at which the temperature is reduced (the annealing schedule) and T_{Start} is the initial temperature of the system.
4. Choose a new configuration for the system that is a 'near' neighbour to the current configuration $\{s'_i\}$ and calculate the new value of the objective function $V(\{s'_i\})$.
5. Compute the change in value of the objective function $\Delta V = V(s') - V(s)$. This may be positive or negative depending upon whether the new configuration improves upon the current global minimum.
6. If $\Delta V < 0$ then the new configuration is energetically favourable and we accept the new configuration as the global minimum.
7. If $\Delta V \geq 0$ then the new configuration is not energetically favourable and we reject it. However if $u < \mathrm{e}^{-\Delta V/T}$ where $u \sim U(0, 1)$ is a uniformly distributed random variable, then we accept the new configuration irrespective of the fact that it is not energetically favourable. This step enables the system to jump randomly from its current minimum into another state (with the probability of jumping being dependent upon the current temperature).
8. Repeat steps 4–7 for $\alpha = 1, \ldots, N_{\text{Sims}}$.
9. Repeat steps 3–8 until $T_k < T_{\text{Stop}}$ where T_{Stop} is the terminal temperature (subject to the constraint $T_{\text{Stop}} \geq 0$).

The goal of simulated annealing is to evolve a system from an arbitrary initial state to a final state with the minimum possible energy (subject to the specified constraints). Simulated annealing is particularly well suited to the case where the number of degrees of freedom of the system is large. The simulated annealing algorithm is remarkably simple and easy to implement. The difficulties arise due to the fact that each problem is different and therefore a certain amount of experimentation (and expertise) is necessary to determine suitable annealing schedules, changes in configuration and number of Monte Carlo simulations performed at each temperature. When approaching a simulated annealing problem the following general guidelines are recommended.

1. Identify the independent variables and constraints amongst them. Identify maximal or most likely ranges for the possible values the independent variables can take.
2. There is no 'correct' cost function for any particular problem and some experimentation may be required to identify the best candidate. If possible perform an initial analysis of the properties of the cost function, for example plot it keeping non-critical independent variables constant, understand its limiting behaviour. The objective is to understand better the gross behaviour of the function (e.g. does it clearly have multiple local minima, or a single global minimum). If the cost function turns out to be a monotonically varying function of the independent variables, then a simpler optimisation scheme such as a gradient follower may in fact suffice.
3. Initially begin the algorithm with a fairly high starting temperature and rapid cooling schedule. Perform some tests to assess the impact of the cooling schedule upon the solution obtained. Gradually decrease the cooling schedule and starting temperature once the 'freezing' properties of the problem become more clearly understood.
4. Initially start with quite gross changes in system configuration for each Monte Carlo simulation at a particular temperature (this allows the system essentially to behave randomly). Gradually refine this through experimentation.
5. Simulated annealing may quickly identify a region where a global extremum occurs, but then take a long time to obtain the actual extremum (due to 'critical slowing down'). Consider interspersing use of simulated annealing with a simpler technique, such as a gradient descent algorithm (which rapidly searches out a local minimum), to obtain greater accuracy.

As a simple exercise the reader should consider how to calculate the global minima of a function like $f(x) = \mathrm{e}^{-\lambda x} \sin \omega x$ when $x \in [0, 10]$. Of course the minima can be

Table B.1 *Results of the different algorithms for calculating the global minima of the test functions*

	Random sampling		Gradient following		Simulated annealing	
	x	$f(x)$	x	$f(x)$	x	$f(x)$
Case A	2.33	−0.79	8.62	−0.42	2.33	−0.79
Case B	4.70	−0.95	11.00	−0.90	4.70	−0.95
Case C	2.35	−0.98	8.64	−0.92	8.64	−0.92

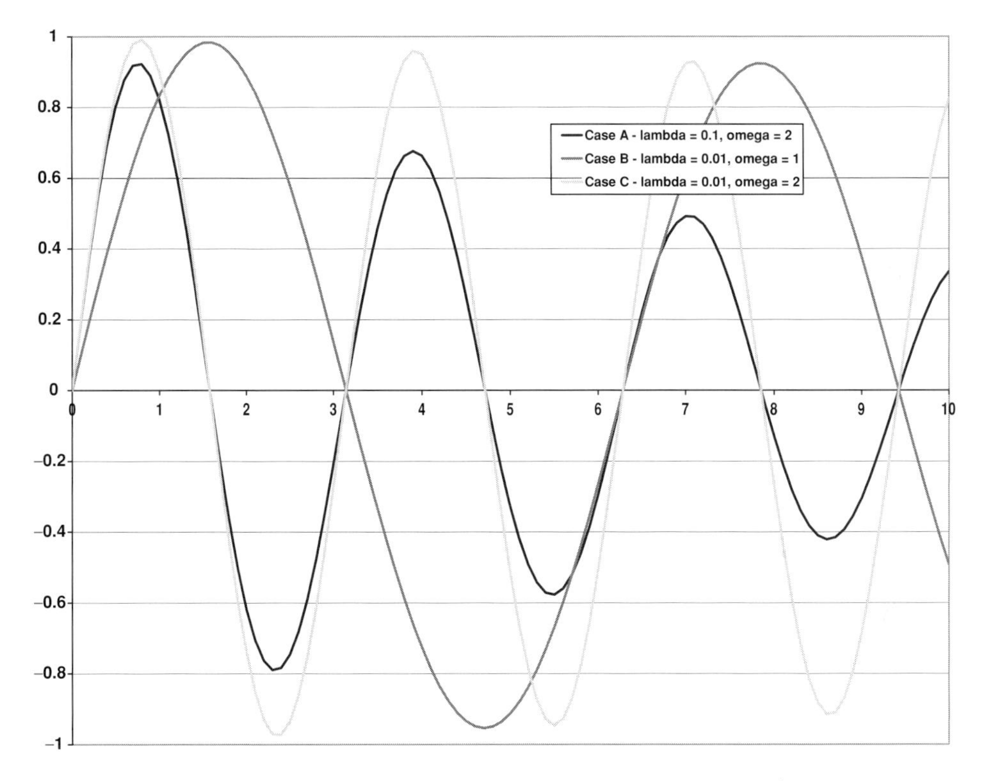

Figure B.1 The three different parameterisations of the function $f(x) = e^{-\lambda x} \sin \omega x$.

computed analytically for this one-dimensional example. Sampling x at random 10 000 times also does a good job of identifying the true minima. However, a gradient based approach where the algorithm simply moves downhill from its initial position can easily return a local minimum. The performance of the simulated annealing algorithm is a little more subtle. Consider the three cases A, B and C with parameters as shown in Figure B.1. Table B.1 shows the results of applying the three different minimisation strategies. In all cases the initial starting value for the algorithm was $x_0 = 8$ (deliberately chosen so as to be close to a local minimum, but far away from the true global minimum). In all cases random sampling does a very good job of finding the global minimum (because the problem is one-dimensional). The gradient follower, however, finds only a local minimum

Table B.2 *Impact of annealing schedule on the performance of the algorithm for case C (near degenerate global minimum)*

Annealing schedule	Simulated annealing	
	x	$f(x)$
0.909	8.64	−0.92
0.990	5.50	−0.95
0.999	2.35	−0.98

because the starting value of $x_0 = 8$ is poor. In cases A and B simulated annealing does a very good job of finding the global minimum. However, in case C it only finds a local minimum. This is because case C has almost degenerate global minimum since $\lambda \to 0$. The random sampling method is unaffected by this since the function parameters have no influence on where the algorithm samples from. But the simulated annealing algorithm is not able to explore adequately the macro shape of the function. This is because the annealing schedule is too rapid. Table B.2 reports the performance of the simulated annealing algorithm as the annealing schedule is slowed down. Because the system spends more time at higher temperatures there is now a greater chance for the algorithm to jump out of the local minimum that it finds itself in. As the annealing schedule is reduced the algorithm eventually finds the true global minimum. This example highlights an important practical point when employing simulated annealing as a stochastic search technique. Specifically, a poor choice of parameters can lead to poor performance. Although simulated annealing is an intuitive and powerful technique, it requires the analyst to have a good understanding of the problem they are trying to solve. In the above example it was easy to see when the algorithm performed poorly. A real problem, however, may not be so simple to interpret. It is advised before using simulated annealing that a good understanding of the actual problem is obtained. Simulated annealing is a beautiful and powerful method, but it requires some expertise to get the best out of it.

References

H. Albrecher, S. A. Ladoucette and W. Schoutens, A Generic One-factor Levy Model for Pricing Synthetic CDOs, www.defaultrisk.com, May 2006.

C. Alexander, *Market Models: A Guide to Financial Data Analysis*, Wiley, 2001.

C. Alexander and S. Narayanan, Option pricing with normal mixture returns, *ISMA Centre Discussion Papers In Finance*, December 2001.

K. A. Allman, *Modeling Structured Finance Cashflows with Microsoft Excel*, Wiley, 2007.

L. Andersen, Portfolio Losses in Factor Models: Term Structures and Intertemporal Loss Dependence, www.defaultrisk.com, 2006.

L. Andersen and J. Sidenius, Extensions to the Gaussian copula: random recovery and random factor loadings, *Journal of Credit Risk*, **1**, 29–70, 2004.

L. Andersen, J. Sidenius and S. Basu, All your hedges in one basket, *Risk*, November 2003.

A. Arvanitis, J. Gregory and J.-P. Laurent, Building models for credit spreads, *Journal of Derivatives*, Spring, 27–43, 1999.

P. Baheti, R. Mashal, M. Naldi and L. Schloegl, Synthetic CDO of CDO's: squaring the delta hedged equity trade, *Lehman Brothers Quantitative Credit Research Quarterly*, June 2004.

P. Baheti, R. Mashal, M. Naldi and L. Schloegl, Squaring factor copula models, *Risk Magazine*, June 2005.

M. Baxter, Dynamic Modelling of Single-name Credits and CDO Tranches, www.defaultrisk.com, March 2006.

M. Baxter, Gamma process dynamic modelling of credit, *Risk Magazine*, October 2007.

M. Baxter and A. Rennie, *Financial Calculus*, Cambridge University Press, 1996.

N. Bennani, The Forward Loss Model: A Dynamic Term Structure Approach for the Pricing of Portfolio Credit Derivatives, www.defaultrisk.com, November 2005.

N. Bennani, A Note on Markov Functional Loss Models, www.defaultrisk.com, November 2006.

T. Bielecki and M. Rutkowski, *Credit Risk: Modeling, Valuation and Hedging*, Springer, 2002.

T. Bielecki, M. Jeanblanc and M. Rutkowski, Hedging of Basket Credit Derivatives in Credit Default Swap Markets, www.defaultrisk.com, 2006a.

T. Bielecki, M. Jeanblanc and M. Rutkowski, Hedging of Credit Derivatives in Models with Totally Unexpected Default, www.defaultrisk.com, 2006b.

F. Black and J. Cox, Valuing corporate securities – some effects of bond indenture provisions, *Journal of Finance*, **31**, 351–367, 1976.

C. Bluhm, CDO Modeling: Techniques, Examples and Applications, www.defaultrisk.com, December 2003.

C. Bluhm and L. Overbeck, *Structured Credit Portfolio Analysis, Baskets and CDO's*, Chapman & Hall/CRC, 2007.

R. Bookstaber, *A Demon of Our Own Design*, Wiley, 2007.

J.-P. Bouchaud and M. Potters, *Theory of Financial Risks – From Statistical Physics to Risk Management*, Cambridge University Press, 2000.

D. Brigo and E. Errais, A Correlation Bridge Between Structural Models and Reduced Form Models for Multiname Credit Derivatives, www.defaultrisk.com, 2005.

D. Brigo and F. Mercurio, *Interest Rate Models – Theory and Practice*, Springer, 2001.

D. Brigo, A. Pallavicini and R. Torresetti, Calibration of CDO Tranches with the Dynamical Generalised Poisson Loss Model, www.damianobrigo.it, May 2007.

R. L. Burden and J. D. Faires, *Numerical Analysis*, 4th edition, PWS–Kent, 1989.

X. Burtschell, J. Gregory and J.-P. Laurent, A Comparative Analysis of CDO Pricing Models, www.defaultrisk.com, April 2005.

X. Burtschell, J. Gregory and J.-P. Laurent, Beyond Gaussian Copula – Stochastic and Local Correlation, www.defaultrisk.com, January 2007.

G. Chacko, A. Sjoman, H. Motohashi and V. Dessain, *Credit Derivatives – A Primer on Credit Risk, Modeling and Instruments*, Wharton School Publishing, 2006.

G. Chaplin, *Credit Derivatives – Risk Management, Trading and Investing*, Wiley Finance, 2005.

Z. Chen and P. Glasserman, Fast Pricing of Basket Default Swaps, www.defaultrisk.com, 2006.

U. Cherubini and G. Della Lunga, *Structured Finance – The Object Oriented Approach*, Wiley Finance, 2007.

U. Cherubini, E. Luciano and W. Vecchiato, *Copula Methods in Finance*, Wiley, 2004.

F. Claudio, A comparative analysis of correlation skew modelling techniques for CDO index tranches, *MSc Thesis*, Kings College London, 2006.

J. Cox, J. Ingersoll and J. Ross, A theory of the term structure of interest rates, *Econometrica*, **53**, 385–407, 1985.

CreditMetrics, *Technical Document*, 1997.

K. Crider, Recent events in the credit correlation market and their impact on hedging, May 2005.

S. Dalton, *Excel add-in Development in C/C++ – Applications in Finance*, Wiley Finance, 2005.

S. Das, *Credit Derivatives CDO's & Structured Credit Products*, 3rd edition, Wiley Finance, 2005.

S. Das and P. Tufano, Pricing credit sensitive debt when interest rates, credit ratings and credit spreads are stochastic, *Journal of Financial Engineering*, **5**, 161–198, 1996.

S. R. Das, D. Duffie, N. Kapadia and L. Saito, Common Failings: How Corporate Defaults Are Correlated, www.defaultrisk.com.

W. Davies, A. Batchvarov and A. Davletova, An introduction to credit default swaps on ABS, *Merrill Lynch Fixed Income Strategy*, April 2005.

M. Davis and V. Lo, Infectious defaults, *Quantitative Finance*, **1**, 382–387, 2001.

A. Debuysscher and M. Szego, The Fourier Transform Method – Technical Document, Moodys Investors Services, www.moodys.com, January 2003.

A. De Servigny and N. Jobst, *The Handbook of Structured Finance*, McGraw-Hill, 2007.

A. De Servigny and O. Renault, Default correlation: empirical evidence, *Working Paper*, Standard and Poors, 2002.

d-fine, Valuation of *N*th to Default Swaps, www.d-fine.co.uk, September 2004.

R. Douglas (Editor), *Credit Derivative Strategies – New Thinking On Managing Risk and Return*, Bloomberg Press, 2007.
D. Duffie and K. Singleton, Modelling term structures of defaultable bonds, *Review of Financial Studies*, **12**, 687–720, 1999.
E. Eberlein, R. Frey and E. August von Hammerstein, Advanced Credit Portfolio Modelling and CDO Pricing, www.defaultrisk.com, September 2007.
P. Embrecht, A. McNeil and D. Strautmann, Correlation and dependency in risk management: properties and pitfalls, *Working Paper*, University of Zurich, 1999.
A. Etheridge, *A Course in Financial Calculus*, Cambridge University Press, 2002.
Fitch, www.fitchratings.com.
C. Flanagan, R. Asato, E. Reardon, C. Muth, G. Schultz, A. Sze, T. Van Voorhis, R. Ahluwalia and T. Ko., Single name CDS of ABS, *JP Morgan Global Structured Finance Research*, March 2005.
C. Flanagan, E. J. Reardon, A. Sbityakov and E. Beinstein, Introducing ABX tranches, *JP Morgan Global Structured Finance Research*, 2007.
R. Frey and J. Backhaus, Dynamic Hedging of Synthetic CDO Tranches with Spread and Contagion Risk, www.defaultrisk.com, 2007.
H. Frydman and T. Schuermann, Credit Rating Dynamics and Markov Mixture Models, www.defaultrisk.com, March 2007.
S. Galiani, Copula functions and their application in pricing and risk managing multiname credit derivative products, *MSc Thesis*, Kings College London, September 2003.
E. Gamma, R. Helm, R. Johnson and J. Vlissides, *Design Patterns – Elements of Reusable Object oriented Software*, Addison-Wesley, 1995.
J. Garcia and S. Goossens, Base Expected Loss Explains Levy Base Correlation Smile, www.defaultrisk.com, July 2007a.
J. Garcia and S. Goossens, Levy Base Correlation Explained, www.defaultrisk.com, August 2007b.
J. Garcia, S. Goossens and W. Schoutens, Let's Jump Together – Pricing of Credit Derivatives from Index Swaptions to CPPI's, www.defaultrisk.com, May 2007a.
J. Garcia, S. Goossens, V. Masol and W. Schoutens, Levy Base Correlation, www.defaultrisk.com, September 2007b.
M. S. Gibson, Understanding the Risk of Synthetic CDO's, July 2004.
K. Giesecke, Credit Risk Modelling and Valuation: An Introduction, www.defaultrisk.com, June 2004.
K. Giesecke, The Correlation Neutral Measure for Portfolio Credit, www.defaultrisk.com, 2007.
K. Gilkes and M. Drexler, Drill-Down Approach for Synthetic CDO Squared Transactions, Standard & Poors Structured Finance, www.standardandpoors.com, December 2003.
P. Glasserman, *Monte Carlo Methods in Financial Engineering*, Springer, 2003.
P. Glasserman and S. Suchintabandid, Correlation Expansions for CDO Pricing, www.defaultrisk.com, October 2006.
A. Greenberg, D. O'Kane and L. Schloegl, LH+: A fast analytical model for CDO hedging and risk management, *Lehman Brothers Quantitative Credit Research*, June 2004.
J. Gregory (Editor), *Credit Derivatives – The Definitive Guide*, Wiley, 2003.
J. Gregory and J.-P. Laurent, I will survive, *Risk*, 103–107, June 2003.
J. Gregory and J.-P, Laurent, In the core of correlation, *Risk*, 87–91, October 2004.
G. Grimmett and D. Stirzaker, *Probability and Random Processes*, 3rd edition, Oxford University Press, 2004.
P. M. Goldbart, N. Goldenfeld and D. Sherrington, *Stealing the Gold: A Celebration of the Pioneering Physics of Sam Edwards*, Oxford Science Publications, 2004.

X. Gou, R. A. Jarrow and C. Menn, A Note on Lando's Formula and Conditional Independence, www.defaultrisk.com, May 2007.

D. Guegan and J. Houdain, CDO Pricing and Factor Models – A New Methodology Using Normal Inverse Gaussian Distributions, www.defaultrisk.com, June 2005.

S. Hager and R. Schobel, A Note on the Correlation Smile, www.defaultrisk.com, December 2006.

J. D. Hamilton, *Time Series Analysis*, Princeton University Press, 1994.

N. Higham, www.maths.manchester.ac.uk/~higham.

N. Higham, Can you count on your correlation matrix, *NAG & Wilmott Finance Seminar*, December 2006.

S. Hooda, Explaining Base Correlation Skew Using NG (Normal Gamma) Process, www.defaultrisk.com, June 2006.

J. Hull, *Options, Futures and Other Derivatives*, 4th edition, Prentice-Hall, 1999.

J. Hull and A. White, Valuation of a CDO and an *n*th to default CDS without Monte Carlo simulation, *Journal of Derivatives*, **12**, 8–23, 2004.

J. Hull and A. White, Valuing Credit Derivatives Using an Implied Copula Approach, www.defaultrisk.com, November 2006a.

J. Hull and A. White, Forwards and European Options on CDO Tranches, www.defaultrisk.com, 2006b.

J. Hull and A. White, Dynamic Models of Portfolio Credit Risk: A Simplified Approach, www.defaultrisk.com, November 2007.

J. Hull, M. Predescu and A. White, The Valuation of Correlation Dependent Credit Derivatives Using a Structural Model, www.defaultrisk.com, March 2005.

S. Inglis and A. Lipton, Factor Models for Credit Correlation, www.defaultrisk.com, June 2007.

I. Iscoe and A. Kreinin, Recursive Valuation of Basket Default Swaps, www.defaultrisk.com.

P. Jackel, *Monte Carlo Methods in Finance*, Wiley, 2002.

K. Jackson and W. Zhang, Valuation of Forward Starting CDO's, www.defaultrisk.com, 2007.

K. Jackson, A. Kreinin and X. Ma, Loss Distribution Evaluation of Synthetic CDO's, www.defaultrisk.com, 2007.

R. A. Jarrow and F. Yu, Counterparty risk and the pricing of defaultable securities, *Journal of Finance*, **53**, 2225–2243, 2001.

R. A. Jarrow, D. Lando and S. Turnbull, A Markov model for the term structure of credit risk spreads, *Review of Financial Studies*, **10**, 481–523, 1997.

R. A. Jarrow, L. Li, M. Mesier and D. R. van Deventer, *CDO Valuation: Fact and Fiction*, Kamakura Corporation, 2007.

D. Joannas and M. Choudhry, *A Primer on Synthetic Collateralised Debt Obligations*, Yieldcurve.publishing, 2003.

M. S. Joshi, *The Concepts and Practice of Mathematical Finance*, Cambridge University Press, 2003.

M. S. Joshi, *C++ Design Concepts and Derivatives Pricing*, Cambridge University Press, 2004a.

M. S. Joshi, Applying Importance Sampling to Pricing Single Tranches of CDO's in a One-factor Li Model, www.quarchome.org, November 2004b.

M. S. Joshi and D. Kainth, Rapid and Accurate Development of Prices and Greeks for *N*th-to-Default Credit Swaps in the Li Model, www.quarchome.org, April 2003.

M. S. Joshi and A. Stacey, IG: a new approach to pricing portfolio credit derivatives, *Risk Magazine*, **19**, July/August 2006.

A. Kakodkar, B. Martin and S. Galiani, *Correlation Trading*, Merrill Lynch, Credit Derivatives, November 2003.

A. Kalemanova, B. Schmid and R. Werner, The normal inverse Gaussian distribution for synthetic CDO Pricing, *Working Paper*, Risklab, Germany, 2005.

M. Kijima and K. Komoribayashi, A Markov chain model for valuing credit risk derivatives, *Journal of Derivatives*, **6**, 97–108, 1998.

C. P. Kindleberger and R. Z. Aliber, *Manias, Panics and Crashes – A History of Financial Crises*, 5th edition, Palgrave Macmillan, 2005.

A. Koenig and B. E. Moo, *Accelerated C++ – Practical Programming by Example*, Addison-Wesley, 2005.

V. Kothari, *Securitization – The Financial Instrument of the Future*, Wiley Finance, 2006.

A. Kreinin and M. Sidelnikova, Regularization algorithms for transition matrices, *Algo Research Quarterly*, **4** (1/2), 25–40, 2001.

D. Lando, On Cox processes and credit risky securities, *Review of Derivatives Research*, **2**, 99–120, 1998.

J.-P. Laurent and J. Gregory, Basket default swaps, CDO's and factor copulas, *Working Paper*, ISFA Actuarial School University of Lyon, October 2003.

J.-P. Laurent, A. Cousin and J.-D. Fermanian, Hedging Default Risks of CDO's in Markovian Contagian Models, www.defaultrisk.com, 2007.

D. Li, On default correlation: a copula approach, *Journal of Fixed Income*, **9**, 43–54, March 2000.

D. Li and M. Liang, CDO Squared Using Gaussian Mixture Model with Transformation of Loss Distribution, www.defaultrisk.com, 2005.

G. Loffler and P. N. Posch, *Credit Risk Modeling using Excel and VBA*, Wiley Finance, 2007.

F. Longstaff and A. Rajan, An Empirical Analysis of the Pricing of Collateralised Debt Obligations, www.defaultrisk.com, April 2006.

F. A. Longstaff and E. Schwartz, Valuing credit derivatives, *Journal of Fixed Income*, **5**, 6–12, 1995.

D. J. Lucas, L. S. Goodman, F. J. Fabozzi and R. J. Manning, *Developments in Collateralized Debt Obligations – New Products and Insights*, Wiley, 2007.

MarkIt, www.markit.com.

L. McGinty and R. Ahluwalia, A model for base correlation calculation, *JP Morgan Credit Derivatives Strategy*, May 2004b.

L. McGinty, E. Beinstein, R. Ahluwalia and M. Watts, Credit correlation: a guide, *JP Morgan Credit Derivatives Strategy*, March 2004a.

R. Merton, On the pricing of corporate debt: the risk structure of interest rates, *Journal of Finance*, **29**, 449–470, 1974.

M. Mezard, G. Parisi and J. A. Virasoro, *Spin Glass Theory and Beyond*, World Scientific Publishing, 1987.

Moodys, www.moodys.com.

T. Moosbrucker, Pricing CDO's with Correlated Variance Gamma Processes, www.defaultrisk.com, 2006.

M. Musiela and M. Rutkowski, *Martingale Methods in Financial Modelling*, Springer, 1998.

R. B. Nelson, *An Introduction to Copulas*, Lecture Notes in Statistics 139, Springer, 1999.

M. Neugebauer, R. Gambel, J. Zelter, R. Hrvatin and M. Gerity, CDO squared: a closer look at correlation, *Fitch Ratings Structured Finance*, www.fitchratings.com, February 2004a.

M. Neugebauer, R. Gambel, C. Hand, R. Gambel, R. Hrvatin and M. Gerity, Analysis of synthetic CDO's of CDO's, *Fitch Ratings Structured Finance*, www.fitchratings.com, September 2004b.

M. Neugebauer S. Hodgson, J. Carter, C. Osako, R. Hrvatin, T. Cunningham and R. Hardee, Understanding and hedging risks in synthetic CDO tranches, *Fitch Ratings Structured Finance*, August 2006.

M. E. J. Newman and G. T. Barkema, *Monte Carlo Methods in Statistical Physics*, Oxford University Press, 1999.

D. O'Kane, Base correlation and a skew model, *Global Derivatives and Risk Management*, 2005.

D. O'Kane and S. Turnbull, Valuation of credit default swaps, *Lehman Brothers Quantitative Credit Research Quarterly*, Q1/Q2, 2003.

B. Oksendal, *Stochastic Differential Equations – An Introduction with Applications*, 5th edition, Springer, 1998.

A. Pain, O. Renault and D. Shelton, Base correlation, the term structure dimension, *Fixed Income Strategy and Analysis*, Citigroup, December 2005.

E. Parcell, Loss Unit Interpolation in the Collateralised Debt Obligation Pricing Model, www.defaultrisk.com, 2006.

E. Parcell and J. Wood, Wiping the Smile off Your Base (Correlation Curve), www.defaultrisk.com, 2007.

N. Patel, Crisis of correlation, *Risk Magazine*, June 2005.

C. Pedersen and S. Sen, Valuation of constant maturity default swaps, *Lehman Brothers Quantitative Credit Research Quarterly*, 30 June 2004.

A. Penaud and J. Selfe, First-to-default swaps, *Wilmott Magazine*, January/February 2003.

A. Petrelli, O. Siu, J. Zhang and V. Kapoor, Optimal Static Hedging of Defaults in CDO's, www.defaultrisk.com, April 2006.

D. Picone, Structuring and Rating Cashflow CDO's with Rating Transition Matrices, www.defaultrisk.com, 2005.

D. Prange and W. Scherer, Correlation Smile Matching with Alpha-Stable Distributions and Fitted Archimedean Copula Models, www.defaultrisk.com, March 2006.

W. H. Press, S. A. Teukolsky, W. T. Vetterling and B. P. Flannery, *Numerical Recipes in C++*, 2nd edition, Cambridge University Press, 2002.

J.-L. Prigent, O. Renault and O. Scaillet, An empirical investigation into credit spread indices, *Journal of Risk*, **3**, 27–55, 2001.

J. Prince, A general review of CDO valuation models, *Global Structured Credit Strategy Paper*, Citigroup, February 2006.

A. Rajan, G. McDermott and R. Roy, *The Structured Credit Handbook*, Wiley Finance, 2007.

R. Rebonato, *Interest-Rate Option Models*, 2nd edition, Wiley, 1998.

R. Rebonato, *Modern Pricing of Interest-Rate Derivatives – The LIBOR Market Model and Beyond*, Princeton University Press, 2002.

R. Rebonato and D. S. Kainth, Tranched Credit Derivatives: Different or Unique? Can Tranched Credit Derivatives Really Be Hedged?, http://www-cfr.jims.cam.ac.uk.

H. Risken, *The Fokker–Planck Equation*, 2nd edition, Springer, 1996.

S. Roman, *Writing Excel Macros*, O'Reilly, 1999.

M. G. Rott and C. P. Fries, Fast and Robust Monte Carlo CDO Sensitivities and Their Efficient Object Oriented Implementation, www.defaultrisk.com, May 2005.

P. Schonbucher, *Credit Derivatives Pricing and Models*, Wiley, 2003.

P. Schonbucher, Portfolio losses and the term structure of loss transition rates: a new methodology for the pricing of portfolio credit derivatives, *Working Paper*, ETH Zurich, 2006.

G. M. Schultz, J. McElravey, S. Whitworth, E. K. Walsh and C. van Heerden, Valuation of subprime ABS credit default swaps, *Wachovia Structured Products Research*, 2007.

D. Shelton, Back to normal, *Global Structured Credit Research*, Citigroup, 2004.

J. Sidenius, On the Term Structure of Loss Distributions – A Forward Model Approach, www.defaultrisk.com, 2006.

J. Sidenius, V. Piterbarg and L. Andersen, A New Framework for Dynamic Credit Portfolio Loss Modelling, www.defaultrisk.com, November 2005.

S&P, www.standardandpoors.com.

M. St Pierre, E. Rousseau, J. Zavattero, O. van Eyseren, A. Arova, D. Pugachevsky, M. Fourny and A. Reyfman, Valuing and Hedging Synthetic CDO Tranches Using Base Correlations, Bear Stearns, May 2004.

E. Tick, *Structured Finance Modeling with Object Oriented VBA*, Wiley Finance, 2007.

R. Torresetti, D. Brigo and A. Pallavicini, Implied Correlation in CDO Tranches: A Paradigm to be Handled with Care, www.ssrn.com, 2006.

R. Torresetti, D. Brigo and A. Pallavicini, Implied Expected Tranched Loss Surface from CDO Data, www.damianobrigo.it, May 2007.

D. Totouom and M. Armstrong, Dynamic copulas and forward starting credit derivatives, *Ecole Nationale Superieure des Mines de Paris*, February 2007.

S. Truck and E. Ozturkmen, Adjustment and Application of Transition Matrices in Credit Risk Models, www.defaultrisk.com, September 2003.

J. Turc, P. Very and D. Benhamou, Pricing CDO's with a smile, *SG Credit Research*, 2005.

O. Vasicek, Probability of loss on loan portfolios, *Working Paper*, KMV Corporation, 1987.

J. Voit, *The Statistical Mechanics of Financial Markets*, 3rd edition, Springer, 2005.

M. Walker, CDO Models – Towards the Next Generation: Incomplete Markets and Term Structure, www.defaultrisk.com, May 2006.

M. B. Walker, Simultaneous Calibration to a Range of Portfolio Credit Derivatives with a Dynamic Discrete-time Multi-step Markov Loss Model, www.defaultrisk.com, August 2007.

D. Wang, S. T. Rachev and F. J. Fabozzi, Pricing of a CDO and a CDS Index: Recent Advances and Future Research, www.defaultrisk.com, 2006.

D. Wang, S. T. Rachev and F. J. Fabozzi, Pricing of Credit Default Swap Index Tranches with One-factor Heavy Tailed Copula Models, www.defaultrisk.com, 2007.

S. Willemann, An Evaluation of the Base Correlation Framework for Synthetic CDO's, www.defaultrisk.com, December 2004.

S. Willemann, Fitting the CDO Correlation Skew: A Tractable Structural Jump Model, www.defaultrisk.com, November 2005.

P. Wilmott, *Paul Wilmott on Quantitative Finance*, Volumes 1 and 2, Wiley, 2000.

G. Xu, Extending Gaussian Copula with Jumps to Match Correlation Smile, www.defaultrisk.com, December 2006.

J. Yang, T. Hurd and X. Zhang, Saddlepoint Approximation Method for Pricing CDO's, www.defaultrisk.com, November 2005.

F. Yu, Correlated defaults in intensity based models, **17** (2), 155–173, 2007.

Index